Lecture Notes in Computer Science 2644
Edited by G. Goos, J. Hartmanis, and J. van Leeuwen

Springer
*Berlin
Heidelberg
New York
Hong Kong
London
Milan
Paris
Tokyo*

Dieter Hogrefe Anthony Wiles (Eds.)

Testing of Communicating Systems

15th IFIP International Conference, TestCom 2003
Sophia Antipolis, France, May 26-28, 2003
Proceedings

 Springer

Series Editors

Gerhard Goos, Karlsruhe University, Germany
Juris Hartmanis, Cornell University, NY, USA
Jan van Leeuwen, Utrecht University, The Netherlands

Volume Editors

Dieter Hogrefe
Universität Göttingen Lehrstuhl für Telematik
Lotzestr. 16-18, 37083 Göttingen, Germany
E-mail: hogrefe@informatik.uni-goettingen.de

Anthony Wiles
ETSI
PEX and Testing Competence Centre (PTCC)
Route des Lucioles, 06921 Sophia Antipolis, France
E-mail: Anthony.Wiles@etsi.fr

Cataloging-in-Publication Data applied for

A catalog record for this book is available from the Library of Congress.

Bibliographic information published by Die Deutsche Bibliothek
Die Deutsche Bibliothek lists this publication in the Deutsche Nationalbibliografie;
detailed bibliographic data is available in the Internet at <http://dnb.ddb.de>.

CR Subject Classification (1998): D.2.5, D.2, C.2

ISSN 0302-9743
ISBN 3-540-40123-7 Springer-Verlag Berlin Heidelberg New York

This work is subject to copyright. All rights are reserved, whether the whole or part of the material is concerned, specifically the rights of translation, reprinting, re-use of illustrations, recitation, broadcasting, reproduction on microfilms or in any other way, and storage in data banks. Duplication of this publication or parts thereof is permitted only under the provisions of the German Copyright Law of September 9, 1965, in its current version, and permission for use must always be obtained from Springer-Verlag. Violations are liable for prosecution under the German Copyright Law.

Springer-Verlag Berlin Heidelberg New York
a member of BertelsmannSpringer Science+Business Media GmbH

http://www.springer.de

©2003 IFIP International Federation for Information Processing, Hofstrasse 3, A-2361 Laxenburg, Austria
Printed in Germany

Typesetting: Camera-ready by author, data conversion by Olgun Computergrafik
Printed on acid-free paper   SPIN: 10930533   06/3142   5 4 3 2 1 0

Preface

This volume contains the proceedings of TestCom 2003, the IFIP TC6/WG 6.1 International Conference on Testing of Communicating Systems, held in Sophia Antipolis, France, during May 26–28, 2003. TestCom denotes a series of international working conferences on testing communicating systems in the data and telecommunication domain. The conference provides a forum for researchers, vendors and users to review, discuss and learn about new approaches, concepts and experiences in the field of testing of communicating systems. This conference in particular focuses on new ways of testing for new-generation networks.

TestCom 2003 is the fifteenth in a series of annual meetings sponsored by IFIP TC6/WG 6.1. The 14 previous meetings were held in Vancouver, Canada (1988), Berlin, Germany (1989), McLean, USA (1990), Leidschendam, The Netherlands (1991), Montreal, Canada (1992), Pau, France (1993), Tokyo, Japan (1994), Evry, France (1995), Darmstadt, Germany (1996), Cheju Island, Korea (1997), Tomsk, Russia (1998), Budapest, Hungary (1999), Ottawa, Canada (2000), and Berlin, Germany (2002).

The scope of the papers presented at TestCom 2003 covers interoperability testing, TTCN-3, automata-based testing, testing of next-generation networks, IP and UMTS, test systems and tools, test specification design and methodology, and industrial experience in testing communication systems.

The TestCom 2003 program consisted of 6 technical sessions, 3 position statement sessions, and a special session on interoperability testing. Three invited speeches gave an overview on actual trends in the testing and telecommunication area and formulated requirements to the testing process.

The proceedings contain the 19 regular papers accepted and presented at the conference. They were selected from 53 submitted papers in a careful selection procedure based on the assessment of three referees for each paper. The proceedings also include the text of the invited talk by Elaine Weyuker, Philippe Cousin, and Ana Cavalli.

The Best Paper Award was given to the authors of "Realizing Distributed TTCN-3 Test Systems with TCI", Ina Schieferdecker and Theofanis Vassiliou-Gioles, affiliated with FhG FOKUS, Germany.

For the first time the PQM Industrial Relevance Award was given to the paper with the most interest to the communications testing industry, selected during the conference by a panel of delegates from within the industry itself.

TestCom 2002 was organized under the auspices of IFIP TC6 by the European Telecommunications Standards Institute, ETSI, and by the University of Göttingen. It was supported by a number of partners including Danet, Davinci Communications, Ericsson, France Telecom, Motorola, Nokia, Testing Tech and Telelogic.

We would like to express our gratitude to the numerous people who contributed to the success of TestCom 2003. The reviewing process was one of the major efforts during the preparation of the conference. It was completed by experts from around the world. The reviewers are listed in these proceedings. We also thank Mr. Hofmann of Springer-Verlag for his constructive support when editing these proceedings. Finally, we would like to thank Mr. Karl Heinz Rosenbrock for making the ETSI facilities available to us and thank the local organizers for the excellent running of the conference, especially Emmanuelle Jouan and Laetitia Taisne.

Our special thanks goes to Tabea Georgi, Kathrin Högg and André Meyer from the University of Lübeck who did the majority of the work in organizing and preparing these proceedings.

May 2003

Dieter Hogrefe
Anthony Wiles

Conference Committees

Conference Chairmen

Dieter Hogrefe, *University of Göttingen, Germany*
Anthony Wiles, *ETSI, France*

Steering Committee

Sam Chanson, *Hong Kong Univ. of Sci. & Tech., China*
Roland Groz, *France Telecom, France*
Guy Leduc, *University of Liege, Belgium*
Alexandre Petrenko, *CRIM, Canada*

Technical Program Committee

B. Baumgarten, *Fraunhofer SIT, Germany*
G. v. Bochmann, *University of Ottawa, Canada*
A.R. Cavalli, *INT, France*
G. Csopaki, *Budapest University of Technology and Economics, Hungary*
S. Dibuz, *Ericsson, Hungary*
A. Gehring, *Deutsche Telekom, Germany*
J. Grabowski, *University of Lübeck, Germany*
R. Hierons, *Brunel University, UK*
T. Higashino, *Osaka University, Japan*
M. Kim, *ICU University, Korea*
H. Koenig, *BTU Cottbus, Germany*
D. Lee, *Bell Labs Research, USA*
G. Maggiore, *TIM, Italy*
O. Monkewich, *OMC International, Canada*
R.L. Probert, *University of Ottawa, Canada*
O. Rafiq, *University of Pau, France*
S. Randall, *PQM Consultants, UK*
I. Schieferdecker, *Fraunhofer FOKUS, Germany*
K. Stöttinger, *Deutsche Telekom, Germany*
K. Suzuki, *Advanced Communication, Japan*
J. Tretmans, *University of Nijmegen, The Netherlands*
A. Ulrich, *Siemens, Germany*
H. Ural, *University of Ottawa, Canada*
M.U. Uyar, *City University of New York, USA*
C. Viho, *IRISA/University of Rennes, France*

C. Willcock, *Nokia, Germany*
J. Wu, *Tsinghua University, China*
N. Yevtushenko, *Tomsk State Universitiy, Russia*

Additional Reviewers

B. Bao, *Bell Labs Research, China*
S. Barbin, *IRISA/University of Rennes I, France*
J. Barta, *Ericsson, Hungary*
G. Batori, *BUTE, Hungary*
E. Bayse, *Institut National des Télécommunications (INT), France*
S. Boroday, *CRIM, Canada*
A. Carolina Minaburo, *RSM/ENST-Bretagne, France*
A.W.H. Chang, *BUTE, Hungary*
S.C. Cheung, *Hong Kong Univ. of Sci. & Tech., China*
C. Chi, *Bell Labs Research, China*
J. Corral, *RSM/ENST-Bretagne, France*
Z.R. Dai, *University of Lübeck, Germany*
A. Duale, *IBM, USA*
M. Fecko, *Telcordia Technologies, Inc., USA*
H. Hallal, *CRIM, Canada*
R. Hao, *Bell Labs Research, China*
T. Hasegawa, *KDDI R&D Labs., Japan*
O. Henniger, *Fraunhofer SIT, Germany*
L. Hevizi, *Ericsson, Hungary*
M. Higuchi, *Kinki University, Japan*
J. Hosszu, *Ericsson, Hungary*
D. Huang, *Bell Labs Research, China*
A. Idoue, *KDDI R&D Labs, Japan*
S. Kang, *Information and Communications University, Republic of Korea*
P. Koopman, *University of Nijmegen, The Netherlands*
I. Koufareva, *Tomsk State University, Russia*
G. Kovacs, *BUTE, Hungary*
P. Krémer, *Ericsson, Hungary*
J. Le Huo, *McGill University, Canada*
D. Le Viet, *BUTE, Hungary*
J. Ma, *Bell Labs Research, China*
A. Mederreg, *Institut National des Télécommunications (INT), France*
M. Meier, *BTU Cottbus, Germany*
T. Mori, *Osaka University, Japan*
H. Neukirchen, *University of Lübeck, Germany*
M. Núñez, *Universidad Complutense de Madrid, Spain*
T. Ogishi *KDDI R&D Labs, Japan*
Z. Pap, *BUTE, Hungary*
A. Pietschker, *Siemens AG, Germany*

S. Prokopenko, *Tomsk State University, Russia*
Z. Rethati, *BUTE, Hungary*
D. Ross, *RSM/ENST-Bretagne, France*
S. Seol, *Information and Communications University, Republic of Korea*
T. Szabo, *BUTE, Hungary*
V. Trennkaev, *Tomsk State University, Russia*
D. Vieira, *Institut National des Télécommunications (INT), France*
E. Vieira, *Institut National des Télécommunications (INT), France*
M. Zuehlke, *BTU Cottbus, Germany*

Organization Committee

Emmanuelle Jouan, *ETSI, France*
Marie-Noëlle Girard, *ETSI, France*
Laetitia Taisne, *ETSI, France*
Tabea Georgi, *University of Lübeck, Germany*
Kathrin Högg, *University of Lübeck, Germany*
André Meyer, *University of Lübeck, Germany*

Partners

Table of Contents

Keynote Speech I

Prediction = Power 1
   Elaine J. Weyuker (AT&T Labs - Research)

Session I   Next Generation Networks, IP and UMTS

UMTS Terminal Testing: A Practical Perspective 10
   Olaf Bergengruen

Testing of 3G 1xEV-DV Stack – A Case Study 20
   Ira Acharya and Hemendra Kumar Singh

Testing SIP Call Flows Using XML Protocol Templates 33
   M. Ranganathan, Olivier Deruelle, and Doug Montgomery

Towards Modeling and Testing of IP Routing Protocols 49
   Jianping Wu, Zhongjie Li, and Xia Yin

Session II   TTCN-3

An Intuitive TTCN-3 Data Presentation Format 63
   Roland Gecse and Sarolta Dibuz

The UML 2.0 Testing Profile and Its Relation to TTCN-3 79
   Ina Schieferdecker, Zhen Ru Dai, Jens Grabowski, and Axel Rennoch

Realizing Distributed TTCN-3 Test Systems with TCI 95
   Ina Schieferdecker and Theofanis Vassiliou-Gioles

T_{IMED}TTCN-3 Based Graphical Real-Time Test Specification 110
   Zhen Ru Dai, Jens Grabowski, and Helmut Neukirchen

Keynote Speech II

Interoperability Events Complementing Conformance Testing Activities .. 128
   Philippe Cousin (ETSI Plugtests^{TM} Service)

Session III   Automata-Based Methodology

Testing Transition Systems with Input and Output Testers 129
   Alexandre Petrenko, Nina Yevtushenko, and Jia Le Huo

Generating Checking Sequences for a Distributed Test Architecture ... 146
   Hasan Ural and Craig Williams

Conformance of Distributed Systems 163
    Maximilian Frey and Bernd-Holger Schlingloff

An Automata-Based Approach to Property Testing in Event Traces 180
    *Hesham Hallal, Sergiy Boroday, Andreas Ulrich,
    and Alexandre Petrenko*

Fault Diagnosis in Extended Finite State Machines...................... 197
    *Khaled El-Fakih, Svetlana Prokopenko, Nina Yevtushenko,
    and Gregor v. Bochmann*

A Guided Method for Testing Timed Input Output Automata 211
    Abdeslam En-Nouaary and Rachida Dssouli

Session IV   Interoperability Testing

Interoperability Testing Based on a Fault Model for a System
of Communicating FSMs... 226
    Vadim Trenkaev, Myungchul Kim, and Soonuk Seol

Framework and Model for Automated Interoperability Test
and Its Application to ROHC 243
    Sarolta Dibuz and Péter Krémer

Keynote Speech III

TestNet: Let's Test Together!... 258
    *Ana Cavalli (INT/Testnet), Edgardo Montes de Oca,
    and Manuel Núñez*

Session V   Test Design, Tools and Methodology

An Open Framework for Managed Regression Testing 265
    Naina Mittal and Ira Acharya

TUB-TCI – An Architecture for Dynamic Deployment
of Test Components 279
    Markus Lepper, Baltasar Trancón y Widemann, and Jacob Wieland

Fast Testing of Critical Properties through Passive Testing 295
    José Antonio Arnedo, Ana Cavalli, and Manuel Núñez

Author Index 311

Prediction = Power

Elaine J. Weyuker

AT&T Labs - Research
180 Park Avenue, Florham Park, NJ 07932
weyuker@research.att.com

Abstract. An argument is made that predictive metrics provide a very powerful means for organizations to assess characteristics of their software systems and allow them to make critical decisions based on the value computed. Five different predictors are discussed aimed at different stages of the software lifecycle ranging from a metric that is based on an architecture review which is done at the earliest stages of development, before even low-level design has begun, to one designed to predict the risk of releasing a system in its current form. Other predictors discussed include the identification of characteristics of files that are likely to be particularly fault-prone, a metric to help a tester charged with regression testing to determine whether or not a particular selective regression testing algorithm is likely to be cost effective to run on a given software system and test suite, and a metric to help determine whether a system is likely to be able to handle a significantly increased workload while maintaining acceptable performance levels.

Keywords: Architecture review, fault-prone, metrics, prediction, regression testing, risk, scalability.

1 Introduction

If we had a crystal ball and knew somehow that certain files in a large software system were going to be particularly problematic in some way, think about what that would allow us to do. If the nature of the problems were incorrect behavior, we could focus our functional testing resources there, increasing the chances of identifying and removing the problems, thereby having a more dependable system than we would otherwise have. It would also probably mean that we could do the testing more economically because we could target our testing efforts to just those problematic files.

Similarly, if the nature of the problems were related to system performance, we could focus our performance testing efforts to identify potential bottlenecks so that workloads could be balanced or distributed more appropriately. In this way, users would never have to wait in long queues or experience unacceptable delays. Undoubtedly you would have happier customers, and likely more customers.

What other sorts of things would we like to be able to predict? What if you were releasing a new product? Initially it would likely have a small user base. Typically when you do performance testing, you test a system's behavior for the initial expected workload, and also a slightly enhanced one. But what if the product were to take off and you now had one or more orders of magnitude more customers? Could you predict how the system would behave under those circumstances and determine when and how many additional servers would be needed? Could you predict whether there was a point at which adding more servers would actually lead to decreased rather than increased throughput, and therefore the system might have to be re-architected? If we could plan for the increased workload by predicting when these significantly increased workloads were likely to occur, we could order new hardware before performance became unacceptable or begin re-architecture so that the customers never saw any negative impact of the increased workload.

When creating a new product, there is often a tradeoff between being first or early to market and improving the dependability of the system. Is it better to get there early and get a large share of the market, or offer the product later with a higher level of reliability than would be possible were you to release the software now? What are the risks associated with each of these scenarios? If we could predict the risk, or expected loss, associated with the release of a software system, we could make informed decisions about whether or not it is wise to release the product in its current state, understanding fully the consequences of whichever decision is made.

Being able to predict these and other related system characteristics can an have enormous impact on the observed dependability of a software system, the cost to deliver the system, and even whether or not the system can be viably produced. In this paper we will examine what sorts of characteristics can be reasonably predicted and how these predictions might be made. We also provide pointers to the literature describing empirical studies used to determine the predictors of interest, as well as studies that show applicability of these predictors for large, industrial software systems as well as the usefulness of these predictors in practice.

In Section 2, we look at predicting the likely quality of the ultimate software system produced by considering the results of architecture reviews that are done well before any implementation has begun. Section 3 examines ways of determining which files in a software system are likely to be particularly fault-prone, and therefore good candidates for concentrating testing resources. In Section 4, we discuss a way of predicting whether or not it is likely to be cost-effective to use a selective regression testing algorithm to try to minimize the regression test suite. If the cost of running the algorithm is high, or the reduction in test suite size is low, it may be more cost-effective to simply rerun as much of the regression test suite as possible rather than using some of those scarce resources of fine-grained analysis with little payoff. Section 5 describes ways of predicting when a system's workload is likely to increase significantly so that steps can be taken to prepare for this situation, thereby assuring that customers always experience acceptable performance, while in Section 6, we discuss how risk can be predicted so that projects can make informed decisions about whether or not it is safe to release a software system in its current state.

2 Architecture Reviews

Architecture is defined as "the blueprint of a software/hardware system. The architecture describes the components that make up the system, the mechanisms for communication between the processes, the information content or messages to be transmitted between processes, the input into the system, the output from the system, and the hardware that the system will run on." [12]

Architecture reviews are performed very early in a software system's lifecycle. [12, 1, 7] It is a standard part of the quality assessment process at AT&T and many other organizations that produce software systems that need to be highly dependable and always available. As soon as the system's requirements and high-level design have been completed, a review is done to assure that the architecture is complete, and that low-level design can begin. It is part of the standard wisdom of the software engineering community that the earlier in the lifecycle problems are identified, the easier and cheaper they are to rectify, and these reviews strive to help identify problems at a very early stage so that they do not negatively impact the system.

The goal of an architecture review, then, is to assess the quality and completeness of the proposed architecture. An obvious extension, therefore, is to use this assessment as a way of predicting the likely quality of the ultimately implemented system, and determining its likelihood of failure. Of course, assuring that the architecture is complete and sound does *not* guarantee that problems will not enter the system during the low-level design, coding, or even testing of the system, but it does assure that the foundation on which these later stages depend is itself appropriate.

For these reasons, Avritzer and Weyuker proposed a way of predicting likely project success based on the results of an architecture review. They developed a questionnaire-based metric that computed a score and provided five ranges indicating whether a project was at low, moderate-low, moderate, moderate-high, or high risk of failure. They used the results of 50 industrial architecture audits performed over a period of two years to create the questionnaire, and then selected seven projects for which they computed the metric. They compared the metric's assessment with the assessment done by several senior personnel familiar with the systems, which at that point had all been in production. They found that the informal and formal assessments matched for six of the seven systems. [3, 4] The seventh project was assessed by the metric as being at moderate risk of failure, while knowledgeable personnel rated it as an excellent system, with very few defects identified in the field. It turned out that the review done for this project was not a true architecture review, but a review done at an even earlier stage, known as a discovery review. Although this sort of review is similar to an architecture review, it is done before the architecture is complete. Its goal is to help a project make decisions and weigh benefits and risks of potential architectural decisions. Therefore, it was appropriate at this stage that certain portions of the architecture were incomplete, and did not indicate potential problems. The results of this case study indicated that it was worthwhile applying this metric for all projects once they had completed an architecture review.

For this reason, Weyuker extended this work and did a larger empirical study limiting consideration to projects during architecture reviews (rather than including the results from projects' discovery reviews.) [19] One of the limitations of the Avritzer-Weyuker metric, however, was that it was difficult and time-consuming to compute because it involved a careful assessment of whether or not the project suffered from any of fifty-four of the most severe problems identified most frequently in the original set of 50 systems studied. For each of the problems, a score had to be assigned indicating the severity of the problem in this instance. Therefore, Weyuker proposed a highly simplified metric that did not require that a questionnaire be completed, and performed a new case study using thirty-six systems for which the results of architecture reviews were present. Of the 36 projects assessed, only one seemed to be at greater risk of failure than the simple metric indicated. It was therefore recommended that projects use this metric to predict their likelihood of success based on the results of their architecture review. In this way, if there is a prediction of a moderate to high risk of failure, there should be sufficient time to modify the architecture to correct problems or complete missing portions so that the project does not move forward with low-level design until a re-review and re-application of the metric indicates that the project has a low risk of failure.

3 Fault-Proneness

Testing software systems for correctness and functionality is an essential and expensive process. It is often estimated that testing and related activities consume as much or more resources as the implementation of the system. Even if that is an exaggeration, there is no doubt that testing a large industrial software system takes a substantial amount of time and money, and that the consequences of doing a poor or inadequate job of testing can be catastrophic. But it is also difficult to do a comprehensive job of testing. For many systems, the input domain is enormous and it is impossible to execute even a very small percentage of the possible inputs during testing. Therefore it is essential that there be a systematic approach to testing and that there be some way of prioritizing potential test cases. Similarly, a large software system may be comprised of many thousands of files, and a decision has to be made which of these files should be most thoroughly exercised, and which can be only lightly tested.

For this reason, if we could identify which files were likely to be most fault-prone, we could give testers a way to allocate testing resources and to prioritize their testing activities. With this goal in mind, a number of research teams have performed empirical studies in recent years to try to identify properties of files that make them particularly fault-prone. [2, 6, 8–10, 13–15] Adams [2], Fenton and Ohlsson [8], Munson and Khoshgoftaar [14], and Ostrand and Weyuker [15] have all observed seeing a very uneven distribution of faults among files within a system. Generally, a very small number of files accounted for the vast majority of the faults. If those files could be identified, then testing effort could be concentrated there, leading to highly dependable systems, with less testing expense.

Given that there tends to be a very uneven distribution of faults among files, a variety of file characteristics have been examined in these empirical studies as possible identifying factors. Some of these characteristics include file size, whether a file is new or appeared in an earlier release, the age of the file, whether earlier versions of the file contained a large number of faults, whether a large number of faults were identified during earlier stages of development, and the number of changes made to a file. Some initial indicators have been identified, but significantly more carefully-done case studies must be prepared.

In a recent empirical study using 13 releases of an industrial inventory tracking system, Ostrand and Weyuker [15] found that among the predictors they considered, the best predictors of fault-proneness involved considering whether the file had been newly-written, and whether it had had a particularly large number of faults in the previous release. Both were shown be promising predictors of high fault densities in the current release.

In a more recent study, they found that if they refined the earlier question about whether the file was new or old, and distinguished between old files that had been changed and those that had not been changed, they found that new files did not always have a higher average fault density than files that appeared in earlier releases. They found, instead, that in more than half of the releases, files that had been changed since the prior release had higher average fault densities than new files, although, for every release, old files that remained unchanged from the prior release always had lower average fault densities than either newly-written files or old, changed files.

Their new research also found that the age of a file was not a particularly good predictor of fault-proneness. For every release studied, they found some cases for which newer files had lower average fault densities than older ones. They therefore concluded that age does not seem to be a good predictor of fault-proneness.

Finally, they examined the question of whether as a system matured, the average fault density of later releases was lower than earlier releases. They did observe a general downward trend, although this value did not decrease monotonically. Hopefully, many other research groups will continue to explore characteristics of files that are particularly fault-prone. If we can learn to make predictions of this nature with any degree of accuracy, there should be an enormous payoff for any organization concerned about producing dependable software systems at reasonable costs.

4 Regression Testing

Whenever a software system has been changed, whether the purpose of the change is to fix defects, enhance the functionality, change the functionality, or any other purpose, there is always the possibility that the fixes or changes have introduced new defects into the product. For this reason, whenever a change is made to the software, the system must be retested. This is known as *regression testing* and consists of rerunning previously-run test cases to make sure that those test cases that performed correctly before the changes, still perform correctly. Since test suites for industrial software systems are often very large, and therefore it is impractical to rerun the entire test suite, *selective regression testing* algorithms have been proposed in order to have an efficient way to select a relatively small subset of the entire test suite. The goal is to select just those test cases to rerun that are relevant to the changes that have been made. By doing this, it is hoped that a significantly smaller subset of test cases can be identified, leading to substantial savings of both time and cost.

Examination of the regression testing literature indicated that several of the proposed algorithms are likely to be computationally very expensive to run, and yield test suites that were almost as large as the full test suite. In the worst case, a great deal of the limited regression testing resources would have been spent selecting the subset, only to learn that the algorithm requires the entire test suite, or nearly the entire test suite, to be rerun. In that case, the test organization is in worse shape than they were initially, since they did not have enough resources to rerun the entire test suite, spent a large portion of their budget learning that they really *should* rerun the entire test suite, and therefore are now able to rerun an even smaller fraction than they would have been able to run in the first place.

This observation led Rosenblum and Weyuker to develop predictors that are computationally very efficient, and indicate to the user whether or not a given selective regression testing strategy is likely to be cost-efficient to use for a given software system and test suite. [16, 17] Additional studies of the use of this predictor appear in [11].

Thus, by using a predictor, a regression testing organization can decide whether or not it is likely to be beneficial to spend part of its limited resources to select an efficient subset of a large test suite, rather than assuming that this is a wise use of resources.

5 Scalability

When we build a new software system, we often have to make a prediction of what the customer base is likely to be. Once the system is complete, we then do performance testing and assess how the system behaves under workloads expected in both the near term and once the product has become more established. But what if the product becomes wildly successful and suddenly has to be able to handle a workload that is several times larger or even orders or magnitude larger than the current workload? Of course, this is every businesses' dream, but if the system cannot handle this increased workload while providing acceptable levels of performance to its customers, it can easily become a nightmare. In addition, many projects use costly custom-designed hardware that must be budgeted for and ordered well in advance of when it will be deployed. On the one hand, a project does not want to order expense equipment on the off-chance that things will go well and it will be needed, but they also do not want to be caught without sufficient capacity if it is needed.

For this reason, Weyuker and Avritzer considered the question of how a project could predict whether or not its software would scale under significantly increased workloads. [18] They defined a new metric, the Performance Non-Scalability Likelihood (PNL), which was designed to be used in conjunction with a workload-based performance testing technique [5].

This metric makes an assessment of the expected loss in performance under varying workloads by distinguishing between system states that provide acceptable performance behavior, and those that don't. The metric incorporates two factors:

- The probability that the system is in a given state (which is determined by the workload characterization used for performance testing).
- The degree to which the system fails to meet acceptable performance while in that state.

A case study was presented that demonstrated how to apply the PNL metric to a large, complex, industrial software system, including the steps needed to model the system, the type of data collected, and the actual computation of the metric. A description of the implications of the computation and experiences of the project of using the information derived from the application of the metric to plan for additional capacity was also presented. This prediction allowed the project to seamlessly rebalance their workloads, identify a potential bottleneck, and deploy additional capacity so that users never encountered unacceptable performance, even though the workload increased substantially over the period of study.

6 Risk

Software risk can be defined to be the expected loss due to failures caused by faults remaining in the software. This is a very important characteristic of software systems because, if we could predict the risk associated with releasing a system in its current form, we could determine whether or not it is safe to do so. In Reference [20], Weyuker discussed the limitations of using some definitions of risk that were proposed in earlier research. The problems centered around either a requirement for information that could not generally be determined, or a requirement that more data be collected than is practically feasible for a large industrial system with a large, complex input domain. Another class of limitation is the fact that many of the risk definitions do not use relevant information that *is* available. For this reason, Weyuker introduced a predictive metric that could be used to determine the likely risk associated with a software system in an industrial environment, thereby helping the development team determine whether or not it is safe to release their software.

The metric incorporated information about the degree to which the software had been tested, as well as the how the software behaved (i.e. whether or not it fails). This metric assures that testers are not rewarded for doing a poor job of testing, by treating unexecuted test cases as if they had been run and failed. For this reason, as the amount of testing increases, the assessed risk typically goes down, because in a typical system, one expects to see a very small percentage of test cases fail. Similarly, if a fault is encountered and a failure occurs, the assessed risk will also decrease once the fault has been corrected. Another feature of this predictive metric is that it includes a mechanism to incorporate the severity of a failure. If the failure is a minor or cosmetic one, it will clearly impact the perceived reliability of the system.

7 Conclusions

We have discussed the importance and value of being able to predict characteristics of software systems. In this way, development and maintenance organizations can determine cost-effective ways of allocating scarce resources such as testing personnel and equipment or laboratory space, and determining whether it is safe to proceed with a proposed architecture or release a system. Each of these predictors have been assessed or developed with case studies performed on large industrial software systems which show both the usefulness of these predictors as well as the feasibility of using these predictors for large systems. Pointers to the literature describing these case studies are provided. Our conclusion is that being able to predict these and related characteristics of software system represent a very powerful mechanism for creating highly dependable systems.

References

1. G. Abowd, L. Bass, P. Clements, R. Kazman, L. Northrup, and A. Zaremski. Recommended best industrial practice for software architecture evaluation. Technical Report CMU/SEI 96-TR-025, available at
   http://www.sei.cmu.edu/products/publications, Jan, 1997.
2. E.N. Adams. Optimizing Preventive Service of Software Products. *IBM J. Res. Develop.*, Vol28, No1, Jan 1984, pp.2-14.
3. A. Avritzer and E.J. Weyuker. Investigating Metrics for Architectural Assessment, *Proc. IEEE/Fifth International Symposium on Software Metrics (Metrics98)*, Bethesda, Md., Nov. 1998, pp.4-10.
4. A. Avritzer and E.J. Weyuker. Metrics to Assess the Likelihood of Project Success Based on Architecture Reviews. *Empirical Software Eng. Journal*, Vol. 4, No. 3, Sept. 1999, pp.197-213.
5. A. Avritzer, J. Kondek, D. Liu, and E.J. Weyuker. Software Performance Testing Based on Workload Characterization. *Proc. ACM/Third International Workshop on Software and Performance (WOSP2002)*, Rome, Italy, July 2002.
6. V.R. Basili and B.T. Perricone. Software Errors and Complexity: An Empirical Investigation. *Communications of the ACM*, Vol27, No1, Jan 1984, pp.42-52.
7. L. Bass, P. Clements, and R. Kazman. Software architecture in practice. Addison Wesley, 1998.
8. N.E. Fenton and N. Ohlsson. Quantitative Analysis of Faults and Failures in a Complex Software System. *IEEE Trans. on Software Engineering*, Vol26, No8, Aug 2000, pp.797-814.

9. T.L. Graves, A.F. Karr, J.S. Marron, and H. Siy. Predicting Fault Incidence Using Software Change History. *IEEE Trans. on Software Engineering*, Vol 26, No. 7, July 2000, pp. 653-661.
10. L. Hatton. Reexamining the Fault Density - Component Size Connection. *IEEE Software*, March/April 1997, pp.89-97.
11. M.J. Harrold, D. Rosenblum, G. Rothermel, and E.J. Weyuker. Empirical Studies of a Prediction Model for Regression Test Selection. *IEEE Trans. on Software Engineering*, Vol 27, No 3, March 2001, pp.248-263.
12. J.P. Holtman. Best current practices: software architecture validation. AT&T, March, 1991.
13. K-H. Moller and D.J. Paulish. An Empirical Investigation of Software Fault Distribution. *Proc. IEEE First Internation Software Metrics Symposium*, Baltimore, Md., May 21-22, 1993, pp.82-90.
14. J.C. Munson and T.M. Khoshgoftaar. The Detection of Fault-Prone Programs. *IEEE Trans. on Software Engineering*, Vol18, No5, May 1992, pp.423-433.
15. T. Ostrand and E.J. Weyuker. The Distribution of Faults in a Large Industrial Software System. *Proc. ACM/International Symposium on Software Testing and Analysis (ISSTA2002)*, Rome, Italy, July 2002, pp.55-64.
16. D.S. Rosenblum and E.J. Weyuker. Predicting the Cost-Effectiveness of Regression Testing Strategies, *Proc. ACM Foundations of Software Engineering Conf (FSE4)*, Oct 1996, pp.118-126.
17. D.S. Rosenblum and E.J. Weyuker. Using Coverage Information to Predict the Cost-Effectiveness of Regression Testing Strategies, *IEEE Trans. on Software Engineering*, March, 1997, pp. 146-156.
18. E.J. Weyuker and A. Avritzer. A Metric to Predict Software Scalability. *Proc. IEEE/Eighth International Symposium on Software Metrics (Metrics2002)*, Ottawa, Canada, June 2002, pp.152-158.
19. E.J. Weyuker. Predicting Project Risk from Architecture Reviews. *Proc. IEEE/Sixth International Symposium on Software Metrics (Metrics99)*, Boca Raton, Fla, Nov. 1999, pp.82-90.
20. E.J. Weyuker. Difficulties Measuring Software Risk in an Industrial Environment. *Proceedings IEEE/International Conference on Dependable Systems and Networks (DSN2001)*, Goteberg, Sweden, July 2001, pp. 15-24.

UMTS Terminal Testing: A Practical Perspective

Olaf Bergengruen

Optimay GmbH – Agere Systems, Orleansstrasse 4, D-81669 Munich, Germany
Olaf.Bergengruen@Optimay.com

Abstract. This paper presents a framework based on the 3GPP test model for a Virtual Test Environment which is the main tool used by wireless-engineers for development and for testing the complete UMTS terminal software. The main requirements to the system are ease of deployment on all engineering workstations, support the design of test scenarios at a high abstraction level and, when running the test scenarios, the system shall perform much faster than in a real run in order to efficiently exercise and debug the relevant functionalities. Furthermore, the Virtual Test Environment is used to run complete test suites within a couple of hours in order to guarantee the quality of the software when it is delivered to the customer.

1 Introduction

The third generation mobile communication technology is currently being developed world-wide. Components and sub-systems will be produced by different companies (e.g. chipsets, SIM cards, RF units, base stations, IP routers, services) and will be eventually assembled into a running system, the UMTS system (Universal Mobile Communication System). The verification that all these pieces work together, i.e. that equipment or unit X interoperates with equipment or unit Y, is called interoperability testing which is very time consuming and is often the bottleneck during system development. In order to facilitate interoperability testing and to guarantee (to a certain degree of confidence) conformance to the specifications, a set of standard world-wide agreed test suites have been developed.

A key aspect for the success of UMTS will be efficient and thorough testing at all levels of development.

Furthermore, due to the increase in complexity of the next generation wireless standards, it is expected that testing effort will rise comparatively more than the effort required for the proper software and hardware development. For this reason it is important to build advanced testing and debugging tools to aid development and validation. We will see test systems with different requirements, purposes and prices for testing RF-units, for verifying software modules, for regression testing, for Type Approval and for production testing.

In this paper we will concentrate on virtual testing, i.e. pure software testing on top of simulated hardware components running on standard workstations or laptops. The system presented here is the main environment used by wireless-engineers at Optimay - Agere Systems to develop the complete Mobile Station software. The system is also used to verify that all (or almost all) functionality delivered to our customers is running properly. For this purpose we periodically (once a day) run the complete test suites on each customer build. This is called regression testing which is an automatic process requiring one to two hours to complete on a powerful workstation dedicated to this task.

The basic idea of virtual testing is to exercise the 'real' Mobile Station software within a simulated environment consisting of a virtual Test System connected to the Mobile Station via a emulated physical layer. This means that at the lowest level, the hardware drivers (controlling the RF units, SIM card, keypad, serial lines, etc.) do not access the proper hardware but the software emulation which was developed for this purpose.

Of course, most Mobile Station manufacturers use test environments for in-house development. But for a simulation environment to be really useful, it needs to be designed with following basic requirements in mind: The system shall be installed and run on any standard workstation; test scenarios shall be easily adapted or created describing the signalling flow and also the cell configurations (like power levels, Sys Infos and neighbour cells); and finally the test environment needs to be much faster than real time. A wireless-engineer will immediately complain if he or she needs more than, say, one minute to run the test scenario and hit the line of code within the Mobile Station software where the breakpoint has been set.

An additional challenge imposed to a test environment results from the fact that a UMTS terminals comprises also other Radio Access Technologies. The market requires mobile terminals to support GSM/GPRS besides UMTS to connect to the Core Network, and also short range protocols like USB, Bluetooth or IEEE 802.11 for the connection to terminal equipment like laptops. Of course, we need to test each stack independently of the others, but even more important and by far more difficult is to verify the interworking between the different components (e.g. verify that the Mobile Station does not crash during inter-RAT handover from UMTS to GSM/GPRS or when Bluetooth is activated for a data connection). In a later section in this paper we will show how we handle different Radio Access Technologies within the Virtual Test Environment.

We start in the next section discussing the scope and main requirements related to conformance testing for a GSM/GPRS/UMTS terminal. But please note that this is not a theoretical paper presenting best possible testing methodologies but an overview resulting from our practical experiences implementing Optimay's test environment which was very much constrained by a turbulent history.

2 Mobile Station Testing / Logical View

From the Mobile Station conformance testing point of view, we need to test basically three interfaces (see Fig. 1): the Um interface to the GSM/GPRS network (GERAN), the Uu interface (the UMTS air interface) to the UTRAN network and the Cu interface to the USIM/SIM card.

The conformance specification for GSM/GPRS terminals is specified in GSM TS 11.10 (or 3GPP TS 51.010). Although this specification is already around 4000 pages

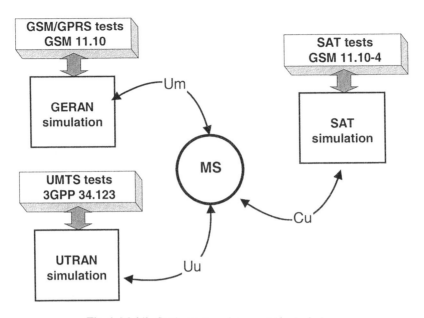

Fig. 1. Mobile Station test environment / logical view

in size, it is in no way sufficient to attain a high degree of confidence in the quality of the Mobile Station software (one of our engineers estimated that GSM 11.10 covers at most 25% of the MS functionality). Thus, we need in addition to GSM TS 11.10 also a substantial number of tests derived from more tricky scenarios found in field testing. And finally, we need performance tests for measuring and optimizing memory and cpu usage. Performance tests are not part of conformance testing nor of Final Type Approval (FTA) procedures but are needed to guarantee and sustain the quality of the overall product.

For the UMTS protocol, conformance tests are specified in 3GPP TS 34.123-1 (the prose specification). Part 3 (3GPP TS 34.123-3) of the specification provides a Test Model and test cases written in TTCN (Tree and Tabular Combined Notation) which are being developed by ETSI Task Force MCC 160. The reason for providing test cases in prose and in TTCN code arises from the difficulties experienced in the past with GSM validation which based on the prose conformance specification. A substantial effort has been invested by test equipment manufacturers, mobile manufacturers and test houses as well in developing proprietary and incompatible test case implementations. Since the prose specification needs a great deal of (human) interpretation and is ambiguous at critical points, inconsistencies and confusion arised resulting for example in a Mobile Station passing a test case from one implementation and failing the same test in an other implementation, and no one being able to tell which is the correct behaviour. For these reasons, there is today consensus in the 3GPP community that an exact test case description in a language like TTCN will simplify conformance testing and save implementation efforts. It is envisaged that eventually the TTCN code will be the mandatory conformance specification and the prose just a scenario description for a better understanding.

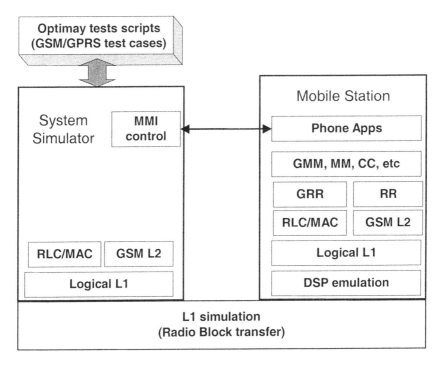

Fig. 2. GSM/GPRS test environment

Finally, the Cu interface enables applications running on the SIM card to interact in a standard way with the Mobile Station and have access to all functionality provided by the Mobile Equipment including call establishment and SMS or mail transfer (access to the ME functionality may happen even without user notice, an ideal situation for developers of 'virus applications' running on SIM cards). To test the SIM Application Toolkit functionality we need a SIM simulation and means to describe the signalling exchanged between ME and SIM in order to implement the conformance tests for SIM Application Toolkit as specified in GSM 11.10-4.

Before we discuss the 3GPP test model, we present in the next section an overview on our GSM/GPRS Test Environment which has been developed over the last 12 years for virtual testing of GSM/GPRS terminals.

3   GSM/GPRS Test Environment

In the good old GSM days, the environment for virtual testing (see Fig. 2) consisted of the System Simulator (SS) connected to the Mobile Station (MS) over a simulated air interface (the lower tester). In order to control the MS and simulate user keystrokes, a simulated MMI interface is connected to the MS (the upper tester) which allows basic user actions like turning on the MS and initiating a call.

The simulated 'air interface' is basically responsible for transferring uplink and downlink radio blocks. Additionally, it handles simulated radio conditions as set by the test scripts which results in proper signals from the emulated DSP code to the protocol stack like measurement report indications and 'bad block' decoding indications. Finally, the simulated 'air interface' has been extended to enable timer synchronization: timer ticks are transferred from the MS to the SS which need to be acknowledged by the SS before the MS continues with the next radio block. In this way, the timer systems from SS and MS are tightly coupled.

The synchronization between the MS and SS timer systems has two main advantages. Firstly, it facilitates developing and debugging the software. Setting a breakpoint in the MS code results in stopping all MS tasks, including the timer ticks, which in turn results in freezing the System Simulator timers, test scripts and other SS processes. This feature is essential for wireless engineers to understand the system behaviour in complex scenarios by inspecting data structures and states in a running system.

Secondly, we can enormously improve the simulation efficiency: since the MS has full control on the timer ticks, it can issue timer ticks when there is nothing else to do (note that the SS also completed the jobs scheduled for the current frame, since the MS waits for the SS tick acknowledgment). In particular, when the MS is in idle mode (i.e. it is reading only paging blocks corresponding to its paging group and in sleep mode during other frames), then we accelerate timer ticks in order to jump from the frame corresponding to one paging group to the next one. Note that the virtual time flow is not directly related to the workstation time on which the simulation is running.

The results of the mechanism explained above is a substantial reduction of simulation time. This is particularly the case for time consuming tests like those related to MM periodic location area update procedures (GSM TS 11.10 clause 26.7) which require more than 30 minutes 'real' execution time. These tests are run in a couple of seconds on our virtual test environment.

Due to historical reasons, our GSM Test Environment was not carefully designed and it was coded basically using the 'quick and dirty' approach, a methodology which is often used in issues related to testing when marketing people and customers expect results and need within a couple of weeks an environment to demonstrate some functionality. Furthermore, since the test environment was viewed as an internal tool which was not to be sold and produce immediate 'cash flow', there was not much interest at the higher management level to put valuable resources on its development, and so the system was patched over the years to upgrade to the current functionality supported by the Mobile Station (I suppose that this is not an uncommon situation in the industry).

Nevertheless, it turned out that the main ideas mentioned above resulted in an extremely successful Virtual Test Environment. Currently we have around 1300 test cases comprising the GSM 11.10 conformance specification and also proprietary tests which verify functionality not thoroughly tested via GSM 11.10. The execution of these tests in a row is called regression testing which is an essential component of Quality Assurance. A full regression test report is required each time the software is delivered to the customer.

The complete test suites need around one hour execution time on a powerful workstation (actually the test suite is executed many times for different versions of the software build for different customers supporting customer specific functionality).

In addition to running protocol tests, we use the same Virtual Test Environment for developing MMI and SIM application toolkit (SAT) functionality. For SAT, we extended the simulation with an additional serial interface connected to a SIM simulation which in turn is controlled by SAT test scripts basically designed to run the tests specified in GSM 11.10-4.

In the next section we present the 3GPP Test Model and we will discuss the issues involved in designing an overall GSM/GPRS/UMTS virtual test environment.

4 The 3GPP Test Model

Mobile equipment manufacturers, test equipment manufacturers, test houses and network providers are all concerned with conformance testing. Thus, all these parties need a common reference and language for discussing test scenarios, designing test cases and building test equipment. The common reference or Test Model shall clearly describe the complete system involving the Mobile Station (referred as UE, User Equipment, in the 3GPP specifications), the System Simulator and all relevant interfaces. Fig. 3 shows the model adopted by 3GPP for signalling tests (see 3GPP TS 34.123-3). The System Simulator basically implements the PHY, MAC and RLC layers, and provides an API compliant to the 3GPP interface as specified in TS 34.123-3. The behaviour of the RRC layer and NAS (Non-Access Stratum including GMM, MM, CC and SM) is explicitly coded in the TTCN test scripts.

A **test case** running on top of the System Simulator consists basically of following steps:

- Configure each cell within the System Simulator
  - Configure PHY / L1
  - Configure MAC
  - Configure RLC
  - Set power levels for the different cells
- Schedule and send System Information Blocks
- Bring the UE into initial state according to the test case (as an example, the UE is brought into Idle Updated state)
  - Perform Location Update procedure
  - Perform GPRS Attach procedure
- Test case body
  - **Stimulate the UE** (for example the SS sends an IDENTITY REQUEST message to the UE, or the SS changes power levels, or the SS commands the UE via e-MMI to establish a call)
  - **Verify responses** sent by the UE and issue a PASS, FAIL or INCONCLUSIVE verdict
- Test case postamble
  - Complete signalling to bring the UE into a stable state

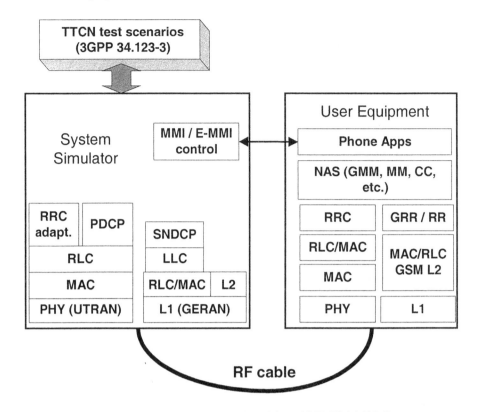

Fig. 3. The 3GPP Test Model (adapted from 3GPP TS 34.123-3)

The Test Model proposed by 3GPP would be the first choice for a model underlying an implementation of a virtual test environment, not only because it is clear structured with well defined interfaces, but also because it facilitates the re-use of software modules and test cases implemented by third parties.

Unfortunately, the 3GPP test model shown above is not sufficient for modeling a complete test environment needed for testing all software components running on a Mobile Station. As already stated, the UMTS stack is only one component in a multi-mode terminal which also includes a mature GSM/GPRS stack and other functionalities like MMI and STK which need also to be validated. Furthermore, all these components have been developed over the last years using proprietary development tools and running on different environments. For example, we developed a scripting language and wrote about 1300 test cases in that language for the GSM/GPRS protocols. These tests comprise the core components of our regression tests which are run on a daily basis.

Thus, the engineering challenge is now how to build a test environment involving the different protocols, the different testers with their test suites and possibly debugging and tracing tools.

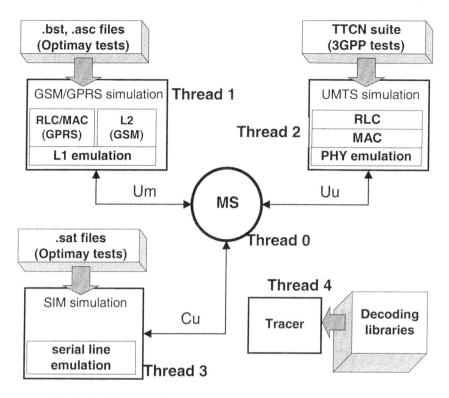

Fig. 4. Architecture of the GSM/GPRS/UMTS Virtual Test Environment

5 UMTS/GSM/GPRS Virtual Test Environment

Our implementation of the Virtual Test Environment (Fig. 4) consists of the proper Mobile Station under test (or User Equipment in 3GPP parlance) and three Testers: the GSM/GPRS System Simulator, the UMTS System Simulator, and the SIM Tester. We decided to apply, as far as possible, the 'golden rule' not to touch running software nor validated test cases unless really necessary. Thus, we encapsulated each Tester into a separated thread implemented as a DLL library. Each Tester thread interfaces with the User Equipment (UE) only via shared buffers protected by semaphores.

The SIM Tester sends and receives SAT command/responses over the simulated serial SIM interface which in turn uses the shared buffers for data transfer. The GERAN and UTRAN Testers use the shared buffers to transfer radio blocks and emulated PHY command/responses.

Additionally, a Tracer thread is responsible for decoding, tracing and 'pretty print' into nice trace logs. The Tracer software is identical to that running on a workstation when testing and tracing the hardware. The Tracer basically fetches trace messages sent by the UE or by the Testers, decodes them using appropriate libraries and outputs into a file. The decoding libraries contain functions for decoding GSM/GPRS messages (GSM TS 04.08), for decoding the UMTS ASN.1 messages (3GPP TS 25.331) and also for tracing proprietary messages used for debugging including interlayer signalling, memory usage on the hardware and DSP states.

There is still one essential component missing in this overview: the synchronization of the threads and, in particular, the coupling of the different timer systems. Each sub-system maintains and manages its own timer system. We need the capability to set breakpoints at any line of code (either at the UE side, at the Testers or even within test scripts) and freeze the whole system when the breakpoint is reached. Also, as explained in a previous section, since the UE is the 'tick master', it can accelerate during a virtual run the rate of its timer ticks sent to the Testers (in particular when the UE is in idle mode only decoding Sys Infos and monitoring neighbour cells). This mechanism greatly improves simulation time. For example, test case 8.3.1.3 Cell update / periodical cell update in CELL_FACH (3GPP TS 34.123-1 version 4.2.0) which at step 13 requires the SS to wait 720 minutes (12 hours!) can be executed in a couple of seconds in our Virtual Test Environment. Of course, this is not a particularly useful test case and hopefully will soon be removed from the test suite.

In order to achieve this tied timer synchronization, and following the ideas developed for GSM/GPRS, we set the UE to issue a timer tick when completing the current frame and then to wait for the tick acknowledgment from each Tester before it goes on to process the next frame. (The code needed for timer synchronization is minimal, but it needs to be thoroughly designed, otherwise it will never work).

6 Future Work

Fortunately for us engineers, there is still plenty of work to be accomplished.

Firstly, we need to add new interfaces to include the W-LAN (IEEE 802.11) and/or Bluetooth stacks in order to run and test a complete TCP/IP application connected to the Mobile Terminal via one of these interfaces. Further, we would like to run 'real' WAP test cases on the Virtual Test Environment which will require a server application running on top of the protocol stacks at the System Simulator side. We still need to investigate the coding effort involved in adapting and connecting TCP/IP and WAP protocols at the System Simulator side for this purpose.

Another issue is related to tools facilitating automatic or semi-automatic design of test cases out of critical scenarios found during field testing. As already mentioned, some software bugs are only found during field testing in tricky situations like Mobility Management or handover procedures at country borders where different providers use different configurations for their PLMNs (Public Land Mobile Networks). It is of high value to exactly reproduce these scenarios within the Virtual Test Environment. For this purpose we need tools supporting the engineers when inspecting the trace files produced during field testing in order to extract base station settings, the System Information broadcasted by the different base stations and the signalling which eventually lead to an erroneous behaviour at the Mobile Station.

Finally, for reducing the simulation time we will distribute the simulation load on different workstations. Several interesting topics on distributed computing need to be investigated like the estimation of memory and cpu potential on idle workstations and the segmentation and distribution of 'simulation patches' over the network.

7 Summary

In this article we presented a practical view of a Virtual Test Environment used in every day work by wireless-engineers developing UMTS Terminal software. Since the environment is software only (i.e. it does not need any UMTS specific hardware) and runs on any standard workstation, engineers can use their preferred developing tools, compilers and debuggers while focusing on the functionality they are currently developing. After selecting test scenarios and setting breakpoints within the Mobile Station software (or also on test scenarios), the system is started and within a couple of seconds it will stop (freezing all timers) at the requested code location for inspecting variables, system states, checking memory usage and so on.

The system is also used for regression testing prior to the delivery of the software to our customers. This is of utmost importance when software modules developed by different groups at different sites and time zones are integrated into a single image which loaded into the Mobile Station hardware will literally execute in your hand.

As one of our engineers remarked 'We will not survive without a proper regression test system verifying at any time all aspects of the phone functionality'. We, as engineers, need to make this point clear also to the company's management, so that the test environment itself is part of the long term company strategy.

References

1. Technical Specification 3GPP 23.010, General UMTS Architecture
2. Technical Specification 3GPP 34.123-1, User Equipment (UE) conformance specification, Part 1: Protocol conformance specification
3. Technical Specification 3GPP 34.123-2, User Equipment (UE) conformance specification, Part 2: Implementation Conformance Statement (ICS) proforma specification
4. Technical Specification 3GPP 34.123-3, User Equipment (UE) conformance specification, Part 3: Abstract Test Suites (ATS)
5. Technical Specification GSM 11.10-4, SIM Application Toolkit conformance specification
6. Technical Specification GSM 11.14, SIM Application Toolkit

Testing of 3G 1xEV-DV Stack – A Case Study

Ira Acharya and Hemendra Kumar Singh

Tata Consultancy Services, D-4, Sector-3, Noida-201301, India
{iraa,hemendra_singh}@delhi.tcs.co.in

Abstract. Due to immense competition in the market, mobile equipment vendors and service providers are faced with the challenge of delivering solutions early, way ahead of their competitors. Time-to-market pressures necessitate a carefully worked out test strategy for verifying and validating the correctness of mobile communication solutions. This paper outlines the various challenges that are faced in the protocol testing of mobile communication products. An evaluation of formal languages such as SDL and TTCN in the design, development and testing phases of such products is also included in the paper. The strategy employed in the testing of a 3G 1xEV-DV Base Station stack has been covered as a case study.

Keywords. Testing, conformance, simulation, test automation, test scripts, testing challenges, IUT, formalism, formal description techniques, 1xEV-DV, EV-DV, DV, protocol stack, 3G, reference implementation, SDL, TTCN, MSC, 3GPP2, queues, timers, task, behavior, LAC, MAC, RLP, Signaling, L2, L3, logging, video conferencing, packet data channel, PDCHCF, Message Integrity.

1 Introduction

With the increase in the sophistication and complexity associated with mobile computing, the challenges faced in the verification and validation of mobile communication components becomes manifold.

This paper describes briefly some of the challenges that are confronted in the testing of components of radio access networks in third generation networks. This paper discusses the challenges in the context of testing a 3G-reference stack for a 1xEV-DV Base Station. The stack development had been undertaken while the standards were still evolving, due to which the risk of having to undertake major rework was very high. As a direct consequence of this, the design, development and testing of the stack required a carefully worked out strategy to be able to cope with changing specifications.

2 Outlining the Challenge

The difficulties that were faced in the testing of the 3G stack solution, by virtue of it being an aggressively evolving technology are listed below:

- Non-availability of handsets or related equipment: One of the major problems encountered in testing of the built product was that handsets against which the de-

veloped product needed to be tested were not available. This was handled by use of the complementary protocol stack method, which is described in the subsequent sections of the paper.
- Conformance to specifications: Testing the implementation against its specification was a very challenging and critical task. The specifications are large and complex, which only increased the intricacies of the problem. TTCN was used for testing conformance of the stack.
- Platform independent protocol: The target segment for this kind of protocol development included mobile manufacturers who use their own custom hardware, therefore the best interest laid in providing a stack solution which would be portable across platforms. SDL was chosen as the design and development platform to handle this. The usage of an SDT provides a mechanism to generate code for most of the contemporary RTOS's prevalent in the industry today.
- Automated Testing Environment: To reduce the testing time, an automated testing environment was desired through the usage of tools. TTCN supporting tools and the SmarTEST tool developed in-house were used for automated testing.
- Evolving Specifications: The most challenging of all issues and unique when compared to the regular testing issues encountered was the non-availability of a frozen standard against which the stack components could be tested. There was the need to be able to cater to changing test conditions and changing Implementations Under Test (IUT). To deal with this issue SDL was used as the design and development platform, as it aids in the rapid incorporation of changes without impacting the rest of the system. SDL tools facilitate testing at the model level itself. They enable the testing of continuously changing systems.
- Time-to-market pressures: To handle the pressures associated with delivering an early solution owing to stiff competition in the market. This resulted in a reduced development lifecycle, and a reduced timeframe for the testing activity. This was again one of the factors, which influenced the decision of choosing SDL as the development platform.

3   The Role of Formalism

Formal Description Techniques (FDT) were used in the design, development and testing of the stack. FDT's are increasingly used in the industry while working with complex communication protocols [15]. In the tight schedules for completion of the work, they helped in guaranteeing syntactically and semantically un-ambiguous formal descriptions of the protocol as well as interoperable and compatible implementations of the protocol so that it can have value to the potential customer.

Specification and Description Language (SDL), [6] was used for the design of the system primarily because of the fact that the SDL system can be tested at an early stage in the design phase, as the built system is completely specified in SDL.

Another formal language introduced by the OSI standardization committees 'Tree and Tabular Combined Notation (TTCN)' was used for describing conformance test cases [7, 8]. Abstract Test Suites (ATS) were created using TTCN. The ATS were a description of tests to be executed for the system. The tests were described using a black-box model, i.e. only control and observe using the available external interfaces. The advantage of using TTCN for testing included the possibility of specifying constraints of complex data types, to react to alternative results of use cases and to define the expected results [9]. Interworking between SDL and TTCN is shown in Figure 1.

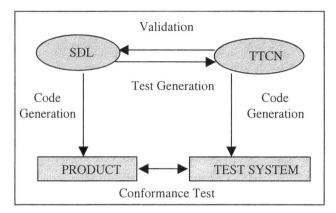

Fig. 1. Interworking between SDL and TTCN

TTCN is extensively used for conformance testing of the system. Conformance testing verifies whether an implementation performs according to the stated standard/specification/environment [10]. This involves the development of Test Suites, a collection of test cases that specify objectives, operating conditions, inputs and expected outputs, in order to evaluate an implementation.

Among the other formal description techniques available include Estelle and Lotos developed by ISO. Estelle is based on an extended finite state machine model and Lotos is based on a calculus of communicating systems [12,13].

4 Testing a 3G 1xEV-DV Stack

1xEV-DV is a CDMA based third generation standard in the wireless value chain and has been standardized within 3rd Generation Partnership Project 2 (3GPP2). It provides for high peak data rates (3.072 Mbps on the forward link and 451.2 Kbps on the reverse link) and promises the enabling of real-time applications such as videoconferencing, voice over IP, 3G-multimedia and telemedicine.

3GPP2 had taken up the 1xEV-DV Standards Development activities. Working Group (WG) 5 of the Technical Specification Group – CDMA 2000 (TSG-C) was entrusted with the responsibility of identifying a framework proposal for the standard. Within WG5, two groups (out of a total of eight) L3NQS and 1XTREME submitted framework proposals for the standard. L3NQS constituted Lucent, LG, LSI, Qualcomm, Nortel and Samsung whereas 1XTREME constituted each of Nokia, Motorola, Philips Semiconductors and Texas Instruments. A harmonized version of the proposals was sought in the light of which Nokia submitted a proposal on a way to move forward during the October 2001 TSG-C meetings. This resulted in a consensus on a harmonized framework for the 1xEV-DV Standard without objection. Thereafter the

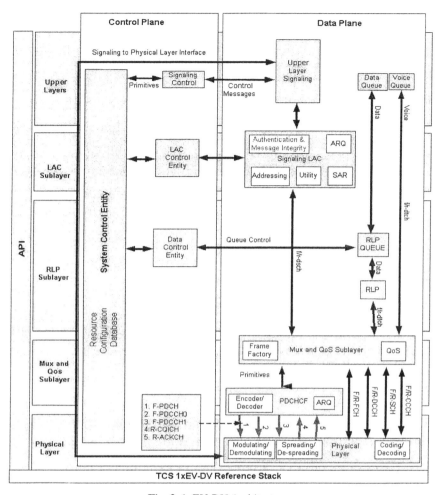

Fig. 2. 1xEV-DV Architecture

Working Groups 2 and 3 in 3GPP2 took up the text development activity and a released standard was available by the end of May 2002 as release 'C' of IS-2000 [1-5].

The 1xEV-DV Base Station stack consists of Upper Layer Signaling, Link Access Control (LAC), Medium Access Control (MAC), Radio Link Protocol (RLP) and the Physical Layer. An additional entity called the Forward Packet Data Channel Control Function (F-PDCHCF) also exists in a 1xEV-DV Base Station stack. The architecture of the 1xEV-DV Base Station stack is depicted in Figure 2.

In order to facilitate testing of the stack, every artifact of the protocol stack was treated as an entity. The artifact may be a layer or sublayer of a stack or it may be the entire Mobile or Base Station. The definition of every artifact as an entity results in it possessing the following attributes:

- Channels: These are the communication paths between processes contained in different entities. It is through channels that entities are notified of events. The be-

havior of an entity is defined solely in terms of its dynamic reactions to events on its input channels, and the production of events on its output channels.
- State: The state of the entity captures its current status at a particular moment in time. An entity undergoes state change on the occurrence of events which are communicated to it through the channels
- Parameters: These are the variables whose values at any point of time form a part of the abstract state of the entity
- Processes: These are the threads of computation that act on the events arriving on the input channels of the entity, or that act upon timers. As a result of the computation performed within processes, events may be generated on the output channels.
- Components: These are again entities, which form sub-entities of the entities behavior.

The entity framework is shown in Figure 3.

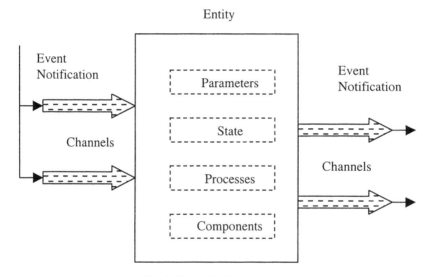

Fig. 3. The Entity Framework

With this definition of an entity SDL was chosen for the design, development and testing of the stack. SDL provides the concept of blocks, processes and channels, which merge very well, with the entity concept. SDL also has a high degree of testability owing to its formalism for parallelism, interfaces, communication and time. The powerful simulation facilities of the tool enabled early testing of the developed system. This allowed an early assessment of the performance of the system as well as that of the required resources. Decision in favor of SDL was also taken due to the following additional reasons:

- Continuously evolving specifications necessitated an easy way of modifying the design model
- It has architecture that reflects the standard
- SDL tools e.g. Telelogic Tau enable code generation for nearly all the commonly used platforms such as VxWorks, Win32, Nucleus, OSE Delta and QNX. This provides a way of dealing with the challenge of delivering portable solutions

- Studies suggested that the time for development of a call automation solution were reduced by 50% [16]

In the SDL system, Architecture is graphically represented with the help of Blocks while Processes represent Behavior. Communications between various processes are handled with the help of Signals and Channels. SDL has its own set of Abstract Data Types to handle data and functions.

The entities such as Signaling, LAC, MAC and Physical Layer of the 1xEV-DV stack were represented by blocks in the SDL system. The various sub-layers such as Authentication, Automatic Repeat Request (ARQ), Addressing, Utility, Message Integrity and Segmentation and Reassembly (SAR) within LAC were represented by processes. Each process undergoes various state transitions during its life cycle depending on the primitive being transmitted from other processes. Primitives between processes were handled in the form of signals sent over the channels in the system. State machines within processes handled the transition between different states. The system level diagram for the 1xEV-DV reference implementation is shown in Figure 4.

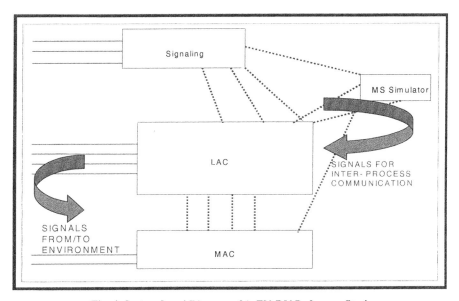

Fig. 4. System Level Diagram of 1xEV-DV Reference Stack

A Reference Implementation (RI) of the 1xEV-DV Base Station stack was developed which included Layer 3 and Layer 2 of the stack [11]. In the absence of actual hardware, Layer 1 was simulated using a simulator tool. The reference implementation was tested independently. The various stages in the workflow of testing of the reference stack are described in the following sections. Also discussed are some of the distinctive techniques adopted to test the system.

Task Testing: At its heart a mobile communication system is a collection of tasks/processes working together to accomplish system function. Each task is often a deterministic program able to execute separately from, and concurrently with, other tasks. In the SDL system each process, analogous to a task, was tested independently using a number of techniques such as simulation and logging. In the absence of implementations for the adjacent layers, the simulation facilities were exploited. Input signals to the process were provided through the connecting channels.

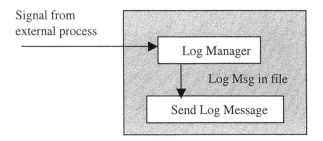

Fig. 5. Logging Framework

Logging was extensively used for testing at various stages. A Log Manager entity in the SDL system was responsible for logging all data received in the form of signals to a file. The output in the file was then examined to verify the functional correctness of that layer/sublayer. The logging framework is depicted in Figure 5.

The Log Manager collected all the parameters to be logged in the form of a signal from the external entity. The Log Manager then directed the message with all the parameters to the 'sendLogMsg' function, which would put the information in the appropriate log file. For instance, if the message to be logged were the error message then it would go in the error log file and if the message contained debug information it would be logged to the debug log file. The logging in SDL is depicted by in Figure 6.

Behavior Testing: Test scenarios covering the different kinds of input signals to a process were designed in order to test the behavior of the various entities. Environment functions for interaction between the outside world and the SDL system were handled in the SDL system. Test cases corresponding to input signals from the environment were executed to verify the correctness of the functionality of entities that received inputs from environment.

Finite State machines (FSM) represented the behavior of each process. The FSMs were tested using Message Sequence Charts (MSC) in SDL. MSC was also used to show interactions between components. MSC's provided a clear description of system communication in the form of message flows. The behavior of the process was expressed in terms of a number of states and inputs, and the outputs and state transitions produced by the arrival of a given input in a given state.

Inter-task Testing: A technique used to carry out inter-task testing, was to store the output from the intermediate processing stages i.e. the outputs from the intermediate entities of the stack were stored into a file and made available to the next entity in sequence. This enabled use of the same parent test data for all of the entities in the stack. This mechanism of testing is depicted in Figure 7.

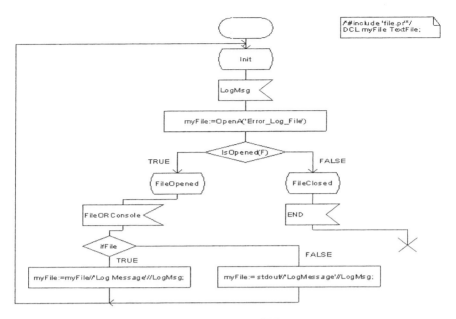

Fig. 6. Logging in SDL

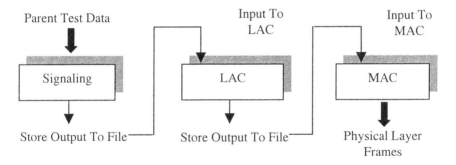

Fig. 7. Inter-task Testing

Tasks that communicate via FIFO queues were tested to uncover errors in the associated processes. Mobile communication system designed with multitasking approach use various mechanism such as semaphores, mailboxes, message systems and FIFO queues for inter task communication. In the stack, queues were used at the boundaries of tasks such as LAC and MAC. The queues were tested extensively with different conditions and inputs to ensure their correctness.

Once errors in individual tasks and in system behavior were isolated, time-related testing was performed. The expiry of timers such as those indicating the sending of the next Physical Layer frame (every 20 ms) was simulated by the manual sending of signals from the environment. These were used to trigger the sending of a Physical Layer frame from the MAC layer.

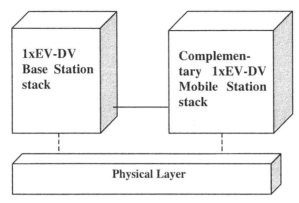

Fig. 8. Complementary Stack for 1xEV-DV

System Testing: SDL provides the feature of simulation using MSCs through which the flow of data can be easily traced. The simulation feature proved to be an invaluable tool, allowing easy verification of message interactions via MSCs. COM files (File containing all the input signals) were generated for the 1xEV-DV system and various scenarios were tested. The excitation in the SDL simulator was done by explicitly feeding test values into system.

In the absence of a Mobile Station implementation, a complementary protocol stack was developed. The complementary protocol stack bypasses the Physical Layer completely and eliminates hardware dependencies while testing the protocol stack and the application layers. It could test the adequacy of system timer values while sending and receiving messages. It could analyze the messages and point out non-conformance to protocol standards.

The complementary stack and 1xEV-DV stack runs on same system. The SW on the system has an alternate module, which hides the routing information from the upper layers and route the messages/signals to the complementary stack. During a normal product run, these messages are to be routed to the Physical layer through the interface provided by the hardware. The complementary stack is shown in Figure 8.

Conformance Testing: In order to verify the functional correctness of each entity in the stack against its protocol specification generic protocol conformance test architecture was designed. TTCN was used extensively to carry out the conformance testing activity.

The Telelogic Tau TTCN tool was employed to create the Test System. Each entity to be tested was configured as the IUT. The test architecture is depicted in Figure 9.

The LT and the UT are the Lower and Upper Testers for the implementation under test (IUT). The Points of Control of Observation (PCO's) are the points in the abstract interface where the IUT can be stimulated and the responses can be inspected. The communication between the LT/UT and the IUT is through Abstract Service Primitives (ASP's). An ASP may or may not contain a Protocol Data Unit (PDU).

The SDL system was simulated using TTCN-SDL co-simulator. The simulation allowed execution of a TTCN test suite in the host environment. For the conformance testing at the system level, the IUT was the 1xEV-DV system. The TTCN-SDL co-simulator was connected with the SDL simulator using the '*start-itex*' command in the SDL simulator window followed by a '*go-forever*' command to execute the SDL system. Then the test cases were executed. The logs were also generated after the tests were carried out.

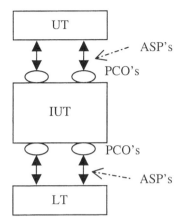

Fig. 9. General Protocol Conformance Test Architecture

Protocol standards bodies such as 3GPP and 3GPP2 have defined standardized conformance test suites, which usually consist of a large number of test cases. The successful execution of these test cases provides a reasonable assurance that the tested implementation follows all the rules defined by the protocol. The TTCN scripts were developed using Signaling Conformance Tests for CDMA 2000 Spread Spectrum Systems [17], since 3GPP2 conformance test cases for 1xEV-DV were not released. Specific tests were designed for testing the F-PDCHCF and the Message Integrity sublayer, which are new in the 1xEV-DV standard. The message flow between the Mobile and Base Station was verified by using the logging framework and MSC's in SDL. A sample MSC for a Mobile Originated 1xEV-DV call is shown in Figure 10 and 11.

In the absence of a Physical Layer implementation, the 1xEV-DV stack was integrated with a tool providing Layer 1 simulation. The tool provided a solution for designing, modeling and simulating analog and digital signals. All the features of physical layer such as modulation, demodulation, encoding, decoding etc. were implemented in the form of blocks. The PDU from the MAC sub layer was presented as an input to the simulator and the simulated data was visualized with the help of various charts.

5   Usage of Testing Tools

In order to provide for test automation, the testing tool SmarTEST was used to automate the functional and regression testing of the stack.

The framework in which SmarTEST can be used for test automation is depicted in Figure 12. Scripts executed the logic of a test case. The execution of a test case

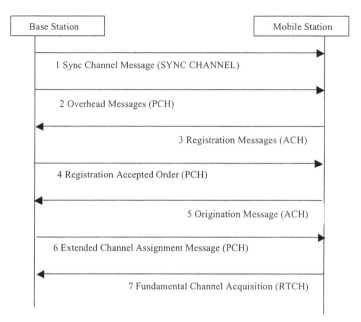

Fig. 10. MSC for a Mobile Originated Call–1

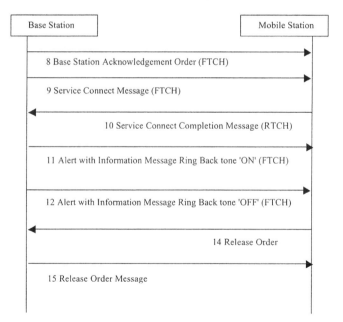

Fig. 11. MSC for a Mobile Originated Call–2

resulted in the generation of Mobile Originated (MO) or Mobile Terminated (MT) calls as desired. Each test case contained the logic required to initiate MO or MT calls with different scenarios. Based on the results of call execution, the script decided the

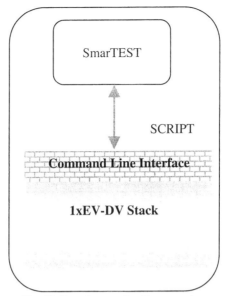

Fig. 12. Test Automation Framework

success or failure of the test. The necessary parameters required for executing the calls were set-up using the SmarTEST GUI, and routed to the stack with the help of scripts. The test cases are managed and automatically executed through the SmarTEST engine. APIs provided on top of the 1xEV-DV stack, enabled the test cases to be driven through its Command Line Interface. Scripts have the ability to execute the logic of the test case and decide the status of success or failure of the test case. The system test cases were managed and automatically executed through the SmarTEST engine. APIs provided on top of the 1xEV-DV stack, enabled the system test cases to be driven through its CLI.

6 Conclusion

Testing of mobile communication protocols is a very complex exercise. With time-to-market becoming a very important business success factor for vendors of mobile communication solutions, the testing activity becomes all the more critical. This challenge is magnified in the absence of hardware. The testing strategy needs to be planned and worked out carefully while the stack development is under progress. Usage of techniques such as complementary protocol stacks and logging can provide a low-cost and flexible solution in the testing of mobile communication protocols. SDL tools enable faster, portable, testable and maintainable protocol stack development.

References

1. C.S0001 Introduction to cdma2000 Standards for Spread Spectrum Systems, Release C 1.0, 3GPP2, May 2002

2. C.S0002 Physical Layer Standard for cdma2000 Spread Spectrum Systems, Release C 1.0, 3GPP2, May 2002
3. C.S0003 Medium Access Control Standard for cdma2000 Spread Spectrum Systems, Release C 1.0, 3GPP2, May 2002
4. C.S0004 Signaling Link Access Control Standard for cdma2000 Spread Spectrum Systems, Release C 1.0, 3GPP2, May 2002
5. C.S0005 Upper Layer Signaling Standard for cdma2000 Spread Spectrum Systems, Release C 1.0, 3GPP2, May 2002
6. Specification and Description Language – 2000, Z.100, ITU-T, November 2000
7. ITU-T Recommendation X.292, OSI conformance testing methodology and framework for protocol Recommendations for ITU-T applications – The Tree and Tabular Combined Notation (TTCN), 1998
8. ISO/IEC 9646-3, Information technology – Open systems interconnection – Conformance testing methodology and framework – Part 3: The Tree and Tabular Combined Notation (TTCN), 1998
9. Performance Analysis of Communication Systems formally specified in SDL - Martin Steppler, ACM Publication, 1998
10. Improving conformance and interoperability testing- Jame D. Kindrick, John A. Sauter, Robert S. Matthews, StandardView Vol. 4, No. 1, March/1996
11. On Testing Hierarchies for Protocols, Deepinder P. Sidhu, Senior Member, IEEE, Howard Motteler, A4ember, IEEE, and Raghu Vallurupalli, 590, IEEE/ACM TRANSACTIONS ON NETWORKING, VOL. 1, NO. 5, OCTOBER 1993
12. Formal Specification Based Conformance Testing, Behqet Sarikaya, Concordia University, Gregor u. Bochmonn, Michel Maksud, Jean-Marc Serre, Universitd de Montreal, ACM Publication, 1986
13. A Test Design Methodology for Protocol Testing, B. Sarikaya. G. Bochmann. and E. Cerny, IEEE Transactions on Software Engineeritlg, Vol. 13, No. 5, March 1987.
14. Roger S. Pressman: Software Engineering – A practitioner's approach
15. Protocol testing: Review of methods and relevance for software testing, Gregor v. Bochmann and Alexandre Petrenko,109, ACM 1994
16. Case Study by Vocalis Group plc. & published by Telelogic
17. Signaling Conformance Tests for cdma2000 Spread Spectrum Systems,3GPP2 TSG-C, OCT 2001.

Testing SIP Call Flows Using XML Protocol Templates

M. Ranganathan, Olivier Deruelle, and Doug Montgomery

Advanced Networking Technologies Division
National Institute of Standards and Technology
100 Bureau Drive, Gaithersburg, MD 20899, USA
{mranga,deruelle,dougm}@antd.nist.gov
http://w3.antd.nist.gov

Abstract. A Session Initiation Protocol (SIP) Call Flow is a causal sequence of messages that is exchanged between interacting SIP entities. We present a novel test system for SIP based on the notion of XML *Protocol Templates*, of SIP call flows. These templates can be pattern matched against incoming messages and augmented with general purpose code to implement specific protocol responses. This architecture allows test systems to be easily scripted, modified and composed. We describe these techniques in the construction of a SIP web-based interoperability tester (SIP-WIT) and comment on their potential more general use for scripting SIP services.

1 Introduction

The Session Initiation Protocol (SIP) [8] is a signaling protocol for setting up and terminating sessions for internet telephony, presence, conferencing and instant messaging. The SIP specification has been through a series of changes since the original RFC [9] was issued. Building comprehensive test tools and protocol stacks that both maintain backward compatibility and incorporate the latest specification is a challenging task. In this paper, we present a test system based on an XML-based pattern of a SIP Call Flow which accomplishes the goal of multi-level testing of SIP-enabled applications.

There are two types of components in a SIP-enabled network. Interior components act as signaling relay points. Examples of interior components include back-to-back user agents (B2BUA) and proxy servers. End components are signaling termination points and this is where the end-user application logic usually resides. Such applications include IP phone user agents (UA), chat clients, instant messaging and presence clients and other SIP-enabled user software. Such applications are usually built on a SIP protocol stack. Figure 1 shows a conceptual layering and heirarchical structure of a SIP stack and its relationship to a SIP application.

The lowest layer of the protocol is the *Message Layer* which reads messages off the network and parses them to present to the higher layers. Certain SIP applications, such as stateless Proxy Servers are built directly on top of

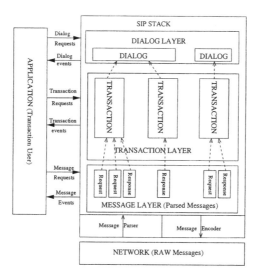

Fig. 1. A SIP application typically consists of the a Transaction User (TU) part where the application logic resides and *Transaction* and *Dialog Layers* which are implemented by a SIP Stack. The SIP Stack interacts with the application using a call/event interface.

the Message Layer. However, most SIP applications rely on the presence of a *Transaction Layer* which conceptually resides on top of the Message Layer. The SIP specification defines a *SIP Transaction* as a SIP Request, the provisional Responses generated by the Request followed by a final Response. The Message Layer presents the Transaction Layer with a stream of parsed messages. These messages can be SIP Requests or SIP Responses. The Transaction Layer is responsible for correlating outgoing Requests with incoming Responses and performing retransmissions of Requests as needed. A SIP Transaction is created as a result of an incoming SIP Request (*Server Transaction*) or as a result of an outgoing SIP Request (*Client Transaction*). A Client Transaction is completed when a final Response to a Request is received and a Server Transaction is completed when the final Response to the Server Transaction is sent out. A given SIP message is part of exactly one Transaction. SIP can run over both reliable and unreliable transports. The Transaction Layer is responsible for re-transmitting SIP Messages as needed.

Some applications such as stateful proxy servers and user agents rely on the establishment of *Dialog*s. A SIP Dialog is a peer to peer association between communicating SIP applications and is established by particular Dialog initiating Transactions. For example, a successful INVITE Transaction results in the creation of a SIP Dialog. Conceptually, the Dialog layer resides on top of the Transaction Layer and a given Transaction is part of exactly one Dialog.

Finally at the highest layer we have the notion of a *SIP Call*, which is identified by a globally unique *Call-ID*. A SIP call can consist of multiple Dialogs. All the Dialogs of the Call have the same Call-ID.

In a correctly functioning SIP application and stack, given a SIP message, the stack can identify to which Call, Transaction and Dialog it belongs to without needing to maintain connection state. Thus although applications and stacks may maintain persistent data structures associated with these abstractions, SIP is often called a stateless protocol.

SIP extensions are under development in a variety of different application domains (for example, instant messaging and networked appliance control), with different Request methods, headers and associated semantics. Still, the concepts of Message, Transaction, Dialog and Call are common to all the domains in which SIP is applied. In all these domains, a SIP application can be envisioned as a state machine that is transitioned on the arrival of messages, creation and completion of Transactions and creation and destruction of Dialogs.

2 Testing the SIP Protocol

Testing a SIP application can be decomposed in roughly the same fashion as the protocol itself. That is, an application may be tested at the Message Layer, Transaction Layer or Dialog Layer. We examine the issues of testing at the various layers in this section.

2.1 Testing at the Message Layer

The most obvious protocol errors are caused by incorrectly formatted SIP messages that do not conform to the specified grammar for URLs and protocol headers or by improperly functioning message parsers. These errors are easily discovered by using a parser that conforms to the specification. While constructing an ad-hoc parser for SIP headers is not difficult, there are pitfalls. The SIP grammar incorporates rules from from various RFCs that define specifications for mail, internet host names, URLs and HTTP. The resultant composite grammar is quite large, context sensitive and easily leads to parser implementation errors. For example, spaces are generally not relevant except in certain cases (for example the Request-Line, Status-Line and URI grammar definitions) where the RFC specifies strictly how many spaces are expected. Another common source of errors which is also an artifact of grammar composition arises from the fact that different sets of characters are legal in different portions of a message. The evolution of the protocol specification through the various revisions has also lead to some issues. For example, SIP URL addresses can appear in SIP messages in various headers. SIP URL addresses appearing in such headers are generally enclosed between pair of <> delimiters except the early RFC did not require this. There were also some early drafts that had context-sensitive disambiguating rules about whether to associate a parameter with a SIP URL or header.

A tool that tests for correct header parsing and formatting must itself be correct in parsing and formatting headers and conform to the SIP RFCs. A good way to achieve this is to use a parser generator. We elaborate on the techniques we have adopted in Section 5.

2.2 Testing at the Transaction Layer

Testing at the Transaction layer can be accomplished by generating messages that will establish and terminate Transactions. The latest SIP RFC (RFC 3261) specifies a robust way of Transaction identification but the earlier RFC (RFC 2543) had some ambiguities. Since interoperability with legacy equipment is often important, a test system must be able to generate both legacy and non-legacy scenarios.

Transaction timeout can be tested by delaying the SIP Response for a Transaction. Transaction matching can be tested by generating a spurious Response, or Responses, with fields left out or mis-specified at various points in the protocol operation. A test system should also be able to generate stray messages that do not correspond to a Transaction and are expected to be rejected by the receiving stack.

Such tests are, for the most part straightforward but there are a few tricky cases. For example, most Transactions are just Request, Response sequences and do not make sense to abort while in progress. However, long running Transactions such as an INVITE Transaction may be aborted by sending the Server side of the Transaction a CANCEL message while the Transaction is in progress. However, the CANCEL only be processed at a certain point in the protocol operation and further operations that reference the canceled Transaction should result in a TRANSACTION NOT FOUND error. A test system must thus be able to generate such messages at specific points in the protocol operation to generate such erroneous conditions.

2.3 Testing at the Dialog Layer

Testing at the Dialog layer can be accomplished by setting up and terminating Dialogs. Crucial to the identification of the Dialog is the *tag* parameters of the SIP Message. The *From* header *tag* parameter identifies one end (the *Client* side) of the peer to peer association and the *To* header *tag* parameter identifies the other end of the peer association (the *Server* side). *Tags* are assigned in a pseudo-random fashion within the context of a Call. *Tags* were not mandatory in the earlier SIP specification. For legacy support, however, applications may support mechanism defined in the older RFC which specified another algorithm for Dialog identification. To test a stack at the Dialog layer, a test system should be able to generate Requests and Responses for established and spurious Dialogs both for legacy and current systems. Our experience with SIP implementations indicates that some common errors include incorrect assignment of *tag* parameters in the headers that identify Dialogs. Testing at the Dialog layer should also include testing for Dialog termination, which is accomplished by sending a BYE message, which can be issued by either side of the Dialog.

2.4 Call Flow Testing

Testing at the level of a SIP Call requires testing the causal sequence of SIP messages, transactions and Dialogs required to establish and release calls. Such sequences of exchanges are described as SIP *Call Flows*. Clearly call flow testing includes all the other layers outlined above, since a Call cannot be set up and terminated without correctly parsing and formatting messages or correctly establishing and terminating Transactions or Dialogs. Thus a test tool that can test at the level of a Call Flow needs to have facilities to test at the other levels as well.

Our approach to test at this level centers around defining XML tags and attributes to define the causal, event-driven behavior of a SIP end point participating in a Call Flow. We call this XML definition a *Protocol Template*. We then construct an event-driven state machine that interprets the Protcol Template to implement the Call Flow. A customizable User Agent which we call a *Responder* takes the Protocol Template as input and and generates the state machine and necessary synchronization actions for running the Call Flow. In the sections that follow, we further detail the design and use of these XML protocol templates as a basis for scripting a web-based SIP Call Flow tester.

3   SIP Testing with Protocol Templates

Our design goal was define a set of XML tags that can be used to represent a Call Flow as a finite state machine. The reason for choosing this approach was twofold. First, a popular way to debug SIP components is to participate in interoperability test events where the predominant mode of testing involves creation of simple signaling scenarios between components under development. We sought to duplicate this approach to testing in our automated test environment. Second we observe that there is currently no standardized way of expressing Call Flows. SIP-related Internet Drafts and RFCs specify Call Flows using sequence diagrams which are informal and subject to errors in interpretation. By choosing an XML representation for Call Flows and by widespread adoption of the conventions we propose, we hope to reduce interpretation errors in the future.

Figure 2 shows the overall conceptual view of our test system. It consists of a scripting layer (*Event Engine*) built on top of our NIST-SIP stack [12]. The Event Engine constructs one or more state machines after reading an XML file (*Protocol Template*) representing one or more call flows. The Protocol Template may be customized by adding code (*Service Script*) whose functions are invoked at specific points in the state machine operation. The Service Script can be inserted directly into the Protocol Template or specified externally as a JAVA [1] [2] class. We elaborate further on this scheme in section 4.

---

[1] The identification of specific software / hardware products or trademarked names in this paper is done soley for the purpose of adequately describing our work. Such identification is not intended to imply recommendation or endorsement by the National Institute of Standards and Technology, nor imply that the products or names idetentified are necessarily the best avialable for the purpose.

[2] JAVA and JAIN are trademarks of SUN Micro Systems.

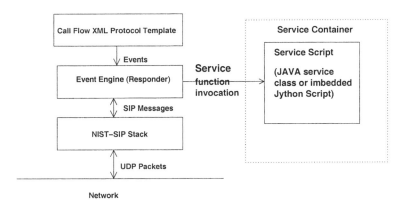

Fig. 2. Scripting Architecture: The Event Engine takes a Protocol Template as an input and constructs a state machine from it. The Protocol Template can invoke Service Script functions at specific transition points in its execution.

4 Protocol Template Programming Model

In this section we elaborate further on the XML representation for defining Protcol Templates. The hierarchy of XML tags that define a Protocol Template shown in Figure 3. Our programming model closely mirrors the layering of the SIP protocol. A **CALL_FLOW** corresponds to a SIP Call Flow and consists of a set of **DIALOG**s. A **DIALOG** is a specification for a finite state machine (FSM) which is represented by a set of XML tags and attributes. This state machine is instantiated when the corresponding SIP Dialog is created and defines the signaling behavior of a SIP-enabled end point. All messages within a Dialog have the same call identifier (Call-Id). Once the Dialog is established, the messages must also have the same *From* and *To* tags. A *Dialog* and its associated instantiation of the state machine defined by the **DIALOG** tag is created when a SIP Request with a previously unseen *Call-Id* or *tag* parameter arrives or is sent out by the system. Subsequently, all matching of messages occurs in the context of the created **DIALOG** so there is a one to one correspondence between an established SIP Dialog and the state machine defined by the matching **DIALOG** that is created as a consequence of the SIP Dialog being established. Because there is a single SIP Dialog for a given SIP message, an outgoing or incoming message can be uniquely associated with at most one instance of State Machine.

Each node of FSM is represented by an **TRANSITION** tag. Each **TRANSITION** tag consists of an optional nested **TRIGGER_MESSAGE** tag and an optional set of **GENERATE** tags. A **TRANSITION** tag can be nested inside of a **CLIENT_TRANSACTION** or **SERVER_TRANSACTION** tag (henceforth referred to generically as a **TRANSACTION**). The **TRANSITION** tags are nodes in the protocol state machine that can be triggered by message arrivals that match the nested **TRIGGER_MESSAGE** and are activated by a boolean combination of events specified in the *enablingEvent* at-

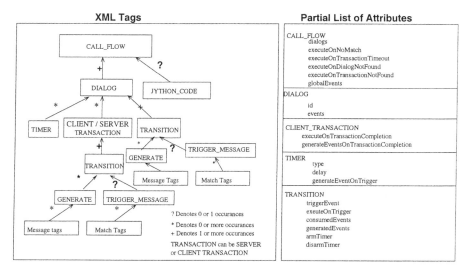

Fig. 3. Hierarchical Arrangement of Tags corresponds to the hierarchy of the SIP protocol. A partial listing of attributes for the XML tags is shown on the right (see text for explanation). The complete DTD is available from [12].

tribute. An *event* is a globally scoped or locally scoped counter (explained below) that is initialized to 0. If an **TRANSITION** node is nested in a **TRANSACTION** tag, then the messages that trigger it must also be part of the enclosing **TRANSACTION**. This provides a way of catching protocol errors related to unmatched Transactions. Note that this is not a static textual match specification because a transaction matching in SIP depends on dynamically generated header parameters. Once a Transaction is created it is associated with an instance of **TRANSACTION** node and **TRANSITION** tags within this node are used to generate state machine transitions in the associated **DIALOG** state machine. When an **TRANSITION** node is activated, it can generate events, activate timers, disable timers or call a Service Script function and optionally generate outgoing messages from the incoming message. The function to be invoked on **TRANSITION** node activation or Transaction completion is specified by the *executeOnTrigger* attribute and *executionOnTransactionCompletion* attributes respectively. The new message to be generated is represented by an optional **GENERATE** tag. The **GENERATE** tag can specify a list of editing rules to be used when generating an outgoing message from the incoming message (assuming that there is one). The **TRIGGER_MESSAGE** is a template that matches incoming messages and can *fire* the **TRANSITION** node if it is activated. If the node is not yet activated because the *enablingEvent* has not yet been satisfied, the fact that the trigger has been seen is noted. If the enabling condition occurs in the future, the node is activated at that time so the order of enabling conditon and trigger message arrival is not relevant. The **TRANSITION** node can be activated by any boolean event expression specified by its *enablingEvent* attribute. If no *enablingEvent* is specified, then the

TRANSITION node is assumed to be always *enabled* and may be triggered by an optionally specified **TRIGGER_MESSAGE**. Initial nodes are specified by the absence of a **TRIGGER_MESSAGE and** the absence of an *enablingEvent* attribute. Initial nodes may be used to start the interaction as soon as the state machine is initialized.

The Figure 3 also shows a partial list of additional attributes. The $generate_*$ attributes specifies a list of events to be generated. The $execute_*$ attributes specifies functions to be executed when specified events occur. The $consume_*$ attributes specify a list of events to be consumed when the specified event occurs. The *enablingEvent* attribute is a boolean expression on the *event* state variables. Events can be scoped either locally (visible only to a **DIALOG** state machine) or globally (visible across the entire **CALL_FLOW**). Global events can enable **TRANSITION** nodes in other **DIALOG**s. Local events are scoped within the **DIALOG** where they occur. The functions invoked from the attributes run the context of a separate class (either a JAVA class or an instance of a *Jython* [6] interpreter) which we call the *Service Class*. The same instance of this class is used for each service call within a Dialog.

Figure 4 shows an **TRANSITION** node *expectOK* from a UAC Call flow. This node is is enabled by the event *INVITEsent*. When the node is activated by this event, an incoming **SIP_RESPONSE** with *statusCode* of 200 can fire the **TRANSITION** node and cause the *OKReceived_ACKsent()* function to be invoked. The firing of the node arms the *byeTimer* timer and generates the *OKreceived_ACKsent* event and generates an outgoing *ACK* message. This message has its *From* header derived form an *agentId* "callee". An **AGENT** is just a way of specifying a list of attributes that are specific to the user that is being called. It allows us to customize a small portion of the code without altering the entire XML File. **AGENT**s may be bound to Registry entries (see section 5).

The service script can directly communicate with the event engine by generating events, starting and stopping timers etc, the same way as can be done with the attributes of the **TRANSITION** tags, allowing for finer grained control at the expense of clarity.

5 Implementation

In order to test for message formatting, our implementation [12] uses the ANTLR [16] parser generator to generate a parser for the SIP grammar. Conversion of the published EBNF to a format that is accepted by popular tools such as YACC and LEX is not straightforward. Advanced tools such as ANTLR make the task easier by allowing for closure on terminals as well as non-terminals, multiple lexical analyzers and the ability to switch between lexical analyzers during parsing and grammar inheritance; however, one must still resort to manual use of as *Syntactic* and *Semantic Predicates* to work through some ambiguities present in the grammar.

Figure 5 depicts the logic of processing incoming messages using our protocol templates. A *Call-Id* header identifies the call for the incoming message. This,

```
<CLIENT_TRANSACTION
     onTransactionCompletion="onCompletion"
  />
  <TRANSITION
     nodeId            = "expectOK"
     enablingEvent     = "INVITEsent"
     executeOnTrigger  = "OKreceived_ACKsent"
     generatedEvent    = "OKreceived_ACKsent"
     armTimer= "byeTimer"
  >
     <TRIGGER_MESSAGE>
        <SIP_RESPONSE>
           <STATUS_LINE
              statusCode = "200"
           />
        </SIP_RESPONSE>
     </TRIGGER_MESSAGE>
     <GENERATE
        retransmit="false"
        removeContent="true"
     >
        <SIP_REQUEST>
           <REQUEST_LINE
              method = "ACK"
              agentId = "callee"
           />
        </SIP_REQUEST>
     </GENERATE>
  </TRANSITION>
</CLIENT_TRANSACTION>
```

```
<JYTHON_CODE>

  def OKreceived_ACKsent():
     print "OK received and Snet an ACK"

  def onCompletion():
     print "Client transaction is complete"

</JYTHON_CODE>
```

```
<AGENTS>
<AGENT
   agentId="caller"
   requestURI="JitterVik@myhome.org"
/>
<AGENT
   agentId = "callee"
   registryEntry="0"
/>
</AGENTS>
```

Fig. 4. A Call Flow is represented as a set of TRANSACTIONs Events TIMERs and triggered TRANSITION nodes. TRANSITIONs are triggered by events and messages can generate outgoing messages and events to trigger other Expect Nodes. An AGENT is a short hand way of referring to a user identity. An AGENT entry can refer to a registry entry in the Proxy server. This is bound at run time to an actual value (see Section 5).

along with the *From* and *To tag* parameters of the incoming request are used to retrieve an instantiated **DIALOG** template for the call. If no template is found for the incoming call, then the one is created by looking for a **DIALOG** that can be instantiated. This is done by searching for a ready node. A node is *ready* if there are no outstanding events or messages for the node which prevent it from being enabled. A start node is one for which there is no *enablingEvent* tag and no **TRIGGER_MESSAGE** nested tag. Our stack and parser are written entirely in JAVA and we use introspection and inheritance to implement pattern matching facilities.

```
Let TRANSITION set be the set of TRANSITION nodes
Apply incoming messsage to all nodes in TRANSITION
Let READY denote a set of TRANSITION nodes that are enabled
Let NON-READY denote a set of TRANSITION nodes that are not yet enabled

While the READY set is not empty
    mark each TRANSITION node in the READY set unexamined
    for each unexamined TRANSITION node in the READY set:
        mark it examined
        if the incoming message matches TRIGGER_MESSAGE nested tag:
            Apply generated events to all NON-READY TRANSITION nodes set
            and move the enabled TRANSITION nodes to the READY set
        If this message resulted in a transaction completion
            Apply completion events to all NON-READY EXPECT nodes set
            and move the enabled TRANSITION nodes to the READY set
        Start any timers that are specifed by the armTimer attribute
        Stop any timers that are specifed by the disarmTimer attribute
        Generate outgoing message list from nested GENERATE tag if it exists
        Invoke the executeOnTrigger method if specifed
        If this is a transaction completion:
            invoke the executeOnTransactionCompletion method if specifed in
            the enclosing TRANSACTION tag
        Send out the outgoing messages
```

Fig. 5. Event Processing Loop implemented by the Responder. The responder reads the XML Protocol Template Specification and constructs a Finite State Machine for the test before running this algorithm. Similar processing occurs on Timer generated events.

Timers can be used to delay sending responses or to send multiple requests or responses. We keep a list of *Timer* records which is scanned periodically for ready timers. When a timer fires, it can generate events. Timer events are always global and timers may be enabled or disabled from **TRANSIITION** nodes.

The evaluation of the boolean expressions that enable the **TRANSITION** nodes is done using an embedded *Jython* control interpreter (different from the one that is used to evaluate the service scripts).

We have used the Protocol Template idea to prototoype and deploy a SIP web-based interoperability tester (SIP-WIT [11]). Figure 6 depicts the implementation architecture of SIP-WIT as comprising three main components: the test proxy, a Responder Event Engine and a Trace Visualizer. The proxy has a XML-based SIP Message pattern matching facility (not described here) that can be used to invoke external tools (including additional Responder instances) while the test is in progress. Both the test proxy and the Responder generate detailed message logs. The proxy uses the *Record-Route* header to ensure that it is in the signaling path for the entire Dialog. While the test is in progress or after the test is complete, the client can visualize the signaling exchanges using the trace viewer application described in section 5.1.

The entire test system is controlled by a *HTTP Servlet Engine* that acts as a front end. The test system user selects a test case and enters appropriate parameters into a HTML form, which results in an instantiation of a test proxy

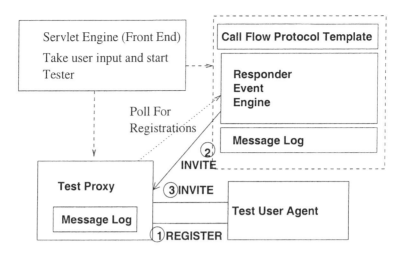

Fig. 6. The Test System: The Servlet Engine is used to start the test components. The Responder takes the Protocol Template as input and constructs a FSM for the test. It polls the test proxy for registrations to synchronize test startup.

and one or more Responders for the test. The trace viewer runs as an applet on the user's browser.

A *SIP Registrar* is a software component in a SIP-enabled network that allows users to register themselves and declare where they may be contacted. Our proxy server implements a registrar funciton and exports its registry entries for access via RPC. The protocol template may have **AGENT** tags which are bound at run time to Registry entries in the proxy. This is convenient when we do not know the identities of the participants of a test a-priori. If such bindings are specified, the *Responder* will poll the Registrar until the binding can be satisfied before it runs the test script. This allows for easy test synchronization and a customization.

5.1 Visualizing the Trace

The Proxy and Responder store their signaling trace for a pre-specified time period and makes it available for viewing by client applications. The traces are accessible via JAVA RMI and are grouped by call identifier. Each trace record has attributes that indicate where the message came from, where it is headed to, transaction identifier and other details that allow for the trace viewer to match Request and Response headers. The stack recognizes a special *NISTExtension* header which allows clients to record status information in SIP messages that are extracted and provided as part of the log file.

The trace viewer application retrieves a trace log from the proxy and can display a message sequence as a set of arcs that pass between stacks. Each stack is identified by IP address and port. The trace is sorted by time and Responses are matched with Requests and color coded appropriately. In order to reduce the

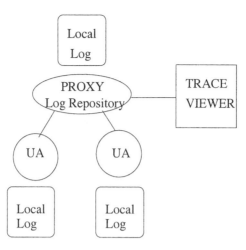

Fig. 7. Signaling Trace Collection : Traces may be collated at a single collection point. The request to fetch a trace is dispatched to slave repositories from the master repository and returned from the master repositry via JAVA RPC.

number of logging-related messages, and have a central collection point, traces are stored locally on each stack and collated on demand. This is done as follows: when a stack is initialized, a remote repository for the trace data may be specified. This is specified as an JAVA RMI URL. When the stack initializes, it registers with the central repository. When a request to retrieve the log comes into the central repository, the central repository dispatches the request to the slave repositories with a log collection request. This request can again be recursively broadcast by the slave repositories to its slaves. The slave repositories respond with the gathered log file. This hierarchical collection structure allows for scalability and decentralization. We have defined an XML syntax for the trace records to allow for possible standardization in future.

In order to avoid clock synchronization problems with merging the traces, we display the traces individually from the point of view of each collection point rather than as a single merged trace. Time stamps for each message are displayed along with the message relative to the beginning of the trace collection time.

In order to aid scalability and clarity, traces are organized by call-Id. Requests and responses are matched by identifying the transaction ID of the request and matching it to the corresponding response. The Figure 8 shows the trace visualization GUI. Each arrow corresponds to a message and the first line of the message is shown along with the arrow.

6 Field Experience

We took our tester implementation to the SIP Interoperability Test Event (SIPIT 11 [18]) where we were able to test against several proxy servers, user agents and IM clients. Our experience with the tester was positive in general but clearly we

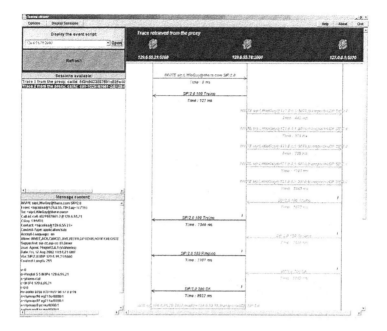

Fig. 8. The Trace Visualizer: This tool can accept traces gathered at the trace repository (which is part of the Proxy) or from an Ethereal trace capture. Trace records are formatted using XML. Arcs are color coded based on transaction and traces are separated by Call-Id.

need to add usability features (we often found ourselves editing configuration files). We were able to test third party call control, instant messaging and simple call flows using our responder and test proxy. We were also able to quickly script tests for extensions that are not part of the test proxy implementation. While further testing is needed, this increases our confidence that the programming model and XML representation are flexible and adequate to handle extensions that we have not yet considered.

An area of concern may be the performance and scalability of the system. When a message is received, the processing engine looks for an available **TRANSITION** node to fire. This currently involves a linear search through all the **TRANSITION** nodes that belong to a **DIALOG**. While this search can be pruned, in practice this turns out not to be a problem because a **DIALOG** usually only consists of a few **TRANSITION**s. The use of *Jython* to evaluate the trigger expressions and enabling conditions does lead to a performance bottleneck. However, *Jython* was only adopted for expedient prototyping purposes and may be replaced in future releases.

6.1 Related Work

As the popularity of the SIP protocol grows, many SIP testers have become commercially available. These are usually geared towards load testing. Load testing involves subjecting SIP components to high signaling loads such as hundreds or thousands of simultaneous calls. Load testing helps in uncovering synchronization bugs and tests scalability. Load testers differ significantly in function and have a different goal than our system.

The ITU-T has adopted TTCN-3 as a basis for building a test suite for SIP [15, 13]. TTCN-3 is a procedural programming language with test-specific extensions which is applicable to a wide range of communication protocols. The test cases thus generated are procedural with an explicit encoding of the protocol to be tested through the TTCN language. On the other hand, our approach is explicitly tailored for SIP testing and our test cases are declarative rather than procedural. Like the TTCN testing approach, we base our tests on pattern matching but in addition we construct an XML description of the protocol call flow to run the test, thus leading to a simpler expression of the test case.

Finally, our approach bears a resemblance to Control XML [17] but functions at a lower level and is explicitly SIP-aware.

7 Conclusions and Future Work

In this paper we presented a protocol template based approach for testing the SIP protocol. Its main advantages are the clean separation between protocol actions and test actions and the specification of entire call flows using the protocol template, which leads to customizable scenario-based protocol testing. We demonstrated the viability of our approach by constructing a SIP web-based interoperability test system and have exercised our system at industry wide interoperability testing events.

Our future work will focus on adding the ability for users to customize their own test scripts by giving them the ability to insert their own service functions to be executed during the execution of the Call Flow. While our initial goal is to expand the capabilities of our test system, the addition of these capabilities will require that we address the two critical issues for any dynamic service creation environment: security and resource control. We plan to use bytecode re-writing techniques to address these issues.

One of the attractive features of SIP is the ability to customize call flows on a per-user basis. For example, users may wish to have the ability to customize call forwarding based on time of day or other considerations. Such customizations possible in a restricted domain using CPL. Here we are suggesting a more general technique which could enhance the programmability currently possible with CPL.

The ideas we have outlined in this paper can be applicable to wider domain than test scripting. Indeed, a SIP Stack is a software component that is aware of protocol state and generates events that can be fielded by a piece of application code. API layers such as JAIN-SIP [2] and JAIN-SIP-Lite [3] define an event model and expose the stack to the application at different layers of event abstraction. The application code is able to express an interest in events at different layers of abstraction via the *Listener* mechanism. However, these models constrain the application to one layer of another. What we have done here is to generalize this so that applications may express an interest in protocol events at any level of abstraction (i.e. at the message layer, transaction layer or dialog layer) in one unified framework. Thus, using an extension of the approach we have defined in this paper, we can go beyond test scripting and define standardized means for expressing dynamic behavior for protocol extensions that are yet to be proposed. More system support may be needed to extend this approach to do this. Exactly what support is needed will be determined by actually building such services. We are working on this idea collaboration with others in industry.

Our test system and the implementation of the mechanisms we have described in the paper are available from [12].

Acknowledgement

This work was sponsored in part by the NIST Advanced Technology Program (ATP) and by the Defense Advanced Research Program (DARPA). NIST-SIP includes the contributions of many people including Christophe Chazeau and Marc Bednarek who where guest researchers on this project at an early stage. An early version of the visualization tool was done as part of a student project at ESIAL conducted by Fabrice Burte, Hugues Moreau, Damien Rigoudy and Damien Rilliard.

References

1. Johnston, A., Donovan, S., Sparks, R., Cunningham, C., Willis, D., Rosenberg, J., Summers, K., Schulzrinne, H.: SIP Call Flows. Note http://www.iptel.org/info/players/ietf/callsignalling/draft-ietf-sipping-call-flowers-00.txt
2. Specification Lead Harris, C. (DynamicSoft Inc.): JAIN SIP 1.0 API. Note http://jcp.org/aboutJava/communityprocess/final/jsr032/
3. Specification Lead Rafferty, C. (Ubiquity Ltd.): JAIN SIP LITE API. Note http://jcp.org/jsr/detail/125.jsp
4. Specification Lead Kristensen, A. (DynamicSoft Inc.): SIP Servlet API. Note http://jcp.org/jsr/detail/116.jsp
5. Lennox, J., Schulzrinne, H.: CPL: A Language for User Control of Internet Telephony Services. Note http://www.ietf.org/internet-drafts/draft-ietf-iptel-cpl-06.txt
6. Hugunin, J., Warsaw, B., van Rossum, G.: Jython: A Python implementation in JAVA. Note http://www.jython.org
7. Lennox, J., Schulzrinne, H., Rosenberg, J.: Common Gateway Interface for SIP. Note http://www.faqs.org/rfcs/rfc3050.html
8. Rosenberg, J., Schulzrinne, H., Camarillo, G., Johnston, A., Peterson, J., Sparks, R., Handley M., Schooler, E.: SIP: Session Initiation Protocol RFC 3261. http://www.ietf.org/rfc/rfc3261.txt
9. Handley, M., Schulzrinne, H., Schooler, E., Rosenberg, J.: SIP: Session Initiation Protocol RFC 2543. Note http://www.ietf.org/rfc/rfc2543.txt

10. Fielding, R., Gettys, J., Mogul, J., Frystyk, H., Masinter, L., Leach, P., Berners-Lee, T.: Hypertext Transfer Protocol–HTTP/1.1 (RFC 2068). Note http://www.ietf.org/rfc/rfc2068.txt
11. NIST Advanced Networking Technologies Division: NIST-SIP Web-based Interoperability Tool (SIP-WIT) Note http://www.antd.nist.gov/sipwit
12. NIST Advanced Networking Technologies Divsion: NIST-SIP Parser and Stack. Note http://www.antd.nist.gov/proj/iptel
13. Wiles, A., Vassiliou-Gioles T., Moseley, S., Mueller, S.: Experiences of Using TTCN-3 for Testing SIP and OSP. Note
http://www.etsi.org/tiphonweb/documents/
Using_TTCN_3_for_Testing_SIP_and_OSPv8.pdf
14. Dahm, M.: Apache Byte Code Engineering Library (BCEL). Note http://www.apache.org
15. Schieferdecker, I., Pietsch, S., Vassilou-Gioles, T.: Systematic Testing of Internet Protocols - First Experiences in Using TTCN-3 For SIP. Note Africom 2001, Capetown, South Africa,
http://www.testingtech.de/technology/Africom2001.PDF
16. Parr, T.: ANTLR parser gnerator. Note http://www.antlr.org
17. Auburn R.J., et al.: Call Control XML. Note http://www.w3.org/TR/ccxml/
18. SIP Interoperability Test Event.: Note http://www.pulver.com/sipit11/

Towards Modeling and Testing of IP Routing Protocols

Jianping Wu, Zhongjie Li, and Xia Yin

Department of Computer Science and Technology, Tsinghua University
Beijing 100084, P.R.China
jianping@cernet.edu.cn
{lzj,yxia}@csnet1.cs.tsinghua.edu.cn

Abstract. Routing protocols are typical distributed systems characterized by dynamic, concurrent and distributed behaviors. As the primary function of routing protocols, routing information processing constitutes the main content of routing protocol conformance testing. However, there isn't a clear model based on which convincing test architecture and test notation can be designed. Also, some important features of routing protocols have not been considered sufficiently in existing test practice. On the other hand, generalized distributed system models, test architectures and test notations usually have limitations when applied directly to specific protocol testing. So it is necessary to study specific cases in order to find a pragmatic and efficient approach. In this paper, an MP-FSM (Finite State Machine with Multiple Ports) model is proposed to describe routing information processing of IP routing protocols. Based on this model and other test requirement particularities, a test architecture called PADTACC (Parallel And Distributed Test Architecture with Centralized-Control) is presented and a test notation called RIPTS (Routing Information Processing Test Script) is defined. We implement the whole approach into a software tester – IRIT (IP Routing Information Tester).

1 Introduction

Routing protocols are core components of the Internet architecture. Therein, the correct running of routing protocols enables connectivity, stability and security of the Internet. In this work we study conformance testing of Internet IP routing protocols including RIP, OSPF and BGP [1].

Inside a router, routing protocols are used to update the routing table dynamically. The main functions of a routing protocol can be divided into two parts: network communication (NC) and routing information processing (RI-Pro). The NC part usually appears in an FSM form functioning to handle low-layer network accessing, establish reliable communication channels for the routing information flows, or detect changes of local network connections. For example, RIP uses UDP to carry its protocol messages. There is no state machine in RIP. OSPF runs on IP, and defines two state machines: the interface machine and the neighbor machine mainly used for Designated Router election, neighboring adjacency establishment and maintenance. BGP has a state machine specifying the procedure how the peers set up, maintain the TCP connections and negotiate parameters. The NC part is relatively easy to model, implement and test. There have been many researches in aspects of protocol modeling, test generation, test system construction, etc [2-5]. Most of them use TTCN (Tree and Tabular Combined Notation) [6] as the test notation to specify an abstract test suite.

The RI-Pro part consists of routing information origination and propagation as well as routing table calculation and update. This part represents the primary function of a routing protocol and should also be the main content of routing protocol conformance testing. There is relatively less work on this part. In reference [5], the authors find that traditional means of testing communication protocols are inappropriate for the testing of routing protocols in the aspects of the test architecture, test notation, etc. They propose some enhancing techniques to improve the test system. SOCRATES [7] is a software test tool exploring automatic network topology generation and probabilistic algorithms, it concentrates on the testing of routing information processing, but it is incapable of testing the inter-area and AS-external routing behaviors of OSPF. As to test systems, currently there are some commercial products, such as HP RouterTester [8], NetCom Systems Smartbits [9], etc. They can be used for both conformance and stress testing of IP routing protocols, focusing on the routing information processing part. They all use self-defined test notation to write test specifications.

Despite these works, there isn't a clear model for routing information processing based on which convincing test architecture and test notation can be designed. Also, some important features of routing protocols have not been considered sufficiently. Some of those features are:

- routing information grouping. Pieces of routing information are packed in one or several packets and come out of the IUT (Implementation Under Test: a software module or system which implements the tested protocol) in a random order. The pieces of routing information contained in one packet appear in a random order too. A general test language like TTCN is clumsy to manipulate these ordering variations. An intelligent "multiple-packets and set matching" mechanism is appropriate.
- complex test configurations. Test configuration includes configuration of the tester and the IUT. Routing protocols such as OSPF have many tunable parameters, options and configuration scenarios. Furthermore, it is common that one test case uses several test configurations. This situation necessitates formalization and automation of test configuration operations.
- complex protocol behaviors. Complex behaviors in routing protocols are difficult to describe in a general test language like TTCN. It is better to program such behaviors by C language. While testing routing information processing, we assume that the network communication part has been tested and is correct. If this part is complex, we'd better implement it by a simulated router which could communicate with the IUT like a real router.

In this paper, we will introduce a formal model of the routing information processing (sect. 2), analyze the specific requirements of routing protocol conformance testing, design an applicable test architecture (Sect. 3) and a flexible test notation and discuss the efficient means of test organization (sect. 4). The complete approach is implemented as a linux-based software tester called IRIT – IP Routing Information Tester (sect. 5). Sect. 6 gives example test topologies of OSPF protocol. Sect. 7 concludes the paper with some discussion on future work.

2 The MP-FSM Model

Fig. 1 shows the skeleton of a routing protocol and its relation with other modules in a router. The NC part lies in the bottom serving the RI-Pro part with low-layer network accessing, reliable communication channels or local connection monitoring. The RI-Pro part is responsible for originating routing information pieces describing local network interfaces and for flooding of the information generated by itself and by others to its neighboring routers (using routing information origination and flooding algorithms). All routing information collected by a router composes its Routing Information Base (RIB). The internal routing table (I-RT) is calculated based on the RIB (using routing table calculation algorithm). In addition, another routing protocol may run on the same router (AS boundary router). Routing management module provides interfaces for routing protocols to modify the system routing table (S-RT). It also manages the route redistribution between different routing protocols. Users configure or monitor the routing protocols via the CLI (Command Line Interface) provided by OAM (Operation and Management). MIBs (Management Information Base) of routing protocols are managed by SNMP module. Both the CLI and MIBs can be used to examine the RIB, I-RT, running statistics and various internal states.

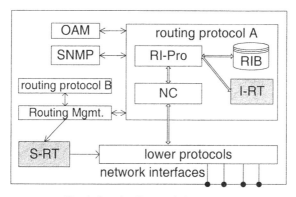

Fig. 1. Routing Protocols in a router

The routing information processing part of routing protocols is described in natural language and there is no clear functional or procedural model. We first give a formal model in definition 1. It is used for later discussion of the test architecture and test notation design.

Definition 1 (MP-FSM model of routing information processing) The routing information processing of IP routing protocols (under a given protocol configuration) can be modeled as an FSM with multiple ports which is defined as an 8-tuple M= $<S, I, O, \Sigma, P, f, g, s_0>$ where

S is the set of states of the routing information processing procedure. A state is defined by the combination of RIB and I-RT: $<R, T>$. S is finite in practice under a given network topology and protocol configuration. s_0 is the initial state with a RIB and I-RT which are generated by this router at startup.

$\Sigma = L \cup E$, is a collection of observable routing information between this router and the environment (neighboring routers). The messages include internal routing information set L and external routing information set E.

P is the set of router's ports. A port is an abstraction of an interface through which all the routing information is communicated among routers. $P=\{1,2,...,r\}$ is the port index. All ports are bidirectional.

I is the set of input actions, $I \subseteq P \times \{recv\} \times \Sigma$, an input action is an input of routing information from a specific port.

O is the set of output actions, $O \subseteq P \times \{send\} \times \Sigma$, an output action is an output of routing information to a specific port.

f is the output function, $S \times I \rightarrow P(O)$, where $P(O)$ is the power set of O. f is the routing information origination and flooding algorithms. The output actions are parallel if they occur in different ports, serial if they occur in the same port. Their sequencing can be arbitrary in either case.

g is the transition function, $S \times I \rightarrow S$. g is the RIB update and routing table calculation algorithm.

A transition can be represented as $s_i \xrightarrow{i/\Omega} s_j$ which states that an MP-FSM, in state s_i, upon an input action i, moves to state s_j and produces a set of output actions Ω ($\subseteq O$). Following this notation we define $f(s_0, i_1 i_2..i_n) = \Omega_1 \Omega_2.. \Omega_n$, $g(s_0, i_1 i_2..i_n) = s_n$, iff. $f(s_{k-1}, i_k) = \Omega_k$, $g(s_{k-1}, i_k) = s_k$, $1 \leq k \leq n$. This is used to describe consecutive input and output actions. Fig. 2 illustrates the MP-FSM model and a state transition scenario where the machine moves from state s_1 to s_2 with input a at port p1, output b at port p2 and output c at port p3 respectively.

3   The PADTACC Test Architecture

Test architecture describes the environment the IUT lies in, generally the connection relations between the IUT and the test system. The behaviors of routing protocols are distributed at several router interfaces interacting with the test system concurrently. Therefore, what we need is a test architecture that supports multiple test points and multiple test processes. This idea immediately results in a simple but widely applied test architecture [10] in which several testers surround the IUT and compose a test system (shown in Fig.3). TTCN-2 [6] improves this test architecture. It introduces the concept of test component (TC) which runs a test process independently. In TTCN-2 test architecture (see Fig.4), there is one Main Test Component (MTC) which creates other test components named Parallel Test Component (PTC). The communication interface between test components is called Coordination Point (CP) and between test component and the IUT Point of Control and Observation (PCO). CP and PCO are both modeled as two FIFO queues: one for each direction of communication. PCOs are connected to the ports of the IUT thru the underlying networks.

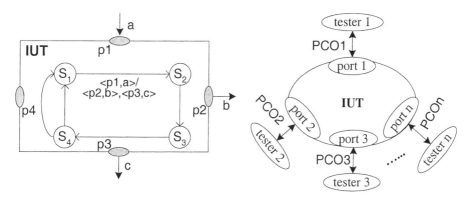

Fig. 2. The MP-FSM Model   **Fig. 3.** A General Distributed Test Architecture

This test architecture is sufficient for use but too generalized. Some modifications can provide a more practical variation. First, we simplify this test architecture by removing CPs between test components. Synchronizations between PTCs will be handled by MTC. As we'll see later, no problems or overhead occur because the necessary synchronizations between PTCs are very simple. Second, we lighten the MTC by cutting its connections to the IUT. Third, because we use linux PCs to implement the test system and a PC usually has limited number of NICs (Network Interface Cards), it will be scalable if we can use any number of PCs. For this purpose, a module named Distributed Communication Utilities (DCU) is added to support distribution of test components on several PCs. The PC on which MTC is located is called "main tester", other PCs are slave testers. PTCs can be distributed on the main tester and slave testers. All the testers sit on an ethernet LAN which is separated from networks connecting the testers with the IUT. We have implemented the DCU using TCP sockets API. Each PTC connects to MTC through a TCP connection, then receives commands and returns test results via this connection.

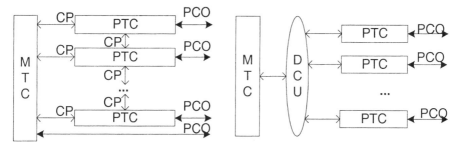

Fig. 4. TTCN-2 Test Architecture   **Fig. 5.** The PADTACC Test Architecture

DCU is also used to access the CLI or SNMP module of the IUT to perform automatic configuration and monitoring functions during the test process, as explained later.

This specialized test architecture, named PADTACC – Parallel And Distributed Test Architecture with Centralized-Control, is shown in Fig.5. The simplification and patching is a reification of TTCN-2 test architecture when applied in routing protocol testing.

4 The RIPTS Test Notation

4.1 Requirements on Test Notation

Flexible and Simple.
In order to test the IUT we need to specify the test actions. Despite the wide applicability of TTCN, it is clumsy to manipulate the ordering randomness of incoming packets or pieces of routing information contained in a packet. Actually, for purpose of testing a specific protocol, a simple script language is enough. In addition, it should be easier to implement and flexible to be customized to meet the special test requirements of routing protocols.

Send, Receive and State Verification Facilities.
FSM-modeled protocols are tested using a transition-cover mechanism [11]. The test notation should provide capability to test a state transition: send to the IUT, receive from the IUT, verify its state.

Formalization and Automation of Test Configurations.
Test configuration includes configuration of the tester and the IUT. It is done before and in a test run because different test cases may use different test configurations. IP, TCP and many other protocols have only a small set of simple configuration tasks, many test cases share one test configuration which is recorded informally and done manually. Routing protocols such as OSPF, however, have many configuration scenarios. Furthermore, it is common that one test case uses several test configurations. This situation necessitates formalization and automation of test configuration operations. We choose to record the test configuration information in the test case and have it done automatically. The test notation should support this improvement. This approach greatly speeds the testing process.

Simulated Router.
In the past, we have overused the test specification language TTCN with a strong wish to keep the test suite pure. We almost walked off our legs until we realized that it is better to program complex protocol behaviors by C language. For example, OSPF neighboring routers have to experience a complex initial synchronization procedure before they enter the normal phase of exchanging further routing information. We implement Simulated Router encapsulating these complex behaviors in C language.

Simulated router (SR) can be viewed as a bipartite routing protocol implementation. The network side, which communicates to the IUT, is compliant with the protocol specification. In case there are state machines defined (in OSPF, BGP), an SR should implement the state machines. If no state machines are defined in the protocol (RIP), SR can be a simple encoder/decoder of the protocol messages. Additional functions such as authentication mechanism, checksum generation and verification, protocol configuration should also be implemented. However, SR need not possess all functions of a real router. For example, it neither originates routing information nor calculates the routing table. All the routing information to be sent to and received from the IUT is specified in the test script. As we will explain next, this is the way we simulate network topologies. The program side, which interfaces with the test script interpreter, acts as a PTC running associated test sequences under the control of MTC. Each network interface of an SR corresponds to one PCO.

Simulated Network Topology.
The nature of routing information processing is to track the dynamic network topologies in order to update the routing table accordingly. Should the tester simulate a network topology by running all containing routers simultaneously? This is unnecessary because the IUT can not see what happen beyond its neighboring routers and it learns the topology through routing information relayed by its neighboring routers. Thus, a tester need only simulate the neighboring routers which hide the simulated network topology behind. Fig. 6 illustrates this principle.

Flood routing information to provide simulated network topologies

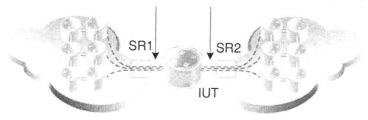

Fig. 6. SR hides the Simulated Network Topologies

4.2 The RIPTS Notation

This section informally presents the syntax, semantics of the RIPTS (Routing Information Processing Test Script) notation and formats of the test case file and test configuration file (see Fig.7). Note "{X}+" means "one or more X", "{X}*" means "zero or more X", "|" means "or", "%" introduces a parameter.

Definition 2 (test suite, test case, test step, test event) A set of test cases for a particular protocol composes a **test suite**. A **test case** is an implementation of a test purpose and comprises several test steps. A **test step** specifies a test configuration and a set of test events run under that configuration. A **test event** is the smallest indivisible unit of a test suite, such as transmission or reception of a packet.

To each test case run, we assign a verdict specifying whether the IUT is correct wrt this test purpose. PASS is assigned if the IUT passes all the test steps in the test case, FAIL otherwise. We say the IUT passes a test step if and only if all the test events in the test step are executed successfully according to their semantics (defined later).

In a test step, **TestCfg** statement designates the current test configuration file and **TestEvents** statement block designates the test events. By embedding test configuration information into test cases, the formerly stated requirement, formalization and automation of test configurations, is satisfied. Putting all test configuration information of a test step in an individual file enables reusability and modularity.

| TestCase %case_index
{
  TestCfg %testcfg_file
  TestEvents
  {
    wait %pco_index %state \|
    %pco_index **send** %ri_file \|
    %pco_index **recv** %ri_file \|
    %pco_index **flush** %ri_file \|
    pause \|
    stop {%pco_index}* \|
    checktable %rib %table
  }+
}+
      Format of a Test Case File | TestCfg %cfg_index
PCO %pco_number
IutCfg %iut_address %iutcfg_file
{
  Tester %tester_address
  {
    SR %srcfg_file %base_pco
  }+
}+

      Format of the %testcfg_file |

Fig. 7. Formats of some RIPTS Files

In a test configuration file, **PCO** statement gives the number of PCOs used by this test step, **IutCfg** statement specifies the IUT configuration where *iut_address* is an IP address of the IUT on the tester-LAN and *iutcfg_file* is the IUT's configuration file. Next statements are used to specify the configurations of testers and simulated routers. Remember that simulated routers act as PTCs and may be distributed on several testers. Each **Tester** statement block gathers all the SRs on one tester whose IP address on the tester-LAN is *tester_address* (used for startup of the SRs). Inside each **Tester** statement block are several **SR** statements which specify *srcfg_file* (the configuration of this SR) and *base_pco* (used to indexing the PCOs associated with this SR). PCO indexing is global. *iutcfg_file* and *srcfg_file* are protocol configuration files. The format of *srcfg_file* depends on SR since it is used by SR. *iutcfg_file* can use the format of *srcfg_file* and serve as the universal configuration script of the IUT. Users either configure the IUT manually according to *iutcfg_file* during the testing or design an automatic configuration mechanism such as using *expect* to access the *telnet virtual terminal* provided by the OAM module of the router under test.

Next, we look into the key construct: test event.

1. **wait %pco_index %state** Informs the relevant PTC (SR) to interact with the IUT until the possible Network Communication FSM reaches the named state. Previously we have written this procedure in TTCN test steps but found it very difficult and error-prone. So, this test event accomplishes the same function of the preamble test steps transferring the IUT to an appropriate starting state in TTCN.
2. **%pco_index send %ri_file** Send all the routing information listed in *ri_file* out of the named PCO to the IUT. It can be used to send many pieces of routing information at one time triggering sequential state transitions. **send** event finishes successfully when all the routing information is sent out.
3. **%pco_index recv %ri_file** Receive routing information from the IUT over the named PCO and see if it matches that specified in *ri_file*. We do not care the ordering of pieces of routing information so it's a multiple-packets and set matching. This test event finishes successfully if we receive all the given routing information within a limited time period, unsuccessfully otherwise. There is a special case when *ri_file* contains no routing information at all. This means no routing information is expected to arrive at this PCO.
4. **%pco_index flush %ri_file** Inform the relevant PTC to flush previously installed routing information via the named PCO. In RIP, the IUT will get a list of route entries with distance 16; in OSPF, the IUT will get a list of MAXAGE LSAs specified in *ri_file*; in BGP, the IUT will get a list of withdrew routes. **flush** is a special **send** event.
5. **pause** The only constructive element in the RIPTS notation, used to separate checking rounds (introduced later) and synchronize PTCs. Also used as a part of user interface implementation: pause the test execution and prompt to proceed. This is useful when we need to configure the IUT manually before go on testing.
6. **stop {%pco_index}*** Shut down the named PCOs. This includes flushing all the routing information previously sent out of named PCOs in *pco_index* list, and then disconnecting with the IUT. An empty *pco_index* list equals an all-PCO list. This test event is similar to the postamble test step returning the IUT to the initial state in TTCN.
7. **checktable %rib %table** *rib* and *table* specify the contents of the RIB and the routing table respectively. They should be verified at the end of each checking round. This state verification is comparable to the UIO verification in formal FSM testing.

Note that test events are not guarded by a timer explicitly. Relevant timers are put in a file together with other test options or parameters and read in a test run. Practice has shown that this approach simplifies writing test cases and enables easy adjustment of test options or parameters.

4.3 Structure of a Test Step

Definition 3 (a checking round) In testing, the tester initiates a sequence of input actions α which makes the IUT move from state s_i to state s_j while originating a sequence of output actions σ. This is called a checking round.

Apparently a checking round is a path of several transitions in the transition diagram of the MP-FSM given by Def. 1. A test step comprises several checking rounds separated by **pause** event. We use // to denote parallel relation, >> sequential relation, | alternative relation. Then the timing structure of a test step can be formalized as follows:

$B = b_1 >> b_2 >> .. >> b_n$, b_i (i=1..n) represents a checking round comprising some parallel test processes:

$b_i = a_{p1} // a_{p2} // .. // a_{pm}$, i=1..n, a test process a_{pk} (k=1..m) is a sequential execution of test events on PCO_k:

a_{pk} = **wait_sync** P_k S_l| **stop** P_k | (P_k **send** U_k >> P_k **recv** U'_k) | (P_k **flush** U_k >> P_k **recv** U'_k) | P_k **recv** U'_k

k=1..m, P_k denotes PCO_k, U_k is the routing information sent out of P_k, U'_k is the routing information received at P_k. S_l is a state of the possible Network Communication FSM.

Thus, a test step is interpreted as a sequential execution of several checking rounds which are parallel execution of several test processes according to above timing relations. In addition, use the **checktable** event at the end of each checking round to verify the end state if necessary.

5 IRIT – IP Routing Information Tester

We implemented all these ideas in a linux-based software tester – IRIT. Fig. 8 shows the software architecture which consists mainly of seven component modules: UI (User Interface), TM (Test Management), TE (Test Execution), DCU (Distributed Communication Utilities), SR (Simulated Router), Test Case Editor (TC Editor), Routing Information Generator (RI-Gen). AC is the abbreviation for Auxiliary Channel. Their roles and functions are presented briefly next.

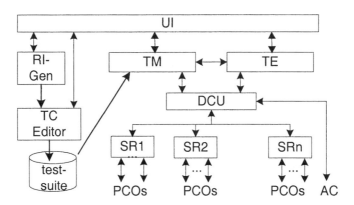

Fig. 8. the Architecture of IRIT

UI is a general designation of all the interfaces between modules and the user (the tester operator). RI-Gen supports manual routing information edit as well as automatic routing information generation. Test Case Editor is used to edit and manage the test case files, test configuration files, protocol configuration files. TM plays an important role in the test organization. It implements the test selection interface by which the user may choose which test cases to run. Also, TM is responsible for the distribution of test configuration files on different testers.

The function of Test Execution is to interpret and execute the test events of a test case. It implements MTC in the PADTACC test architecture, runs the main test process and schedules the PTCs to run parallel test processes. A Simulated Router implements a PTC and its function has been detailed in sect. 4. DCU is used for the communications between test components and used to communicate with the IUT in other ways for other purposes, e.g. for remote configuration of the IUT.

6 Example Test Cases

Fig. 9 illustrates three test topologies. Each corresponds to an OSPFv2 test case. SRx denotes simulated router, Rx for virtual router (that "exists" in the simulated networks). ABR indicates area border router, ASBR for AS border router. Thick short lines labeled with N1, N2, ... are multi-access networks. Lines connect interfaces to networks. OSPF areas are partitioned spatially and denoted by A0, A1 and so on.

TOP2 is used to test summary-LSA origination of the IUT under various area route aggregation and suppression configurations. This test case is typical in that it consists of five test steps with different test configurations. Fig. 10 shows the configurations of the IUT and what summary-LSAs should be originated by the IUT into area A1 and A0 (or A2). TOP3 is used to test that the IUT should originate ASBR-summary-LSA and ASE-LSA into normal areas but not into stub areas. The corresponding test case is shown in Fig.11 with comments. TOP4 is used to test the behaviors of the IUT with a virtual link configured. The test case is omitted here.

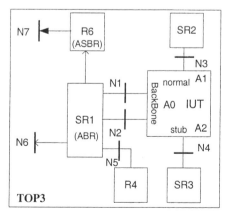
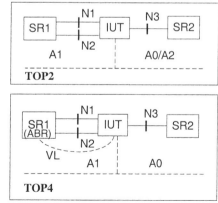

Fig. 9. Example Test Topologies of OSPFv2

N1: 10.1.1.0/24; N2: 10.1.2.0/24; N3: 10.1.3.0/24
IUT interface1: 10.1.1.5, metric=1; interface2: 10.1.2.5, metric=2;
    interface3: 10.1.3.5, metric=3

Test Step	IutCfg (IUT configurations)	summary-LSA (route prefix, metric) the IUT should originate into this area
1	A0: no aggregate	(10.1.1.0/24, 1) (10.1.2.0/24, 2)
	A1: no aggregate	(10.1.3.0/24, 3)
2	A0: 10.0.0.0/12	(10.1.0.0/16, 2)
	A1: 10.1.0.0/16	(10.0.0.0/12, 3)
3	A0: 10.1.3.0/24 suppress	(10.1.0.0/16, 1)
	A1: 10.1.0.0/16; 10.1.2.0/24 suppress	-
4	A0: 10.1.0.0/16 suppress	(10.1.1.0/24, 1)
	A1: 10.1.0.0/16 suppress 10.1.1.0/24	-
5	A1: no aggregate	(10.1.3.0/24, 3)
	A2: stub, default-cost=10	(10.1.1.0/24,1) (10.1.2.0/24, 2) (0.0.0.0/0, 10)

Fig. 10. The TOP2 Test Case Outline

Test Case Statements	Comments
TestCase 3	pco1, pco2 belongs to SR1, pco3 belongs to SR2, pco4 belongs to SR3 as specified in *testcfg1*. A2 is a stub area configured not to import summary-LSAs.
TestCfg testcfg1	
TestEvents	
wait all full	wait completion of synchronization on all PCOs.
pco1 recv top3_a0_noaggr.ol	specify what LSAs are expected on each PCO, .ol files list the LSAs expected.
pco2 recv top3_a0_noaggr.ol	
pco3 recv top3_a1_noaggr.ol	
pco4 recv top3_a2_stub.ol	
pause	signal the end of a checking round
pco1 send top3_total5.ol	SR1 originates a rourter-LSA. It also simulates routes to R6, N6 and N7. The IUT should take N2 as the next hop network to R6, this is reflected in the LSAs originated into area A1.
pco1 recv empty.ol	
pco2 recv top3_total5.ol	
pco3 recv top3_total4.ol	
pco4 recv empty.ol	
stop	stop this test case

Fig. 11. The TOP3 Test Case

7 Summary

The main functions of a routing protocol can be divided into two parts: network communication (NC) and routing information processing (RI-Pro). The RI-Pro part is primary and should be the main content of routing protocol conformance test. This paper develops a pragmatic approach to test RI-Pro functions of IP routing protocols including RIP,BGP,OSPF [1].

An MP-FSM model is proposed to describe routing information processing of IP routing protocols. Some simplification and patching is made to TTCN-2 test architecture, resulting in the Parallel and Distributed Test Architecture with Centralized-Control (PADTACC). Compared to decentralized control, centralized control is simpler to realize the coordination and synchronization of test processes.

As a general test notation, TTCN has some limitations and extensions are not easy. We have designed a simple test notation called Routing Information Processing Test Script (RIPTS). RIPTS has advantages over a general test notation (e.g. TTCN) in the aspects of easy-understanding, extensibility, test efficiency and reliability, etc. Some limitations of TTCN-2 will be overcome in TTCN-3. For example, the newly introduced "external function" allows the use of functions that are written in other languages. It also provides a means to configure and monitor the IUT automatically in a test run.

All these ideas are implemented in a linux-based software tester - IP Routing Information Tester (IRIT). Compared with commercial testers like [8,9], IRIT is low-cost, easy to use. Also, it supports use of any PCs to compose a test system and thus gets rid of the limitation on available network interfaces. Formalized and automatic test configuration is another highlight that achieves faster test speed.

Future work based on IRIT exists in the following issues: stress testing, network topology simulation, routing information test generation. In addition, an effort will be taken to relate our work with TTCN-3 and translate the test cases to TTCN-3.

Acknowledgements

This research is supported by National Natural Science Foundation of China under Grant No. 90104002 and No.60102009. The authors would like to thank the anonymous reviewers for constructive comments on this paper.

References

1. RFC1058, RFC2328, RFC1771. URL = http://www.ietf.org/rfc.html
2. Wang Lirui, Ye Xinming. A Formal Approach to Conformance Testing of OSPF Routing Protocol. Proc. of the 6th Asia-Pacific Conf. on Communications, Page(s): 1286-1290
3. Jun Bi, Jianping Wu. A concurrent TTCN based approach to conformance testing of distributed routing protocol OSPFv2. Proc. of 7th Int. Conf. on Computer Communications and Networks. Page(s): 760 –767, 1998

4. Zhao Yixin, Wu Jianping, Yin Xia, et al. Test of BGP-4 based on the Protocol Integrated Test System. Proc. of 6th Asia-Pacific Conf. on Communications. Korea, 2000, 347-355
5. Wu Jianping, Zhao Yixin, Yin Xia. From Active to Passive: Progress in Testing of Internet Routing Protocols. KLUWER ACADEMIC PUBLISHERS. Proceeding of FORTE/PSTV 2001, Korea. 2001. 101-116
6. ISO/IEC 9646. Information Processing Systems, Open System Interconnection, OSI Conformance Testing Methodology and Framework- Part 3: The tree and tabular combined notation. 1998.
7. HAO R B, LEE D, RAKESH K. et al. Testing IP Routing Protocols - From Probabilistic Algorithms to a Software Tool. KLUWER ACADEMIC PUBLISHERS. Proc FORTE/PSTV2000. Pisa, Italy. 2000, 249-266
8. Agilent Technologies. RouterTester. http://www.agilent.com/comms/RouterTester, 2001
9. NetCom Systems. http://www.netcomsystems.com. 2000
10. Luo, G., Dssouli, R., Bochmann, G.V.etc. Test generation for the distributed test architecture. International Conference on Information Engineering '93, Page(s): 670 -674 vol.2
11. Lee, D., Yannakakis, M. Principles and methods of testing finite state machines-a survey. Proceedings of the IEEE, Volume: 84 Issue: 8, Aug. 1996. Page(s): 1090 –1123
12. Expect Home Page. URL = http://expect.nist.gov

An Intuitive TTCN-3 Data Presentation Format

Roland Gecse and Sarolta Dibuz

Ericsson Hungary, 1037 Budapest, Laborc 1
{Roland.Gecse,Sarolta.Dibuz}@eth.ericsson.se

Abstract. This paper describes the TTCN-3 Data Presentation Format (DPF). DPF is an intuitive graphical notation for representing TTCN-3 Core Language (CL) [1] types and values. The major advantage of using DPF compared to free-text editing is that a DPF implementation ensures a consistent type and template structure by excluding references to unexisting entities while significantly reducing typing work. The result is a shorter test suite development time. DPF covers all excluded parts of Graphical Presentation Format (GFT) [3]. We believe that DPF and GFT together could be the basis for building the ultimate graphical representation of TTCN-3.

1 Introduction

Traditional test suite design starts with the preparation of declarations and constraints parts. This job requires careful analysis of SUT and good understanding of the exact purpose of the test. A load test, for instance, requires less detailed resolution of PDUs as conformance test. The former focuses on whether the SUT can service a huge number of typical requests while the latter targets to prove standards conformity. Another important factor of data part development is the time. Working out type and template definitions takes 20-50% of total test suite development time. The preparation of data part in CL comprises of freetext editing of textual definitions. Unfortunately, existing presentation formats (TFT [2], GFT [3]) do not provide help on this. TFT basically fragments CL code into table fields while GFT targets dynamic behaviour only. The intention with DPF is to facilitate test suite data part development and thereby make test designers' work easier.

A TTCN-3 module can be subdivided by functionality into several sections. Some of these – constant declarations, type definitions, signature declarations, template declarations, module parameter declarations, port and component type definitions – constitute what we refer to as *data part*. DPF operates on data part only – that is where its name comes from. Other sections containing test cases, functions, alt steps as well as the module control part pass through DPF unmodified (Figure 1).

DPF manifests itself in a utility, which provides a graphical representation of data part constructions and assists the user to manipulate these. This GUI provides separate panes for each section of the data part. The sections have been introduced because we deliberately avoided using a distinct graphical symbol for

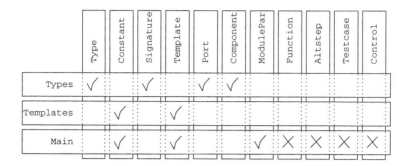

Fig. 1. Mapping of DPF and TTCN-3 parts

each type in order to keep the notation intuitve. This approach resulted that the DPF notation became context sensitive; the same graphical construct can mean different things when appearing at different sections, i.e. a template declaration for a given type, for instance, looks similar to a constant declaration for the same type provided the template consists of specific values only.

The example of Figure 1 shows an imaginary test suite, splitted up into three modules. This example presents how a DPF utility handles TTCN-3 modularity. Each module contains some sections denoted either with a check-mark or a cross depending on whether they are relevant to DPF or not. The module *Main* is the root module of the test suite as it contains the control part governing the test execution. This module includes not only test cases, functions, alt steps (the dynamic part) and module parameter definitions but some additional constants and templates, too. The other two modules appear because the root module imports some or all of their definitions. The module *Types* contains type, communication port and component type definitions and some signature declarations while the module *Templates* holds constants and templates for the types defined in either of the modules.

The GUI provides means for creating or editing adequate entities in every section. A new template can be built, for instance, based on an existing type definition simply by choosing the appropriate type from the type definitions section and assigning values or wildcards to each of its fields. Alternatively, a new type can be defined by extending or reducing an existing type or by merging several types. It could be ensured that no undefined types or templates are referenced, too. Optionally a DPF implementation could perform some basic checks such as if all fields of a template had a value assigned or whether the designer really wanted to create optional set-of types, etc. The fabricated definitions can be saved into a CL module at any time during editing. Regardless of the changes made to the data part DPF keeps dynamic module parts untouched. The only difference between the original CL module and the one that has been processed by DPF is that DPF may reorder some sections.

The rest of the paper presents graphical notations of DPF. The simple type and value notations are described first. It is followed by the structured type constructs. A dedicated section describes the embedded type definition as DPF relieves the TTCN-3 limitation of embedded type definitions. Finally the notation for special CL features like signatures, templates, communication port and component types are introduced. Each section contains some examples together with their CL mapping when appropriate in order to get an impression of DPF.

2 Simple Type and Value

The basis of DPF is the graphical notation for simple type as all structured types are *built*[1] of simple types. Figure 2 presents the format of DPF simple type. The graphical representation consists of a rectangle with a mandatory simple type identifier in the middle. Three placeholders are also included for the optional *field identifier*, *with-attributes* and *value* attributes.

```
[field-identifier]
          simple-type-spec
[with-attributes]          [value]
```

Fig. 2. DPF notation of simple type

The simple types of DPF include all predefined CL simple basic types and basic string types. DPF supports subtyping according to CL 6.2 in [1]. Although the permitted subtyping methods are equivalent DPF subtyping syntax slightly diverges from CL. The mandatory simple-type-spec carries all type properties including subtyping. The three optional attributes add further information to the simple type. These include contextual properties, encoding directives and assigned values.

2.1 Simple Type Specification

It has the following ABNF [4] syntax:

```
simple-type-spec =
    [ simple-type-id ":" ] base-simple-type-id [ subtype-spec ]
```

where simple-type-id and base-simple-type-id shall be unique identifiers within the scope of the module. subtype-spec is any valid CL subtype specification (SubTypeSpec, prod. 44 of CL BNF) of the corresponding base-simple-type-id.

Depending on presence of optional parts simple-type-spec can be a *simple type reference*, a *simple type alias*, a *subtype definition* or an *inline subtype definition*.

---

[1] The actual building methods are introduced at the structured types.

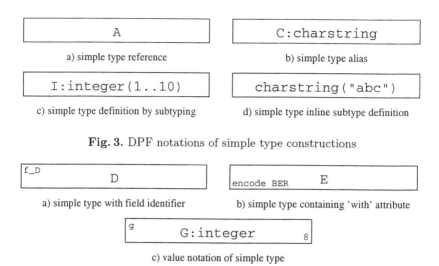

Fig. 3. DPF notations of simple type constructions

Fig. 4. Examples of simple types with attributes

1. The simple type reference (Figure 3/a) consists of a reference to a predefined simple type or a user-defined simple type. Both optional simple-type-id and subtype-spec are absent.
2. Simple type alias (Figure 3/b) is composed of a unique simple type identifier followed by a semicolon and the reference to an existing simple type. This construct is used to create a new simple type (an alias) with the given identifier from a base type. The new alias type has the same value domain as its base type. The CL mapping of the example is type charstring C;.
3. The subtype definition in Figure 3/c constructs a new simple type from an existing simple type by means of permitted subtyping methods of base-simple-type-id. The resulting subtype is considered as a distinct simple type. The CL equivalent of the example is: type integer I (1..10);.
4. The inline subtype definition (Figure 3/d) derives an unidentified new type from another simple type using subtyping. The difference between subtype definition and inline subtype definition is that the latter can not be referenced because it has no identifier.

It shall be noted that items 1 and 4 may occur in structured type definitions only or additional attributes must be added to provide semantics!

2.2 Simple Type Attributes

The DPF representation provides three attributes to describe the context of simple type specifications. The following paragraphs introduce each in detail.

Field Identifier. The optional context specific field-identifier (Figure 4/a) has different meaning in different module parts. It can stand for the identifier of a constant or variable of the current simple type or the field identifier assigned to a structured type element.

CL requires an identifier to be assigned to each structured type element in order to distinguish between them. DPF provides graphical symbols for this purpose that are sufficient in most cases. Consequently identifiers can become superfluous and may be omitted. However, the identifiers must be provided when exporting definitions into CL. Thus, missing field identifiers are automatically generated during the export[2]. The `field-identifier` shall be unique within the scope of the given structured type.

With Attributes. The `with-attributes` field (Figure 4/b) is the location of CL with attributes. The `with-attributes` can be separately set for each simple type in the bottom left corner of the rectangle symbolizing the simple type. The syntax is similar to the format of CL with statement (CL BNF prod.491) except that `WithKeyWord` and the enclosing curly braces are omitted. The content of the `with-attributes` field shall be inserted into the with attributes of the given simple type when exporting DPF into CL.

The Value. The bottom right value field is only used in value notation and must be empty in type definitions. This field represents the value assigned to the type instance specified in field identifier. It holds a specific value or expression, which is valid for the current profile and results in a value adequate to the type defined. The example in Figure 4/c presents a simple type instance g of type G, an `integer` alias, having the value 8 assigned.

3 Structured Type and Value

DPF follows a general technique for building structured types; it defines operations for constructing structured types. These operations are ordering, choice, permutation and repetition. Ordering, which joins types into a sequence is expressed by the record construct. Choice offers type alternatives equivalent to union type of CL. Permutation of element types is represented with unordered set. Finally, type repetition is modeled with quantors.

DPF type definitions are represented on a plane where the imaginary vertical axis specifies ordering while the horizontal shows the potential alternatives. The axes themselves are invisible in DPF. The representation can always read unambiguously in top-down and left-right direction.

The graphical notation of structured types (Figure 5) consists of a tag incorporating the list of element types arranged on the plane according to the introduced principle.

---

[2] A suitable procedure is to generate `field-identifier`s from type identifiers by prefixing with literal `f_` and postfixing with `_n`, where n is the index of occurrences of the given type within the given context.

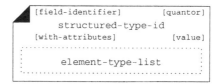

Fig. 5. DPF notation of structured type

The tag contains the structured type identifier (`structured-type-id`) and optionally the `field-identifier`, `quantor`, `with-attributes` and `value`. The meaning of these is similar to the identically named properties of simple types.

Field Identifier. The optional `field-identifier` can be used to assign an identifier to an instance of an embedded type definition. In value notation it is used to identify the appropriate constant, variable or template instance.

Quantor. The quantor is used to express record-of and set-of constructs (see section 4.5) including subtyping facility. The quantor is placed into the upper right corner of the tag. It can only be used when the construct comprises of a single element type! The quantor must appear in value notation but it must be omitted in stuctured type reference. The quantor has the format (ABNF):

quantor   =   min [".." max]   /   "*"   /   "+"

where `min` and `max` are integral numbers such that `max>min`. The shorthands `"*"` = `0..infinity` and `"+"` = `1..infinity` are also allowed.

With Attributes. The optional `with-attributes` has the same syntax and semantics as the with attributes of simple types in section 2.2 except that the with attributes specified for structured type apply for all element types. In case of an element type reference appearing in `element-type-list` has different with attributes this overrides the with attributes of the structured type.

Value. The value field shall only be used in value notation. It may hold either a reference to a structured type value or an expression that evaluates to the given structured type value. The value specified in the tag can be partially superseded by values defined in element list.

3.1 Element Type List

The `element-type-list` determines structured type content. This can include references to simple or complex types as well as embedded type definitions. The graphical format and layout of `element-type-list` depends on the construction used for structured type definition. Section 4 presents the graphical format of each structured type in detail.

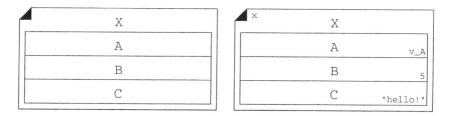

Fig. 6. Example for a record type and value notation

The element-type-list is omitted at structured type reference and structured type alias constructions. The graphical notation of these is identical to their counterparts at the simple types, i.e. simple type reference as shown in Figure 3/a and simple type alias in Figure 3/b.

4 Structured Type Constructions

4.1 Record Type and Value

Fixed ordering of element types is expressed using the record construction. The graphical notation of record type is a chain of element types along the position axis arranged such that rectangles of neighboring element types have joint edges. The record type definition example on Figure 6 is shown together with its value notation. Type X consists of an ordered list of types A, B and C. None of the element types have field names assigned. Note that A, B and C are not necessarily simple types. The value notation to the right assigns A the value v_A by reference, while B and C get the literals 5 and "hello!" respectively. The equivalent CL definitions are:

```
type record X {              const X x := {
    A    f_A,                    f_A := v_A,
    B    f_B,                    f_B := 5,
    C    f_C                     f_C := "hello!"
}                            }
```

4.2 Union Type and Value

The union construct expresses a choice of alternative types. DPF represents type alternatives orthogonal to the position axis. The element types are enlisted similar to the record construct but along the horizontal axis. The example in Figure 7 (left) defines the union Y as an alternative of the three element types A, B and C.

The value notation of union type permits only a single alternative to get value assigned. The not selected alternatives can as well be omitted. The example in Figure 7 (right) shows a union value containing element type B having value 5 assigned. The CL equivalent definitions are:

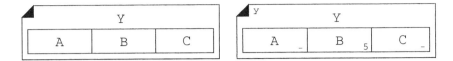

Fig. 7. Example for union type and value

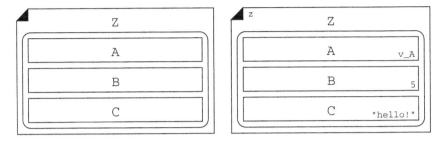

Fig. 8. DPF notation for set type and value

```
type union Y {           const Y y := {
    A    f_A,                f_B := 5
    B    f_B,            }
    C    f_C }
```

4.3 Set Type and Value

Arbitrary ordering of types is expressed with the set construct. The graphical representation of set is formed by enclosing the element types into a rectangular box with rounded corners such that the element types must not touch each other.

The example set type Z in Figure 8 (left) stands for any permutations of element types A, B and C. The value notation for the set type (Figure 8 (right)) assigns value to each element type either literally or by value reference. The equivalent CL definitions are:

```
type set Z {             const Z z := {
    A    f_A,                f_A := v_a,
    B    f_B,                f_B := 5,
    C    f_C                 f_C := "hello!"
}                        }
```

4.4 Optional Elements in Structured Types

Protocol specifications often make use of optional types. Optional types may occur in record and set types. The graphical notation distinguishes optional element types with a dotted border. The element types B and C in the example in Figure 9 are optional in both X0 record and Z0 set types.

The CL equivalent type dinfitions are:

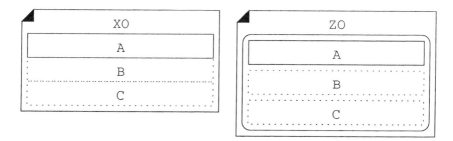

Fig. 9. DPF notation for optional element types in record and set types

```
type record XO {            type set ZO {
    A   f_A,                    A   f_A,
    B   f_B optional,           B   f_B optional,
    C   f_C optional            C   f_C optional
}                           }
```

The value notation of structured types containing optional element types is similar to structured types except that omitted optional element types get the "−" (omit) value assigned. Figure 10 contains example values for both types defined in Figure 9.

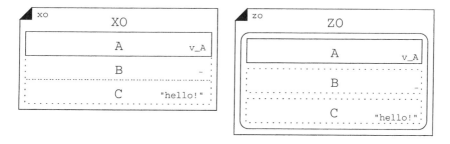

Fig. 10. DPF value notation for optional record and set types

The CL mapping of the constants in Figure 10:

```
const XO xo := {            const ZO zo := {
    f_A := v_A,                 f_A := v_A,
    f_B := omit,                f_B := omit,
    f_C := "hello!"             f_C := "hello!"
}                           }
```

4.5 Record and Set of Types

The quantor inside the tag of record or set types expresses record-of or set-of types. The element type list of set or record types definition must consist of a single element type. Note that the quantor can put length restrictions on set-of and record-of types and thereby express subtype constraint. The set-of type

P of the example in Figure 11 stands for zero or more unordered appearances of element type A. The record-of type R represents an ordered list of 1 to 3 repetitions of type A.

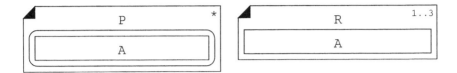

Fig. 11. ITN notation for set-of and record-of types

The equivalent CL definitions are:

```
type set of A P;          type record length(1..3) of A R;
```

The type notation for set-of and record-of types shall contain as many instances of its element type as specified by the quantor. The value of the quantor must always be in accordance with the type definition. The set-of and record-of values of the example in Figure 12 both consist of three element instances.

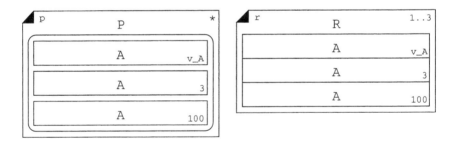

Fig. 12. Value notation for set-of and record-of types

The corresponding CL declarations are:

```
const P p := { v_a, 3, 100 };  const R r := { v_a, 3, 100 };
```

5 Embedded Type Definitions

TTCN-3 forbids embedded type definitions. DPF on the contrary provides graphical notation for embedded type definitions. The only requirement is that even the embedded type shall have an identifier assigned. The element-type-list part of structured type definitions may contain embedded type definitions of both simple and structured types. At embedded typed definition, the type definition to be embedded simply replaces its reference in the element type list.

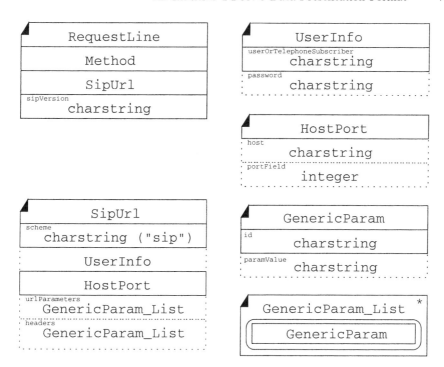

Fig. 13. DPF "flat" definitions for SIP RequestLine

The inner type inherits the field name and value attributes of the reference. The frame format (optional, mandatory) is also kept in the embedded type. When exporting DPF definitions into CL format, all these definitions shall be decomposed into standalone "flat" type definitions.

DPF implementation shall enable editing in flat view (Figure 13), in embedded view (Figure 14) and in mixed view (Figure 16). Furthermore, it shall support conversion between these modes by expanding references into embedded type definitions or vice versa collapsing embedded type definitions into references. The default type view could be the uppermost level flat view. The desired type reference shall then be expanded into an embedded definition at user request. The embedded view of a type could be collapsed when the user finished editing. This way the user could keep overview of the whole type hierarchy.

The example in Figure 13 shows structured type definitions of RequestLine part of the SIP Invite-Request message. The complete definition consists of five tables (except the simple type definitions). The equivalent embedded definition in Figure 14 fits into one table. Although the embedded notation can be harder to read in this printed form it expresses element type hierarchy and gives a better overview of the entire complex type definition.

The embedded graphical representation increases overview of complex data definitions and thereby helps to detect certain unconformities, which may occur in definitions. These include record-of/set-of types referenced in an optional

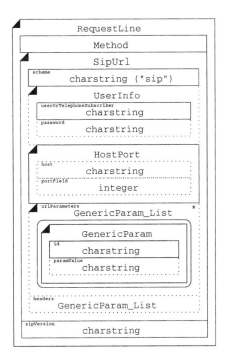
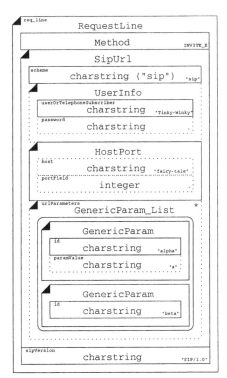

Fig. 14. Embedded DPF view of SIP RequestLine with an example value

field, an optional definition of a record/set consisting of solely optional elements and much more. The GenericParam_List type, which is defined to be a set-of GenericParam type, for instance, appears twice as an optional field of SipUrl. This feature can further contribute to correctly designed declaration parts.

The value notation for embedded type follows the same way as ordinary structured types. Figure 14 (right) presents a dummy value for RequestLine.

6 Recursive Definition

The recursive definition is a special case of embedded definition. The defined type contains a reference to itself. Following the CL guideline DPF does not place any restrictions on recursions therefore it is the user's responsibility to make sure that no endless recursion loop is defined. Figure 15 presents a simple example recursive type definition (left).

```
type record U {              var U u := { f_a := 1, f_U := {
    a   f_a,                     f_a := 3, f_U := {
    U   f_U optional                f_a := 5, f_U := omit } } };
}
```

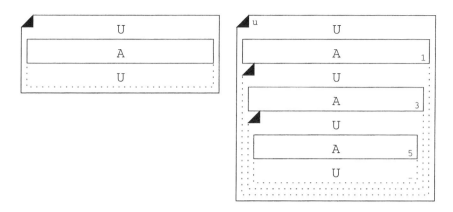

Fig. 15. Example of recursive type definition and value notation

The value notation, however, must not contain recursion. The recursive definitions shall be unfolded similarly to embedded type definitions before getting a value assigned. A matching value for the example recursive type definition is shown in Figure 15 (right).

7 Signature Declaration

The declaration of signatures is a prerequisite of procedure based communication. In DPF, the signature declarations are collected into a separate section. The graphical representation of a signature resembles to structured type definition (Figure 5). The signature name takes the place of structured type identifier. The return type identifier replaces the field identifier while the exceptions go into the top right corner into the position of the quantor. In case of non-blocking signature types the noblock keyword replaces the return type in the top left corner of the tag. The signature parameters appear on the element type list in an ordered fashion. For each item of the parameter list: the parameter type goes into the middle, the parameter name into the top left identifier while the direction indicator keyword into the top right corner. Figure 16 (right) contains the DPF equivalent of the following CL signature declaration:

```
signature IPConnection (inout ProtocolID protocol,
   in IPAddress srcHost, in integer srcPort,
   out IPAddress dstHost, out integer dstPort)
return SuccessIndication exception (ErrorException);
```

8 Template Definition

DPF provides graphical representation for CL templates. The notation resembles to value notation. The major difference is that wildcard symbols expressing

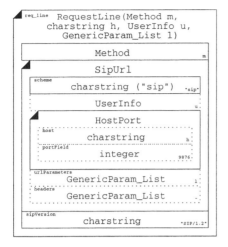
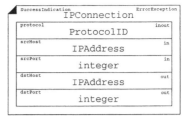

Fig. 16. DPF example for parameterized template definition (left) and signature declaration (right)

matching mechanisms can be used beside specific values and value references inside the bottom right value attribute. Parameterized templates are also available. The formal parameters list must be specified in parentheses behind the template identifier inside the tag. The formal parameters can then be incorporated as references in expressions of the value field.

DPF also supports modified templates. The name of the base template shall appear in the value attribute of the outermost template definition. The GUI shall then fill all elements according to the base template and offer the user to make changes. The implementation keeps track of the changes and creates appropriate modified templates when the module is exported to CL. A benefit of this approach is that the user always see the full template even if she works with modified templates. Optionally DPF implementation could also perform template optimization, which includes the generation of modified and parameterized templates when feasable.

Note that the representation on Figure 16 (left) contains both embedded definitions and references at the same time. This is an example of a mixed view.

9 Component and Port Type Definitions

Component and port types belong to different DPF sections. Both are based upon the graphical notation of the set type (Section 4.3).

DPF supports procedural, message based as well as mixed port types. The structured-type-spec of port type definition contains the name of the given port type. The field identifier consists of the literal procedure, message or mixed signalling the operation method of the port. The element type list consists of messages, procedure signatures or both consequently. The message type defini-

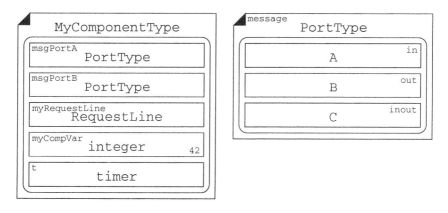

Fig. 17. DPF notation of component and port type definitions

tions and procedure signature declarations come from their respective sections. The keyword in, out or inout appearing on the top right corner determines the direction of the message or signature. Figure 17 (right) shows an example port type definition.

The graphical layout of component type is similar. The component type name belongs to the structured-type-spec. The component type definition must not have any attributes. TTCN-3 components may contain port, timer and variable declarations. The element type list uses DPF value notation. Port declarations consist of a reference to a port type and a port name inside the field identifier. Ports must not have other attributes that the field identifier. Variables are also declared by reference but these may have an initializer present inside the value attribute. Timers are declared using the timer keyword as simple-type-spec. Similarly to variables, timer may also have an initial value set. The example in Figure 17 (left) shows the DPF definition of MyComponentType. The CL equivalent definitions:

```
type component MyComponentType {          type port PortType message
    port PortType msgPortA, msgPortB;     {
    var integer myCompVar := 42;              in    A;
    var RequestLine myRequestLine;            out   B;
    timer t;                                  inout C;
}                                         }
```

10 Conclusion

We presented a new TTCN-3 presentation format, the Data Presentation Format, for intuitive graphical representation of TTCN-3 data. DPF provides graphical notation for all kinds of CL types and values, which is especially useful in large-scale test development where protocol(s) contain several hundreds or thousands of type definitions and templates. A typical DPF implementation enhances test data design and thereby reduces the lead time of test suite development. We hope that DPF gets acceptance from the testing community either as a stand-alone presentation format or as a complementary part of GFT.

References

1. Methods for Testing and Specification (MTS); The Testing and Test Control Notation version 3; Part 1: TTCN-3 Core Language, ETSI ES 201 873-1 V2.2.0 (2002-03).
2. Methods for Testing and Specification (MTS); The Testing and Test Control Notation version 3; Part 2: TTCN-3 Tabular Presentation Format (TFT), ETSI ES 201 873-2 V2.2.0, (2002-03).
3. Methods for Testing and Specification (MTS); The Testing and Test Control Notation version 3; Part 3: TTCN-3 Graphical presentation Format (GFT), ETSI TR 101 873-3 V1.2.1, (2002-05).
4. D. Crocker, Ed., P. Overell: Augmented BNF for Syntax Specifications: ABNF, RFC-2234, Nov. 1997.

The UML 2.0 Testing Profile and Its Relation to TTCN-3

Ina Schieferdecker[1], Zhen Ru Dai[2], Jens Grabowski[2], and Axel Rennoch[1]

[1] Fraunhofer FOKUS, Competence Center for Testing, Interoperability and Performance
Kaiserin-Augusta-Allee 31, D-10589 Berlin
http://www.fokus.fhg.de/tip

[2] University of Lübeck, Institute for Telematics
Ratzeburger Allee 160, D-23538 Lübeck
http://www.itm.mu-luebeck.de

Abstract. UML models focus primarily on the definition of system structure and behaviour, but provide only limited means for describing test objectives and test procedures. However, with the approach towards system engineering with automated code generation, the need for solid conformance testing has increased. In June 2001, an OMG Request For Proposal (RFP) on an UML2.0 Testing Profile (UTP) has been initiated. This RFP solicits proposals for a UML2.0 profile, which enables the specification of tests for structural and behavioural aspects of computational UML models, and which is capable to interoperate with existing test technologies for black box testing. This paper discusses different approaches for testing with UML and discusses the ongoing work of the Testing Profile. Special emphasize is laid on the mapping of UML2.0 testing concepts to the standardized Testing and Test Control Notation (TTCN-3).

1 Introduction

It is well known that the development and implementation of conformance tests is expensive w.r.t. time and money. Several initiatives and efforts have been undertaken to establish an approach to automate - or at least to provide significant support for an automated - test generation. Algorithms have been defined to derive tests from formal system specification given in various notations. Their usage has been demonstrated with sample applications. But today, none of the approaches is widely used in the industrial practise for large applications. One reason may be the difficulty to select the test cases from a (theoretical) unbounded number of tests, which result from test generation algorithms. But often it is simply the lack of a formal system specification of the implementation under test, i.e. the base for the application of the test generation algorithms is missing.

Economic reasons still require computer support for test generation to enhance confidence in software reliability. The traditional distinction between system and test design appears inefficient and is possibly faulty since knowledge transfer between two specifications is needed. Today modern system modelling techniques like UML have an increasing acceptance in the software development community which is much higher than the acceptance of Formal Description Technics in the past (the reason might be that the better tool support nowadays). There is a chance to involve system developers in the test definition process if the modelling language allows to integrate testing related information. Test related information at this early stage means to benefit from the know-how of a system developer and to capture testing related ideas of developers in the test definition process.

In principle, it is possible to start the test derivation with a system model only, i.e. skeletons of the test cases will be generated from the system model. As mentioned before, this process may lead to an unbounded number of tests (e.g. due to an infinite number of test data values). A practical alternative may be the incooperation of test relevant information into the system model, e.g. with annotations provided by the system developer to restrict the number of test cases from the very beginning.

UML technology focuses primarily on the definition of system structure and behaviour and provides limited means for describing test procedures only. With the approach towards system engineering according to model-driven architectures with automated code generation, the need for solid conformance testing, certification and branding has increased. In June 2001, an OMG Request For Proposal (RFP) on an UML Testing Profile has been initiated. It shall provide specification means to define precisely tests for structural (static) and behavioural (dynamic) aspects of systems modelled in UML.

IBM, Ericsson, FOKUS, Motorola, Rational, Softeam and Telelogic formed a consortium to develop that UML Testing Profile (UTP). This profile is based on the concepts of the upcoming version of UML, UML2.0, which is still an ongoing work. The work of UTP is based on recent developments in testing such as TTCN-3 and COTE. It provides mappings to established test environments such as JUnit and TTCN-3.

This paper presents different approaches to UML based testing and discusses the need for test specifications in UML (Section 2). Section 3 presents the recent work on the UML testing profile. In Section 4, the relationship of UTP and TTCN-3 concepts is discussed. In addition, a possible way of mapping UTP specifications to TTCN-3 definitions is presented. The paper concludes with an outlook of the UTP work.

2 Different Approaches on Testing with UML

Several approaches for the integration of testing related information in system developments with UML exist. According to the classical test methodology, two different aspects have to be distinguished: the modelling of system related features (i.e. of the system under test, the SUT), and the definition of the test model features (i.e. of the test system).

In the following, different testing frameworks and practical approaches based on UML systems and tools are discussed. One approach is called *integration testing*. It adopts the UML syntax and uses statechart diagrams for the objects under test. The aim of integration testing is to minimize testing costs, time and effort, i.e. to initially develop customized test drivers, test stubs and test cases and to adapt and rerun them repeatedly for regression testing purposes at each level of integration. Tool support is available e.g. for applications built with the UML modelling tools in Rational Rose.

Other initiatives have developed *UML based test notations*. In , the use of UML to support test development has been investigated to encourage the parallel development of a conformance test suite and a standard system specification. The presented guidelines have been given in the context of TTCN as the target test notation. The suggested test development activities adopt a straight forward approach, i.e. the identification of the independent system components is followed by a definition of the test configuration, test case structure and test cases. UML component or deployment diagrams can be used to represent the test configuration, e.g. in a distributed environment system parts are represented by UML components, points of control and observation (PCOs) by UML interfaces. The test suite structure can be defined in UML with class diagrams whereas any hierarchy of possible test (sub)groups is expressed by a nested class structure. The test behaviour of a test case can be defined using various UML features: UML interaction diagrams, i.e. sequence and collaboration diagrams, and state transition diagrams.

JUnit[19] is a framework for automated unit tests based on Java. Because of its simplicity, JUnit has become popular for *extreme programming* where permanent code integration and code testing are required. JUnit provides an own graphical user interface. A JUnit test defines a testsuite which is composed of several test cases or test suites. A test case contains test methods, a *setup()* method, a *teardown()* method, a *main* method which runs the testcase and an optional *suite()* method which groups test methods in a test suite. At the end of a test run, JUnit reports a *pass, failure* or *error* as its test result. Black box testing can be realized by defining private or protected test methods. JUnit has already been integrated in many case tools, e.g. JBuilder. From a UML model, skeletons for unit tests of individual classes or packages can be recursively generated, including the tested classes and packages. However, the hard part, namely the coding of the dynamic part of the testcases, is not addressed.

The consequent application of UML for the specification of a component-based SUT leads to an interesting approach for the derivation of a *component-based test system*. In , a Test Framework (TFW) is proposed which contains base types of test components. The test purpose independent (i.e. generic) behaviour for these components has been predefined. It covers setup and configuration of a test system, initiation of tests, exchange of coordination messages between test components and collection of test results. The full test system itself is built from the test components. For both test system and SUT the technology independent UML method has been selected e.g. to allow the test system to share the same static information with the SUT. The TFW builds the final test system from a generic test system (GTS) which has one test manager as the main components, one front end, one Main Test Component (MTC) and a set of parallel TCs (PTC). Similar to the Conformance Testing Methodology and Framework (CTMF) , the system behaviour of the GTS comprises test preamble (configuration establishment and test initiation), test body (generic test case) and test postamble (test end and configuration release), whereas the configuration may be hold for several tests in sequence. A user defined test system (UTS) will inherit from the GTS components MTC and TCs and will be refined according to the selected test purposes. During these refinements, both the test case independent and dependent behaviour have to be distinguished. The generic behaviour may be overloaded to locate references to e.g. naming service, timer management, SUT specific initial objects or UTS specific object repositories, or assign a particular order on the PTC start, etc. The testcase dependent behaviour can be defined with sequence diagrams to provide e.g. sequential request/reply pairs at several interfaces. Multiple instances of these behavioural definitions may be used to describe performance tests, too. It has been proposed to define other (alternative) behaviour with separate sequence diagrams (on default events, timeouts, unexpected behaviour to be ignored). An association of such diagrams to the test case dependent behaviour may be possible with the introduction of activity diagrams.

In most cases, the automation of test generation processes is based on a sequence of different tool applications. The tools perform individual tasks within a sequence of manual or (semi-) automatic transformations (e.g. compilation) of refinements (e.g. extractions) in a step-wise approach. An UML based test tool chain has been proposed by the AGEDIS project. The approach is characterized by the goal to reuse/adopt existing system validation and test code generation tools. It starts with a standard UML system model managed by usual UML tools. Further processing of the UML model results (via XML) in the Intermediate Format (IF) which has been defined and chosen to describe the system model in a state machine manner, and to have a suitable input document to existing model checking and test suite generation tools. The resulting Abstract Test Suite (ATS) is provided by the Test Generation with Verification (TGV) tool in the standardized TTCN format. This allows the application of TTCN tools from the telecommunication industry to produce executable test cases in the desired target API language (e.g. C, C++, Java).

The approaches presented above show that different academic and industrial initiatives have already been undertaken to test on the basis of UML. It appears reasonable that testing becomes an issue within UML itself.

3 The UML2.0 Testing Profile

In this section, the ongoing work for the testing profile is presented. The profile is developed in several steps: the concept space combining various testing areas is defined, a metamodel aligned with the UML 2.0 meta-model is developed, an the applicability of the concepts is analyzed via example and a mapping to existing test infrastructures demonstrates the practical use.

The work on UTP was – besides other sources - based on as the only standardized test notation TTCN-3. The initial intent to base the UML Testing Profile on the Graphical Format of TTCN-3 could not be taken directly since additional requirements from software testing together with the alignment with UML required additions and generalizations. Major generalizations in UTP are:

- the separation of test behaviour and test evaluation by introducing a new test component: the *arbiter*. This enables the easy reuse of test behaviour for other testing kinds without changing the test behaviour but just the arbiter. This concept is comparable to the evaluate function of TSSL.
- the integration of the concepts of test control, test group and test case into just *one concept of a test case*, which can be decomposed into several lower level test cases. This enables the easy reuse of test case definitions in various hierarchies. A test suite is then just a top-level test case. This concept is comparable to the test object concept of TSSL.

- the support of *data partitions* not only for observations, but also for stimuli. This allows to describe test cases logically without having the need to define the stimulus data completely but as a set or range of values. This concept is comparable to the concept of Test Data Definitions (TDD) of ADL.

Furthermore, some additions ease the practical use of the UML Testing Profile:

- an *initial test configuration* is used to describe the setup of the test components and the connectivity to the SUT and between each other
- component and deployment diagrams are used to enable the definition of software components realizing a test suite and to describe the requirements regarding test execution on certain nodes in a network.

The different background of the UTP members has led to an intensive discussion on the basic set of terms as a number of topics allow alternative views. The result is explained by describing the actual terminology. It has been agreed to distinguish three major groups of terms:

- *test architecture*, i.e. the elements and their relationship which are involved in a test,
- *test data*, i.e. the structures and meaning of values to be processed in a test, and
- *test behaviour*, which address the observations and activities during a test.

3.1 Test Architecture

The test architecture sub package covers the concepts for specifying test components, the interfaces of and connections between test components and to the SUT. Test components are active entities within the test system which perform the test behavior defined in a test case (see Test Behavior sub package) by using test data as defined in the Test Data sub package.

The *test architecture* is a set of related classes and/or components from which test case specific configurations may be specified. A *test context* groups test cases with the same initial test configuration. The *test configuration* is a collection of parts representing test components and the SUT and the connections between the test components and to the SUT. The test configuration defines both (1) test components and connections when a test case is started and (2) the maximal number of test components and connections during the test execution. A *test component* is an active object within a test system performing a test scenario. A test component has a set of interfaces via which it may communicate with other test components or with the SUT when the respective interfaces are connected. An *arbiter* is a specific test component to evaluate test results and to assign the overall verdict of a test case. There is a default arbiter for functional, conformance testing, which generates *pass, fail, inconc,* and *error* as verdict, where these verdicts are ordered as pass < inconc < fail < error. In addition to test components, utility parts can be used to denote helper and miscellaneous parts to realize a test system, e.g. to contain additional data to be used during testing.

An *interface* is a specification of a set of possible operations/messages which a client may request of/send to a test component or to the SUT or which a server may receive from a test component or SUT. An interface is either procedure-based or message-based. A *connection* is a communication path between two interfaces.

The *system under test (SUT)* is characterised by the set of interfaces via which a real SUT can be controlled and observed during testing. An SUT can be on different abstraction levels: a complete systems, a subsystem, a single component, object or even a class.

An example is given for a bank automaton - an ATM (Fig. 1). The bank automaton offers various interfaces, in particular, a port to the bank network and interfaces to the user to insert and withdraw a bankcard as well as to take the money. The test objective is to check that it is possible to debit the account provided that enough funds are available. The package ATMTest imports the definition from the ATM SUT and defines the test suite ATMtestsuite as well as the classes for the test components HWEmulator and BankEmulator. The test suite is used to define one test case validWithdrawal(). authorizeCard() is an auxillary operation used within the test case.

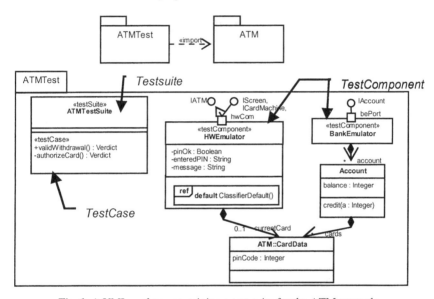

Fig. 1. A UML package containing a test suite for the ATM example

In addition, the test configuration as the internal structure of test suite is given. The test behavior is assigned to the test case and is invoked when the test case is invoked. The test suite uses two test components (Fig. 2Fig. 2): a bank emulator be and a hardware emulator hw. A utility part current represents the bankcard used during the tests. The test components are connected with the SUT via the interfaces atmPort and netCom.

3.2 Test Data

The test data sub package covers the concepts for data sent to the SUT and received from the SUT. Mechanisms in order to change and compare test data are used to enable precise and succinct test specifications. Data can be concrete (i.e. a specific value) or abstract (i.e. a logically described set of values). Logical partitions are used

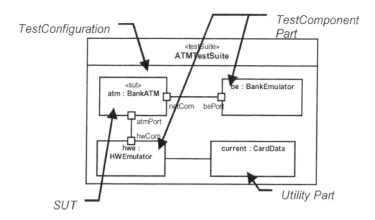

Fig. 2. The test configuration for the ATM example

to define such value sets within test parameters. Coding rules are part of the test specification and denote the encoding and decoding of test data. By means of coding rules, the interfaces of the SUT can be bound to certain encodings such as for CORBA GIOP/IIOP, IDL, ASN.1 PER or XML.

In the ATM example, different messages to and from the SUT are used. They are declared in the class diagram of the ATM such as

```
messageDisplay(in Message:string)
```

In the test behaviour, concrete data is used for example

```
messageDisplay("EnterPIN")
```

Another example is of a data declaration is

```
constant Integer amount {findAccount(current).balance > amount }
```

where `amount` is characterized by the constraint contained in parenthesis, i.e. it has to be less then the balance of the account belonging to the current bankcard.

3.3 Test Behavior

A *test case* is a specification of one case to test the system, including what to test with which input, result, and under which conditions. It uses a concrete technical specification of how the SUT should be tested - the test behaviour. A test case is the implementation of a test objective for a particular test configuration, which is defined by the test behaviour. A test case uses an arbiter to evaluate the outcome of its test behaviour. A test objective is a general description of what should be tested. The test behaviour is the specification of behaviour performed on a given test configuration, i.e. sequences, alternatives, loops and defaults of stimuli to and observations from the SUT. Test behaviours can be defined by any behavioural diagram of UML 2.0, i.e. as interaction diagrams or state machines. There can be a designated main test behaviour for a given test configuration. By invocation, test cases can make use of other test behaviours.

A *verdict* is the outcome of a test case being *pass*, *fail*, *inconc*, or *error* as defined in TTCN-3. Additional verdict information can be used to denote specific test outcomes e.g. for performance tests. Every test component handles a local verdict. Verdict updates are reported to the arbiter for calculation of the overall verdict of the test case. Different schemes to realize an arbiter and the coordination with the test components exist.

A *validation action* is an action to evaluate the status of the execution of a test scenario by assessing the SUT observations and/or additional characteristics/parameters of the SUT. A validation action is performed by a test component and sets the local verdict of that component.

Defaults can be defined on three levels: individually for events in interaction diagrams or for states in state machines, for test components of a specific class or for all test components in a test system, i.e. the *basedefault()*. These defaults are evaluated in sequence – from the event default up to the basedefault.

During the execution of a test case a *test trace* is generated. It contains logs for each action performed during that test case execution and the test result of that test case execution. A log action can be used to store additional information in the test trace.

Fig. 3 (left side) depicts the ATM test case *ValidWithdrawal()*. The objective of the test is to verify that if a user inserts and authorizes a valid card correctly, he is able to withdraw money if he has sufficient funds, i.e. the test case defines a test for a valid withdrawal of money: after authorization of the bankcard (by referencing to the *authorizeCard* operation) the *withdrawal* operation is selected and an *amount* requested, which is smaller than the balance of the account related to the bankcard. This is defined by a logical partition with a constraint on amount (see top on the left side of Fig. 3). The SUT then interacts with the bank emulator *be* to debit the account and delivers the money afterwards. An event specific default *DisplayDefault* is used for the display event in order to handle different display messages specifically. The default behavior is depicted by means of a note notation. The specification of the default is shown in Fig. 3 (right side). Finally, the verdict is set to *pass*.

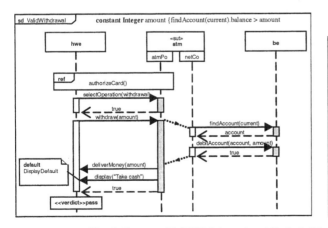
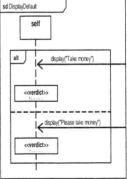

Fig. 3. Test case ValidWithdrawal and Default DisplayDefault

4 The Relation to TTCN-3

Since there was no accepted test notation in UML yet, the UTP request for proposal was an ideal opportunity to bring TTCN-3 in form of GFT to the attention of the UML world. In fact, GFT is the archetype for UTP. UTP uses several concepts being developed in GFT. Still, GFT and UTP differ in several respects: UTP is based on the object oriented paradigm of UML where behaviours are bound to objects only, while GFT is based on the TTCN-3 concept of functions and binding of functions to test components. UTP uses additional diagrams to define e.g. the test architecture, test configuration and test deployment. Test behaviour can be defined as interaction diagrams but also as state machines. While GFT supports dynamic configurations in terms of kind and number of test components and the connectivity to the SUT and between test components[1], UTP uses static configurations where only the number of test components may vary but not the structure of the connections between test components. In addition, UTP has only one FIFO queue per test component, while GFT uses a FIFO queue per test component port.

New concepts in UTP are the arbiter, the validation action, the test trace and the logical partition. According to its definition, an arbiter is a special test component which is responsible for assigning verdicts. Therefore, an arbiter can easily be defined as a test component in TTCN-3 which is created at the beginning of a test-run. In order to overrule the verdict mechanism of TTCN-3, a special verdict type has to be used in addition. Validation action can be realized with external functions. The logical partition of test data for test stimuli is not supported in TTCN-3. Here, a data generation function (as an external function) has to be used in order to select a specific value from that logical partition to be sent to the SUT. A test trace is not specifically part of the TTCN-3 concepts, but can be considered as a test case with just a single sequential execution and, therefore, requires the same specification concepts as test cases.

The verdict handling in GFT is bound to the well-established verdict handling of conformance testing, while UTP uses in addition the ability of user-defined verdicts and the arbitration of verdicts, i.e. the definition of algorithms of when and how verdicts are determined. Additional validation actions can be used to calculate local verdicts of test components by the use of external information from the test execution suite.

Another difference is that of default handling for unexpected or irrelevant behaviour from the SUT: GFT uses function-based defaults which can be dynamically activated and deactivated during test execution, while UTP uses structural defaults, which are bound to the structure of a test system – from test component level down to event/state level – leading to a defaults hierarchy and less dynamic default handling.

UTP supports UML data only, i.e. primitive types (Boolean, String, Integer) and classes, while GFT supports all types available in TTCN-3 such as basic types (integer, char, universal char, float, Boolean, objid, verdicttype), basic string types (bitstring, hexstring, octetstring, charstring, universal charstring), user-defined structured types (record, record of, set, set of, enumerated, union) and anytype. In addition, any imported data like ASN.1 or IDL is supported.

---

[1] In TTCN-3 and hence in GFT, ports can even be connected, reconnected, started, stopped and cleared during test execution, which leads to dynamic test configurations in terms of connectivity between test components and to the SUT.

Table 1. Relation of UTP and TTCN-3 concepts and the principal translation from UTP to TTCN-3

UML Testing Profile	TTCN-3
Test Architecture:	
Package	Module
Test Suite	Group covering all test cases of a test suite Having a specific TSI component type (to access the SUT) Having a specific behavioral function to set up the initial test configuration for this test suite
System Under Test (SUT)	The test system accesses the SUT via the abstract test system interfaces (TSI). The SUT interfaces result in port types used by TSI One additional port is needed to communicate with a user-defined arbiter Potentially additional ports are needed to coordinate/synchronize test components
Interfaces	Port types
Test Components	Component types
Test Configurations	Configuration operations create, start, connect, disconnect, map, unmap, running and done for dynamic test configurations. Behavioral function to set up the initial test configuration
Arbiter	The UTP default arbiter is a TTCN-3 built-in User-defined arbiters are realized by the MTC
Test Data:	
Test Parameter, Test Argument	(Inline) templates are used for both test stimulus and test observations
Coding rules	Encode attribute
Test Behaviour:	
Test Case	Testcase
Test Objective	Not part of TTCN-3, just a comment to a test case definition
Test Behaviour	Functions generated via mapping functions per behavior feature of a test suite Test case behavior resulting from creating test components and starting their behavior, MTC just as a „controller" which also controls the arbiter
Test Trace	Not part of TTCN-3, but could be mapped just as a strict sequential behavioral function
Stimulus	Sending messages and Calling operations Replying to operation invocations (however, raising exceptions is not yet well handled in UML 2.0)
Observation	Rreceiving messages, operation invocations, and operation replies (however, catching exceptions is not yet well handled in UML 2.0)
Default	Altstep and activation/deactivation of the altsteps along the default hierarchy
Coordination	Message exchange between test components.
Verdict	The default arbiter and its verdict handling is an integral part of TTCN-3 For user-defined, a special verdict type and updating the arbiter with set verdicts is needed
Validation Action	External function or data functions resulting in a value of the specific verdict type
Log Action	Log operation

Last but not least, GFT and UTP are on different levels of abstractions: GFT (being part of TTCN-3) is on a detailed test case specification level (i.e. on a level from which executable tests can directly be derived). However, UTP can also be used on more abstract levels by defining just the principal constituents of e.g. a test purpose or of a test case without giving all the details needed to execute the tests. While this is of great advantage in the test design process, additional means have to be taken in order to generate executable tests. For example, the expressiveness of UML 2.0 sequence diagrams allows to describe a whole set of test cases by just one diagram, so that test generation methods have to be applied in order to derive these tests from the diagram.

Overall, UTP is targeted at UML providing selected extensions to the features of GFT/TTCN-3 as well as restricting/omitting other TTCN-3 features. Table 1 compares the UML 2.0 testing profile concepts with existing TTCN-3 testing concepts. All UML Testing Profile concepts have direct correspondence or can be mapped to TTCN-3 testing concepts. A mapping from UTP to TTCN-3 is possible but not the other way around. The principal approach for the mapping to TTCN-3 consists of two major steps: (1) take UTP stereotypes and associations and assign them to TTCN-3 concepts and (2) define procedures how to collect required information for the TTCN-3 modules to be generated.

In the following, an example mapping[2] is provided for the Bank ATM case study described in the previous section (test case in Fig. 3). Two TTCN-3 modules are generated: one for ATM being the SUT (and being defined in a separate UML package) and another module for the ATM test architecture defining the tests for the Bank ATM also in a separate UML package. The module ATM provides all the signatures available at the SUT interfaces, which are used during testing.

```
module ATM {
    //withdraw(amount : Integer): Boolean
    signature withdraw(integer amount) return boolean;
    //isPinCorrect(c : Integer) : Boolean
    signature isPinCorrect(integer c) return boolean;
    //selectOperation(op : OpKind) : Boolean
    signature selectOperation(OpKind op) return boolean;
    … // and so on
}
```

The module for the ATM test architecture ATMTestArchitecture imports all the definitions from the ATM module, defines the group for the ATM test suite, provides port and component type definitions within the group, the function to set up the initial test configuration and finally the test cases. In order to make this mapping more compelling, a user-defined arbiter is assumed in addition and the default handling is made explicitly.

```
module ATMTestArchitecture {
    import from ATM all;
    // utility Account type
    type record Account {
                integer balance,
                charstring number
    }
    //credit(a : Integer)
    signature credit(integer a);
```

---

[2] Please note that UTP will provide an example mapping only as there are several ways to map to TTCN-3. It is not the intend to restrict the mappings to a single one, but rather to show the principles and to leave options for the implementers.

```
            //debit(a : Integer)
            signature debit(integer a);
            // utility accnts : Account [0..*]
            external const Account accnts[0..infinity];
            group ATMSuite {
                ... // all the definitions constituting the tests for ATM
            } // group ATMSuite
    } // module ATMTestArchitecture
```

The required and provided interfaces are reflected in corresponding port definitions `atmPort_PType` and `netCom_PType`, which are then used in the component type definitions `BankEmulator_CType` and `HWEmulator_CType` to constitute the component types for the PTCs:

```
            //required interfaces: IScreen, ICardMachine, IMoneyBox
            //provided interface: IATM
            type port atmPort_PType procedure {
                in display_; //Iscreen
                in ejectCard; //ICardMachine
                in deliverMoney; //IMoneyBox
                in getStatus; // status information
                out withdraw, isPinCorrect,
                    selectOperation, storeCardData; //IATM
                out enterPIN; //to give a PIN
            }
            //required interface: IAccount
            //no provided interface
            type port netCom_PType procedure {
                in debitAccount, findAccount //IAccount
            }
            // test component type BankEmulator
            type component BankEmulator_CType {
              port netCom_PType bePort;
              port Arbiter_PType arbiter; // user defined arbiter
            }
            // test component type HWEmulator
            type component HWEmulator_CType {
              port atmPort_PType hwCom;
              port Arbiter_PType arbiter; // user defined arbiter
            }
```

The following shows the mapping for a user-defined arbiter. A specific type `MyVerdict_Type` together with an arbitration function `Arbitration` is used to calculate the overall verdict during test case execution. The final assessment is given by mapping the user-defined verdicts to the TTCN-3 verdict at the end. This enables e.g. the use of statistical verdicts where e.g. 5% failures lead to fail but less failures to pass. The arbiter is realized by the MTC. It receives verdict update information via a separate port `arbiter`. The arbitrated verdict is stored in a local variable `mv`.

```
            //the arbitration
            type enumerated MyVerdict_Type {
                pass_, fail_, inconc_, none_
            }
            type port Arbiter_PType message {
                inout MyVerdict_Type
            }
            // the MTC is just a controller
            type component MTC_CType {
                port Arbiter_PType arbiter; // user defined arbiter
                var MyVerdict_Type mv:= none_;
            }
```

```
function Arbitration
(BankEmulator_CType be, HWEmulator_CType hwe)
runs on MTC_CType {
    while (be.running or hwe.running) {
        alt {
            [] arbiter.receive(none_) {…}
               [] …}
        }
    }
    if (mv == pass_) { setverdict(pass) }
    else …
}
```

The defaults in the defaults hierarchy are mapped to several altsteps, which will be invoked later along that hierarchy. In this example, an altstep for every component type is defined, i.e. HWEmulator_classifierdefault and BankEmulator_classifierdefault.

```
altstep HWEmulator_classifierdefault()
runs on HWEmulator_CType {
    var charstring s;
    [] hwCom.getcall(getStatus:{}) {
       hwCom.reply(getStatus:{} value true);}
    [] hwCom.getcall(ejectCard:{}) {arbiter.send(fail_);}
    [] hwCom.getcall(display_:{?}) -> param (s) {
            if (s == "Connection lost") {
   arbiter.send(inconc_) } else {arbiter.send(fail_)} }
}
altstep BankEmulator_classifierdefault()
runs on BankEmulator_CType {
    … }
```

The component type for the test system interface SUT_CType is constituted by the ports netCom and atmPort used during testing in the specific test suite. A configuration function ATMSuite_Configuration sets up the initial test configuration and is invoked at first by every test case of that test suite.

```
// SUT
type component SUT_CType {
  port netCom_PType netCom;
  port atmPort_PType atmPort;
}

// setup the configuration
function ATMSuite_Configuration
( in SUT_CType theSUT, in MTC_CType theMTC, inout
    BankEmulator_CType be, inout HWEmulator_CType hwe)
{
 be:=BankEmulator_CType.create;
 map(theSUT:netCom,be:bePort); //map to the SUT
 hwe:=HWEmulator_CType.create;
 map(theSUT:atmPort,hwe:hwCom); //map to the SUT
 connect(theMTC:arbiter,be:arbiter); // arbitration
 connect(theMTC:arbiter,hwe:arbiter); // arbitration
}
```

The validWithdrawal test case uses two PTCs hwe and be each having its own test behaviour, which is defined by behavioural functions validWithdrawal_hwe and validWithdrawal_be as shown below.

```
function validWithdrawal_hwe()
runs on HWEmulator_CType {
    activate(HWEmulator_default());
    activate(HWEmulator_classifierdefault());
    authorizeCard_hwe();
    hwCom.call(selectOperation:{withdrawal}) {
      [] hwCom.getreply(selectOperation:{?} value true)
          {}
    }
    if (amount <= accnt.balance) {
        hwCom.call(withdraw:{amount},nowait);
        hwCom.getcall(deliverMoney: {amount});
        hwCom.getcall(display_:{"Take cash"});
        hwCom.getreply(withdraw:{?} value true);
    }
    else {
        log("not enough money on the card
            to withdraw amount");
        setverdict(inconc);
    }
    setverdict(pass);
}

function validWithdrawal_be()
  runs on BankEmulator_CType {
    activate(BankEmulator_default());
    bePort.getcall(findAccount: {current});
    bePort.reply(findAccount: {current} value accnt);
    bePort.getcall(debitAccount: {accnt,amount});
    bePort.reply(debitAccount: {accnt,amount}
       value true);
    setverdict(pass);
}
```

Finally, the test case can be provided. According to the initial test configuration, two PTCs `hwe` and `be` are used. The configuration is set up with `ATMSuite_Configuration`. The test behaviour on the PTCs is started with `validWithdrawal_hwe` and `validWithdrawal_be`. The arbiter `Arbitration(be,hwe)` controls the correct termination of the test case. This completes the mapping.

```
//+ validWithdrawal() : Verdict
   testcase validWithdrawal_test()
     runs on MTC_CType system SUT_CType {
       var HWEmulator_CType hwe;
       var BankEmulator_CType be; // initial configuration
       ATMSuite_Configuration(system,mtc,be,hwe);
       hwe.start(validWithdrawal_hwe());
       be.start(validWithdrawal_be());
       Arbitration(be,hwe);
   }
```

5 Outlook

UML has been discovered by both software engineers and test developers to specify system and test models in a platform independent manner. With an integrated approach of developing a system and its tests within one framework, tests can be developed more efficiently and economically. Special attention has been given to the OMG's initiative on defining a UML testing profile. It supports independent test laboratories in their work but also the system engineers to perform the test runs by their own.

The status of the basic test concepts and terminology which have been presented in this paper can be regarded as a consensus of different R&D scientists and engineers working in heterogeneous IT fields like object-oriented systems or telecom protocols specifically on testing aspects. Fundamental elements of the UML testing profile's test architecture, test data and test behaviour have been collected and applied exemplarily using UML related class and sequence diagrams. A comparison with the established concepts of TTCN-3 confirms the suitability of the selected definitions in the UML testing profile. The UML testing profile elements can be mapped to TTCN-3 but not vice versa. This mapping allows to base implementations of the UML testing profile on top of TTCN-3 test environments.

At the time of writing this contribution, the work on the UML profile for testing is still ongoing in OMG and no final decisions have been made on the UML extension mechanisms, i.e. stereotypes, constraints or tagged values, selected for the different testing concepts. Due to the dependencies on the UML2.0 release, which is expected only in 2003, it is expected that the final submission of the UML profile for testing will be available mid to end of 2003. Nevertheless the importance of testing with UML has to be elaborated earlier to assist its acceptance.

Acknowledgements

The authors thank the UTP consortium and supporters for the joint work and discussions. Particular thanks go to Paul Baker, Oystein Haugen, Serge Lucio, Johan Nordin, Eric Samuelson and Clay Williams.

References

1. J. Hartmann et al.: UML-Based Integration Testing. ISSTA´00. Portland, Oregon.
2. ETSI: Methods for Testing and Specifications (MTS); Methodological approach to the use of object-orientation in the standards making process. ETSI EG 201 872 (August 2001). Sophia Antipolis (F).
3. ISO/IEC 9646-3: Information Technology - Open Systems Interconnection - Conformance Testing Methodology and Framework (CTMF) - Part 3: The Tree and Tabular Combined Notation (TTCN), edition 2, Dec. 1997.
4. M. Born et al.: Test Framework for Component-Based Systems. ICDCS' 2000 & DS-VV'2000, Taipei (Taiwan), April 2000.
5. ISO/IEC 9646: Information Technology - Open Systems Interconnection - Conformance Testing Methodology and Framework (CTMF).
6. C. Crichton et al.: Using UML for Automatic Test Generation: ASE'2001.
7. A. Cavarra et al.: AGEDIS Language Specification. Project Deliverable 2.2. The AGEDIS project, 2001, http://www.agedis.de.
8. L. Clark et al.: Achieving Cross-Platform Compatibility with Increased Productivity and Quality using the OMG's Model Driven Architecture. Lockheed Martin Corporation, 2001.
9. ETSI ES 201 873-1: The Testing and Test Control Notation version 3; Part 1: TTCN-3 Core Language. V2.1.0 (2001-10), 2001; also an ITU-T standard Z.140.
10. ETSI DES 201 873-3 V2.0.0: The Testing and Test Control Notation version 3; Part3: Graphical Presentation Format for TTCN-3 (GFT). V2.0.0 (2001-11), 2001.

11. J.-C. Fernandez et al.: An experiment in automatic generation of test suites for protocols with verification technology. Science of Computer Programming, 1997. http://citeseer.nj.new.com/2326.html.
12. C. Jard, S. Pickin: COTE – Component Testing using the Unified Modelling Language. - ERCIM News No.48, January 2002.
13. T. Vassiliou-Gioles et al.: Configuration and Execution Support for Distributed Systems.- IWTCS'99, Budapest, Hungary, Sept. 1999.
14. E. Rudolph, J. Grabowski, and P. Graubmann. Towards a Harmonization of UML-Sequence Diagrams and MSC. In R. Dssouli, G. v. Bochmann, and Y. Lahav, editors, SDL'99 - The next Millenium. Elsevier, June 1999.
15. E. Rudolph, I. Schieferdecker, and J. Grabowski. Development of an MSC/UML Test Format. BT'2000 - Formale Beschreibungstechniken für verteilte Systeme. Shaker Verlag, Aachen, June 2000.
16. R. Soley: Model Driven Architecture: An Introduction. http://www.omg.org.
17. The Open Group: ADL 2.0 Translation System, 1998. http://adl.opengroup.org/
18. I. Wilie et al.: UML Action Specification Language (ASL) Reference Guide. Kennedy Carter Ltd., Feb. 2001.
19. R. Hightower, N. Lesiecki: Java Tools for eXtreme Programming, Wiley Computer Publishing, 2002.
20. I.Schieferdecker, J. Grabowski: The Graphical Format of TTCN-3 in the context of MSC and UML. Proceedings of the 3rd Workshop of the SDL Forum Society on SDL and MSC (SAM'2002), Aberystwyth (UK), June, 24 - 26, 2002.
21. UML testing profile home page: http://www.fokus.gmd.de/U2TP/

Realizing Distributed TTCN-3 Test Systems with TCI

Ina Schieferdecker[1] and Theofanis Vassiliou-Gioles[2]

[1] Fraunhofer FOKUS, Competence Center for Testing, Interoperability and Performance
Kaiserin-Augusta-Allee 31, D-10589 Berlin, Germany
schieferdecker@fokus.fhg.de
[2] Testing Technologies IST GmbH, Oranienburger Str. 65, D-10117 Berlin, Germany
vassiliou@testingtech.de

Abstract. Distributed test setups for efficient load, performance, scalability, interworking, and end-to-end tests are gaining importance for the assessment of distributed communicating systems. The Testing and Test Control Notation TTCN-3 provides concepts for component-based distributed test systems in dynamic test configurations, where test components may reside on various network nodes to be near the interfaces of the tested system. The realization of executable TTCN-3 tests on concrete test platforms involves TTCN-3 compilation/interpretation and adaptations to the test platform. The TTCN-3 Control Interfaces TCI define entities, interfaces, types and operations needed to flexibly manage and distribute TTCN-3 based test systems. It complements and completes the TTCN-3 Runtime Interface TRI. This paper discusses the underlying concepts of TCI and demonstrates its use for the realization of a distributed test for the Session Initiation Protocol SIP.

1 Introduction

The Testing and Test Control Notation TTCN-3 is a test specification and implementation language to define test procedures for black-box testing of distributed systems. TTCN-3 allows an easy and efficient description of complex distributed test behavior in terms of sequences, alternatives, loops and parallel stimuli and responses. Stimuli and responses are exchanged at the interfaces of the system under test, which are defined as a collection of ports being either message-based for asynchronous communication or signature-based for synchronous communication. The test system can use any number of test components to perform test procedures in parallel. Likewise to the interfaces of the system under test, the interfaces of the test components are described as ports.

The development of TTCN-3 was forced by key players of the telecommunication industries and science to get a single test notation for nearly all black-box testing needs. Especially the newly introduced support of dynamic distributed test setups, i.e. dynamic creation and termination of test components including dynamic connections between test components and to the system under test (the SUT), enables new applications of TTCN while keeping the mature and stable test concepts. TTCN-3 test specifications are not only a basis for functional and conformance testing, but also for performance, load and scalability tests. Such tests require varying load conditions for the SUT, which can be realized by an ensemble of parallel test components. Since the test system has to be as performant as the system under test, any realistic load for the SUT can be realized in a distributed environment only: the parallel test component have to be distributed and located on remote nodes in a network constituting a distributed test system.

One essential benefit of TTCN-3 is that it enables the specification of tests in a platform independent manner. Hence, TTCN-3 provides the concepts of test components, their creation, their communication links to each other and to the SUT, their execution and termination as such on abstract level only. Means to control the distributed execution of test components and coordination between them are outside TTCN-3.

However, the application of executable tests to a SUT within test campaigns requires the realization and implementation of such distributed test systems in a network of test nodes – at best in a well-defined manner to enable a standardized adaptation for the management, component handling, communication and logging between distributed test nodes. Another aspect of this adaptation is the ability to reuse external encoders/decoders, which are also outside TTCN-3 and just referenced within a test specification.

Well-defined interfaces as a set of operations independent of the target, i.e. independent of the SUT, processing platform, implementation language, etc will enable that code from any TTCN-3 compiler or interpreter supporting and using these interfaces can be executed on any test platform/test device, which supports and uses these interfaces. A first step towards this code independence was done with the TTCN-3 Runtime Interface TRI 24: TRI provides an interface to adapt a TTCN-3 test system to the SUT by providing means to adapt the communication with the SUT as well as to adapt the timer handling. As such, TRI defines a local adaptation to the SUT only. The aspects of test management, component handling (both on local and remote nodes) as well as the type and value handling have not been considered by TRI. These aspects can be summarized as being the adaptation to the test system being either a single test device or a test platform consisting of several test nodes[1]. TRI has to be supplemented by interfaces to enable a well-defined adaptation to the test platform. These interfaces are called the TTCN-3 Control Interfaces TCI.

TCI together with TRI provide a complete solution for a well-defined adaptation to the test system and to the SUT providing maximal flexibility in realizing TTCN-3 test systems. Only recently – at the ETSI MTS meeting, October 2002 – the importance of TRI and TCI for TTCN-3 has been reflected: TRI 4 and TCI 5 were made integral parts of the multi-part standard for TTCN-3.

TCI is currently developed at ETSI and is expected to become together with TRI the future standard interface set for all TTCN-3 test system implementations. It considers previous work in GCI 8 on type and value interfaces and in TSP1+ 7 and its implementation 6 on distributed test systems. However, the approach of TCI on generic distributed test setups is new in several respects:

- it enables the implementation of the new TTCN-3 concepts for dynamic test configurations – different network and platform technologies can be used to realize distribution and communication within the test platform,

---

[1] In the following we will use the term test platform to refer both to a single node as well as a multi node test system.

- it enables the flexible reuse of coders/decoders without predefining any internal type and value representation for the test system implementation and
- it enables the integration of TTCN-3 test systems into existing test management environments and applications by concentrating on TTCN-3 related test management aspects only: this enables more flexibility for test management specific application domain and goes beyond TSP1+ 7

We provide first insights into how various kinds of test systems can be build based on the concepts of TCI: in Section 2, the general entities, interfaces and types of TCI together with a selected uses case on a more complex test components handling are presented and discussed. Afterwards in Section 3, the approach is illustrated with an application example for testing the Session Initiation Protocol SIP 1. Section 4 concludes the paper with a summary and an outlook on the future work on TCI.

2 Overview of the TTCN-3 Control Interfaces

A TTCN-3 test system can be conceptually thought of as a set of interacting entities where each entity corresponds to a particular aspect of functionality in a test system implementation. These entities manage test execution, interpret or execute compiled TTCN-3 code, realize proper communication with the SUT, handle types, values and test components, implement external functions, and handle timer operations.

The part of the test system that deals with interpretation and execution of TTCN-3 modules, i.e., the Executable Test Suite (ETS), is shown as the *TTCN-3 Executable (TE)*. Within the TE individual structural elements can be identified, like Control, Behaviour, Components, Types and Values and Queues. The structural elements within the TE represent functionality that is defined within a TTCN-3 module or by a TTCN-3 specification itself. For example, the structural element "Control" represents the control part within a TTCN-3 module, while the structural element "Queues" represent the requirement on a TTCN-3 Executable that each port of a test component maintains its own port queue. While the first is specified within a TTCN-3 module the later is defined by the TTCN-3 specification. The TE corresponds typically to the executable code produced by a TTCN-3 compiler or a TTCN-3 interpreter. Prior to a test system implementation, the *Abstract Test Suite (ATS)*, i.e., the TTCN-3 modules being the test specification, has been compiled into an executable format - the *Executable Test Suite (ETS)*. The TE can be executed centralized, i.e. on a single test device or distributed, i.e. on different physical test devices. Although the structural entities of the TE implement a complete TTCN-3 module, single structural entities might be distributed over several test devices.

The TE implements a TTCN-3 module on an abstract level. The other entities of a TTCN-3 test make these abstract concepts concrete. For example the abstract concept of sending an event or receiving a timeout cannot be implemented within the TE. The platform adaptors of the test system realize e.g. the encoding of the message and it's sending over concrete physical means or measuring the time and determine when a timer has expired, respectively. The *System Adaptor (SA)* for the communication with the SUT and the *Platform Adaptor (PA)* for the realization of timers and their interaction with the TE are defined in [TRI].

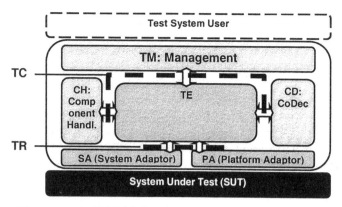

Fig. 1. General Structure of a TTCN-3 Test System. A TTCN-3 test system implementation consists of a part that deals with the interpretation and execution of TTCN-3 module (TE) and parts that either adapt the test system to the System Under Test (SA and PA) or to the test system platform (TM, CD and CH).

The TCI specification defines the interaction between the TE and the *Test Management and Control (TMC)* entities. In TMC, we can distinguish between functionality related to test execution management (TM), component handling (CH), and coding and decoding handling (CD).

The *test management (TM)* entity is responsible for the overall management of a test system. After the test system has been initialised, test execution starts within the TM entity. The entity is responsible for the proper invocation of TTCN-3 modules, i.e., propagating module parameters and/or IXIT information, i.e. implementation extra information for testing, to the TE if necessary. Typically, this entity would also implement a test system user interface. In addition, the TM entity performs test event logging and presentation to the test system user.

As the TE can be distributed among several test devices the *component handling (CH)* is responsible to implement the distribution and communication between the distributed entities. The CH provides the means to synchronize the different entities of the test system being potentially located on several nodes. The general structure of a test system distributed via several nodes is depicted in Figure 2.

On each node, a test execution TE together with system adaptor SA, platform adaptor PA and coder/decoder CD is performed. The entities CH and TM mediate the test management and test component handling between the TEs on each node. There is a special TE that is identified to be the TE that started a test case[2] and that is responsible for calculating the final verdict of that test case. Besides this, all TEs are handled the same.

Communication is in one respect the message or procedure based communication between TTCN-3 components. Therefore, the CH adapts message and procedure based communication of TTCN-3 components to the particular execution platform of the test system. It is aware of connections between TTCN-3 test component communication ports. It is responsible to propagate send request operations from a single

---

[2] Please note that in course of executing a TTCN-3 module there can be at most one test case being executed.

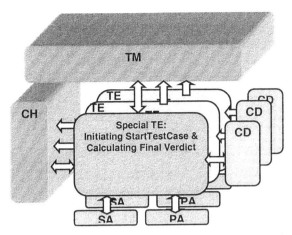

Fig. 2. General Structure of a distributed TTCN-3 Test System. A distributed TTCN-3 test system consists of one CH and TM entity. Each TE is located on a separate node, together with its own SA, PA and CD entities.

TTCN-3 component that resides within a certain TE to the targeted component residing potentially in a different instance of the same TE on a different test device. It then notifies the TE about received test events by enqueueing them in the port queues of the TE.

Procedure based communication operations between TTCN-3 components are also visible at the CH. The CH is responsible to distinguish between the different messages within procedure-based communication (i.e., call, reply, and exception) and to propagate them in the appropriate manner to the targeted component TE. TTCN-3 procedure based communication semantics, i.e., the effect of such operation on TTCN-3 test component execution, are to be handled in the TE.

Furthermore, there is additional test component management communication necessary in order to implement the distribution of test components between several test devices. Component management communication includes the indication of the creation of test components, the starting of execution of a test component, verdict distribution as well as component termination indication. For this the CH does not implement the behaviour of TTCN-3 component but the communication between several components that are implemented within the TE.

The *coding/decoding (CD)* entity is responsible for the encoding and decoding of TTCN-3 values into bitstrings suitable to be sent to the SUT. The TE determines which codecs shall be used and passes the TTCN-3 data to the appropriate codec in order to obtain the encoded data. Received data is decoded in the CD entity by using the appropriate decoder, which translates the received data into TTCN-3 values.

The TCI operations of TM, CH and CD are atomic operations in the calling entity. The called entity, which implements a TCI operation, returns control to the calling entity as soon as its intended effect has been accomplished or if the operation cannot be completed successfully. The called entity is not blocked, so that performant test implementations are enabled.

TCI is defined by a set of abstract types to realize the TTCN-3 type and value system, a set of operations at required and provided subinterfaces of TM, CH and CD, and a set of scenarios to show the use of abstract types and operations. The types and

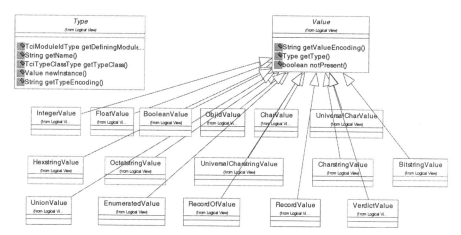

Fig. 3. Hierarchy of abstract values. Each abstract value provides at least the operations of the abstract data type Value and form a hierarchy of abstract values. Operations on abstract values are not shown in this figure.

operations are provided in the OMG IDL 9. Language mappings to Java and C will show the realization of TCI on specific test platforms.

2.1 The Abstract Data Type Model

Abstract data types are use to describe on a high-level which kind of data shall be passed from a calling to a called entity. In addition, the abstract data types are used to define how TTCN-3 data is passed from the TE to a coder that encode a TTCN-3 value representation to a bitstring and from a decoder to the TE to decode a bitstring into a TTCN-3 value representation. For these abstract data type a set of operations are defined in order to process the data by the coder/decoder. The concrete representation of these abstract data types as well as the definition of basic data types like String, boolean are defined in a respective language mappings.

A set of abstract data types builds up the TTCN-3 type and value representation. For every abstract data type a set of operations has been defined in order to define the functionality of the abstract data type. Operations on or with this abstract data type return either a value of this abstract type or a basic type like boolean. The abstract TTCN-3 type and value representation consists of two parts:

- An abstract data type Type that represents all TTCN-3 types in a TTCN-3 module
- Different abstract data types that represent TTCN-3 values, i.e. TTCN-3 values of a given TTCN-3 type. This can be either values of TTCN-3 predefined types or of TTCN-3 user-defined types.

For the abstract data type Type operations to reference predefined and user-defined TTCN-3 data types, and to create and maintain TTCN-3 values are defined. The following figure presents the hierarchy between the abstract data types for TTCN-3 values (short: abstract values):

All TTCN-3 abstract values share a common base abstract data type, the abstract data type Value. For the abstract values that share the common base abstract data type, all operations that are defined on the base data type are implicitly defined for the abstract values, too.

2.2 The Test Management Interface

The TCI Test Management Interface (TM) describes the operations a TE is required to implement and the operations a test management implementation shall provide to the TE. A test management implementation provides overall test management to the test system user. It requires from the TE the presence of operations to start and stop test execution of a TTCN-3 module or of certain test cases in a TTCN-3 module. In turn it provides operations to the TE for resolving module parameter at runtime, logging and the indication of execution termination.

2.3 The Component Handling Interface

A component handling implementation distributes TTCN-3 configuration operations like create, connect and start and inter-component communication like send on a connected port among one or more TTCN-3 Executables participating in a test session.

The basic principle is that CD is not *implementing* any kind TTCN-3 functionality. Instead it will be informed by the TE that for example a test component shall be created. Based on component handling (CH) internal knowledge the request for creation of a test component will be transmitted to another (remote) participating TE. This second (remote) participating TE will create the TTCN-3 component and will provide a handle back to the requesting (local) TE. The requesting (local) TE can now operate on the created test component via this component handle.

Within the operation definitions, the terms local TE and remote TE is used to highlight the fact that a test system implementation might be distributed over several test devices, each of them hosting a complete TE. The terms "local" and "remote" always refer to the interface currently described. For convenience reasons, the term "local" refers always to the TE being either the callee of an operation (for *required* operations) or the caller of an operation (for *provided* operations). While the TE is conceptually considered as being distributed the CH is considered to be non-distributed. This can either be achieved using a centralized architecture or by using a middleware-platform that abstracts from distribution aspects. Although the TE might be distributed over different physical device, there might be configurations where only one, non-distributed TE will participate in a test session. In this case the term "local" and "remote" refer to the same TE instance.

Although all TTCN-3 Executable participating in a test session are equal there is a distinct TE*. This TE* is the TE where a test case has been started explicitly, i.e. the explicit tciStartTestCase() has been processed, or the test control by means of tciStartControl(). The reason for this distinction is, that this TE* is responsible for global verdict calculation and is therefore informed about any test component termination yielding the final local verdict of the terminated test component. Finally, the TE* will notify the test management upon termination of a test case execution with the overall final test case verdict.

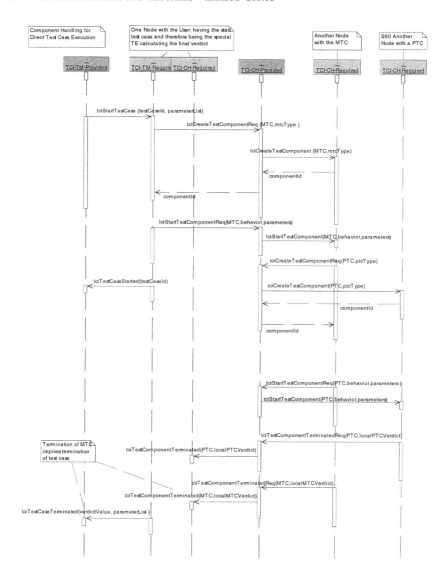

Fig. 4. Test case execution and termination. Test execution will be started at the TCI-TM interface. TM will be informed after termination of test execution together with the final and the actual parameter list.

2.4 The Codec Interface

A codec implementation encodes TTCN-3 values according to the encoding attribute into a bitstring and decodes a bitstring according to decoding hypothesis. To be able to decode a bitstring into a TTCN-3 value, the CD requires certain functionality from the TE. The basic operation required by CD implementation from the TE is the provisioning of value instances, either for basic or structured types. Together with the possibility to query values on their type information the CD provides encoding and decoding functionality to the TE. An example of how this can be achieved will be presented later.

2.5 A Selected Use Case

This section shows a use case on starting a test case directly from the test management user interface. The module containing the test case is selected first. When the test case is started, the main test component is created. Then, the test case behaviour is started on this main test component. Whenever a parallel test component is used within a test case, it is handled the same: the parallel test component is created first: giving a test component create request to the TCI-CH entity, which propagates the test component create to the TE in which the parallel test component shall be created.

The identifier for the created parallel test component is returned. The identifier is then used to start the PTC behaviour of the start operation. When the PTC terminates its execution, a test component terminate request together with the local test verdict is issued in order to inform CH about this termination. The same is done when the main test component terminates. In addition, the termination of the main test component leads to the overall termination of the test case. The test management interface receives in that case the final verdict.

3 A Distributed TTCN-3 Test System Example

To illustrate the construction of a distributed test system, a real-world IP application scenario is presented. The scenario demonstrates the transition from a pure functional test to a scalability test for the Session Initiation Protocol (SIP) [1]. In the given example, the ability of a proxy server to handle simultaneous initiation of several multimedia sessions is analyzed. SIP is an IETF signaling protocol for the establishment, maintenance and tear down of multimedia connections in the Internet using UDP or TCP as underlying transport mechanism.

Figure 5. illustrates a general test configuration for testing a SIP UA Server. Figure 5.a displays a test configuration where a single Parallel Test Component, executing a functional test behavior, interacts with the System Under Test, while Figure 5.b shows the simultaneous execution of the same functional behavior on multiple parallel test components. In both cases, the Master Test Component coordinates the test execution.

3.1 A TTCN-3 ATS Example

A TTCN-3 abstract test suite (ATS) specifying a simple functional test case using the single PTC configuration can be specified as follows:

```
module sipTest {
  testcase functionalTest()
     runs on MyMtcType system MyTSI
  {
     var MyPtcType sipUAc = MyPtcType.create;
     map(sipUAc:S, system:R);
     connect(mtc:C, sipUAc:C);
     sipUAc.start(uaCBehavior(USER));
     C.send(startPrimitive);
     all component.done;
  }
}
```

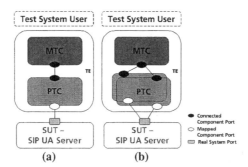

Fig. 5. A general SIP User Agent (UA) Server testing configuration. Parallel test components (PTC) interact with the System Under Test (SUT) at the real the real system port[3]. The parallel test components communicate with the Master Test Component (MTC) using the connected ports. The Master and the Parallel Test Components are executed within the TTCN-3 Executable (TE) while the Test System User manages test execution.

First, the Master Test Component of type MyMtcType creates a Parallel Test Component of type MyPtcType. After mapping the PTC port S to a Test System Interface port R (as defined in MyTSI) and connecting the port C of the MTC and the PTC the behavior uaCBehavior() will be executed. The MTC instructs the PTC to proceed with test case execution by sending the startPrimitive template. The test case will terminate after the PTC has terminated test behavior execution. This example assumes that uaCBehavior() specifies all the necessary behavior in order to asses the SUT.

One aim in a multiple test components scenario is to reuse definitions of a functional test scenario for the definition of e.g. a scalability test. The PTC will execute the same test behavior while the MTC is responsible to set up the test configuration.

In our particular example, MAXNUMBER PTCs shall execute their behavior simultaneously. For this the MTC instructs the PTCs to start the test via the startPrimitive after all participating test components have been created and the functional test behavior uaCBehavior()has been started[4]. MAXNUMBER has been defined as an integer module parameter. The value of MAXNUMBER will be resolved at runtime.

The following TTCN-3 fragment illustrates this scalability test scenario:

```
modulepar { integer MAXNUMBER := 10 ; // 10 is default }

testcase scalabilityTest()
    runs on MyMtcType system MyTSI
```

---

[3] The figures display only the mapped component ports at the PTC for the communication with the SUT. The necessary test system interface port has been omitted for the sake of readability.

[4] We distinguish between the starting of test component, which is achieved using the start operation on a component and the execution of a test, i.e. performing communication with the SUT to asses its validity. From a TTCN-3 point of view the test behavior start after the call of the start operation. From the test logic point of view we assume that the test will start after the reception of the startPrimitive.

```
{
  var integer i;

  for(i:=0; i < MAXNUMBER; i := i + 1) {
    sipUAc[i] = MyPtcType.create;
    map(sipUAc[i]:S, system:R);
    connect(mtc:C, sipUAc[i]:C);
    sipUAc.start(uaCBehavior(USER[i]));
  }

  for(i:=0; i < MAXNUMBER; i := i + 1) {
    C.send(startPrimitive) to sipUAc[i] ;
  }

  all component.done ;
}
```

Typically, a TTCN-3 ATS contains not only the type, data and behavior definitions but also a control part, that relates the execution of test cases. This example assumes that the `scalabilityTest()` will be executed only if the `functionalTest()` was successful, i.e. terminated with the verdict PASS.

3.2 Application of the TCI Operations

Although the definition of language mapping for TCI is still in progress, a Java language mapping, derived from the IDL definitions in TCI is presented here to illustrate the application of the TCI operations. Although the used operations and signature are by no means complete they present a good selection for an insight into possible implementation of TCI operations.

In order to start test execution, the test system user has to instruct the TTCN-3 Executable (TE) to start either the control part of the module or a particular test case. TCI defines for this two different operations, the `tciStartTestCase()` and the `tciStartControl()` operation. Prior to test execution the module has to be selected.

```
public class MyManagement() implements TciTMProvided {
  static TciTMRequired TE = getTE();

  public void main(String[] args) {
    TE.tciSetModule("sipTest");
    if(startTestCase) {
      TE.tciStartTestCase("functionalTest", null);
      waitForTestCaseTermination() ;
      TE.tciStartTestCase("scalabilityTest", null);
      waitForTestCaseTermination() ;
    } else {
      TE.tciStartControl();
      waitForTestControlTermination(); }
  }
  // -- TCI-TM Provided Implementations --
  ...
  public Value tciGetModulePar(String param) {
    if(param.equals("MAXNUMBER")) return determineMAXNUMBER();
    return null; }

  public void tciTestCaseTerminated(VerdictValue
      verdict, TciParameterList list) { ... }

  public void tciTestControlTerminated() { ... }
}
```

Like all following code fragments, this fragment is not complete and shall give only an impression on how the operations exposed by the TE in the *TM Required* interface might be used, and how the TM and the other components provide the functionality needed for correct execution of the TTCN-3 within the TE.

The main method first determines whether the test cases should be executed from within the control part (startTestCase == false) or the test cases shall be started directly. After test execution has been initiated within the TE, the TM (an instance of class MyManagement) has to wait until the TE indicates termination of test execution using the methods provided by the TM.

The TE can be accessed by the TM using a global variable TE, which implements the interface TciTMRequired. The handling of this global knowledge is out of scope of TCI and will not be further discussed.

After test execution has started, the TE will execute the test behavior as defined in the TTCN-3 specification. The TCI Component Handling Interface addresses all issues related to component management and inter-component communication. A possible message exchange for a test component creation is shown in Figure 5.

The following Java fragment shows a possible implementation of the CH:

```
public class MyCH() implements TciCHProvided {

  static TciCHRequired ONE_NODE     = getONE_NODE();
  static TciCHRequired ANOTHER_NODE = getANOTHER_NODE();
  static int noOfComponentsCreated  = 0 ;

  public TriComponentId tciCreateTestComponentReq (
      TciTestComponentKind kind, Type componentType) {
    if(distributed && ((noOfComponentsCreated % 2) == 0) {
      TriComponentId tc =
        ANOTHER_NODE.tciCreateTestComponent(kind,
          componentType); register(kind, tc, ANOTHER_NODE);
      noOfComponentsCreated++;
      return tc ; }
    else {
      TriComponentId tc =
        ONE_NODE.tciCreateTestComponent(kind,
          componentType); register(kind, tc, ONE_NODE);
      noOfComponentsCreated++;
      return tc ; }

  }
}
```

This implementation assumes that there will be only one instance of MyCH (the CH) and that this instance is known to the TE. However, the fragment shows also that CH can have access to multiple instances of TE. The location of this, possible different, instances is not restricted to the same test device. It is an implementation decision on how to retrieve references to the distributed objects.

Based on the variable distributed, every second component will be created on a remote (ANOTHER_NODE) TE implementation. If distributed is false every component will be created on the local TE implementation (ONE_NODE)[5]. As tciComponentCreationReq will be called also when a test component creates another test component (e.g. the MTC a PTC, or a PTC another PTC), this implementation would distribute test components equally on both participating nodes, ONE_NODE and ANOTHER_NODE. CH performs book-keeping by registering the kind, the id and the place a component resides for later usage.

---

[5] It is assumed that the tciCreateComponentReq was called by ONE_NODE.

Starting of test behavior as well as setting up connections between components and performing communication is performed in a comparable way. The following Java fragment can give an impression on how the CH could be extended to provide this functionality.

```java
public void tciConnectReq (
    TriPortId fromPort, TriPortId toPort) {
  TriComponentId fromC=fromPort.getComponentId() ;
  TriComponentId toC=toPort.getComponentId() ;

  // resolve() returns the TciCHRequired instance the
  // remote component resides on
  resolve(fromC).tciConnected(fromPort, toPort) ;
  resolve(toC).tciConnected(toPort, fromPort) ;
  registerConnection(fromPort, toPort) ;
}

public void tciSendConnected (TriPortId sender,
    TriComponentId receiver, Value sendMessage) {

  // retrieveConnection returns the TciCHRequired
  //instance the receiver resides on
  retrieveConnection(sender, receiver).
     tciEnqueueMsgConnected(sender,
         receiver, sendMessage) ;
}
```

As can be seen, the main task of the CH is to route the requested operations to the destination and monitor the setup and tear down of connections. By exposing this TCI-CH interface to the user, the user can implement distribution strategies as required. Whenever data has to be passed from the TE to the remaining test system, i.e. TM, CH or SA and PA (as defined in [4]), data that has been defined abstract within TTCN-3 has to be translated into a concrete representation. In case of communication with the SUT the abstract data has to be encoded according to the encoding rules. Besides the abstract TTCN-3 data types and values as described in previous sections, TCI defines the Codec Interface (TCI-CD) to enable the TE to pass the abstract TTCN-3 data to the appropriate codecs prior to sending data or performing matching operations on received data. For each encoding rule CD provides two operations to the TE, `encode()` and `decode()`. An implementation in Java might look as follows:

```java
public class MyEncodingRule() implements TciCDProvided {
  public TriMessage encode(Value value) {

    // TRI
    TriMessage encodedMsg = new TriMessageImpl();
    return encodedMsg.setEncodedMessage(furtherEncode(value));
  }

  public Value decode(TriMessage message,
      Type decodingHypothesis) {
    byte[] encodedMsg = message.getEncodedMessage() ; //TRI
    return furtherDecode(encodedMsg, decodingHypothesis) ;
  }
}
```

For each encoding rule this functionality has to be provided. In order to be able to encode the TTCN-3 data from a from a `Value` into an encoded message, the TE offers additional functionality to the TM, CH and the CD for the handling of types and values.

The abstract data type `Type` offers some functionality to obtain instances of any TTCN-3 type as defined in a TTCN-3 module. This applies both for predefined and user-defined types. Differentiation between basic types and structured types can be achieved using a type class concept, i.e. different types classes exists for each predefined or subtype TTCN-3 type, (e.g. the type class is BITSTRING if the type represent a TTCN-3 `bitstring` type) and for structured types (e.g. the class is RECORD if the type represents a TTCN-3 record type). Additionally the encoding of a type, according to the TTCN-3 specification can be retrieved. Instantiations of a given type, the Values, can be created by using the `newInstance()` operation that is defined for the abstract data type Type.

Values can be read and manipulated depending on their type. For example for a TTCN-3 `integer` the abstract data type IntegerValue has been defined, with operation like

```
- integer getInt()
- void setInt(in integer integerValue )
```

TTCN-3 `record` values have operations like

```
- Value getField(in String fieldName)
- void   setField(in String fieldName, Value fieldValue)
- String[] getFieldNames()
```

Using this operation as offered by the TE, CD is able to encode and decode values according to the encoding, CH is able to send messages without the need of encoding them, and the TM to provide value for module parameters.

4 Conclusions

This paper discusses the realization of distributed test systems being defined in TTCN-3. Although TTCN-3 does not define distribution schemes or distribution patterns for test components, i.e. distribution is not in the scope of TTCN-3, the pure fact that component-based test systems with dynamic test configurations can be used to define e.g. load and scalability tests on an abstract level requires the ability for distribution: only if a test system is performant enough it can make valid assessments about a test systems. The possibility to distribute test components and to handle their setup, coordination and communication is key here.

The TTCN-3 Control Interfaces TCI are an approach to close this gap and provide together with TRI a complete set of interfaces, entities, types and operations for the adaptation of TTCN-3 tests to the test platform and the tested system. TCI provides means for test management, test component handling and coding/decoding. This paper presents the basic concepts of TCI and discussed its implementation concepts. The potential of TCI are illustrated for a load test of a SIP application.

Future work beyond TCI will consider automated means for test deployment onto test platforms: while TCI provides means for automated test execution, the preparation of the test platform with e.g. all the code needed to perform the tests or the configuration of the SUT and the test devices, is not yet considered. First approaches exist for the deployment of distributed systems in general. These concepts need to be investigated for application in a test context in general and specifically for TTCN-3. An automated and flexible mixture of SUT and test components to check certain test purposes will be in particular interesting for software testing.

References

1. J. Rosenberg, H. Schulzrinne, et al: "SIP: Session Initiation Protocol", Draft IETF SIP RFC 3621, June 2002.
2. S. Schulz, T. Vassiliou-Gioles: "Implementation of TTCN-3 Test Systems using the TRI", IFIP 14th Intern. Conf. on Testing Communicating Systems -TestCom 2002-, Berlin, Germany, March 2002.
3. ETSI ES 201 873 – 1, v2.2.1: "The Testing and Test Control Notation TTCN-3: Core Language ", Oct. 2002.
4. ETSI ES 201 873 – 5, v1.0: "The TTCN-3 Runtime Interface (TRI); Concepts and Definition of the TRI", Oct. 2002, Draft.
5. ETSI DES 201 873 – 6, v1.0: "The TTCN-3 Control Interfaces (TCI); Concepts and Definition of the TCI", Oct. 2002, Draft.
6. T. Vassiliou-Gioles, M. Li, I. Schieferdecker, M. Born, M. Winkler: Configuration and Execution Support for Distributed Tests. - IFIP 12th International Workshop on Testing of Communicating Systems (IWTCS'99), Budapest (Hungary), Sept. 1999.
7. ETSI ES 201 770 V4.2.4: "The Test Synchronization Protocol TSP1+", Sept. 2000.
8. F. Brady and R.M. Baker, "INTOOL/GCI; Generic Compiler/Interpreter interface; GCI Interface Specification", INTOOL CGI/NPL038 (V2.2), Infrastructural Tools for Information Technology and Telecommunications Conformance Testing, Dec. 1996.
9. OMG CORBA v2.2: "The Common Object Request Broker: Architecture and Specification", Section 3, Feb. 1998.

*TIMED*TTCN-3 Based Graphical Real-Time Test Specification

Zhen Ru Dai, Jens Grabowski, and Helmut Neukirchen

Institute for Telematics, University of Lübeck
Ratzeburger Allee 160, D-23538 Lübeck, Germany
{dai,grabowsk,neukirchen}@itm.uni-luebeck.de
Tel.: +49 451 500 3721, Fax: +49 451 500 3722

Abstract. The textual *Testing and Test Control Notation* (TTCN-3) is frequently used in combination with *Message Sequence Chart* (MSC) and the MSC-based *Graphical Presentation Format for TTCN-3* (GFT). Both, MSC and GFT allow an automatic generation of TTCN-3 test case descriptions.
*TIMED*TTCN-3 is an extension of TTCN-3 for testing real-time properties and has been submitted for standardization. For a complete integration of *TIMED*TTCN-3 into the TTCN-3-based testing process, the usage of *TIMED*TTCN-3 in combination with MSC and GFT needs to be established.
This paper presents our approach for graphical real-time test specification based on MSC and *TIMED*GFT, which is our real-time extension of GFT. We explain how MSC can be used for the description of real-time test purposes and define *TIMED*GFT. Our approach includes the automatic generation of *TIMED*TTCN-3 test cases based on MSC test purposes and *TIMED*GFT diagrams.

1 Introduction

The *Testing and Test Control Notation* (TTCN-3) [8, 13] of the *European Telecommunications Standards Institute* (ETSI) becomes more and more popular for test specification and test implementation. Several TTCN-3 tools supporting the design, documentation, compilation and execution of test campaigns are under development or are already commercially available [4, 5, 22, 23]. Trials have shown that TTCN-3 is a modular and flexible test language that can be used effectively for testing numerous applications like, e.g., SIP [17], IPv6 [21], IDL-based interfaces [6] or XML-based interfaces [19]. Like the previous edition of TTCN-3, in the following called TTCN-2++ [7], TTCN-3 focusses on the description of test cases for functional black box testing.

The need for testing non-functional properties has been discussed in literature and several real-time and performance extensions for TTCN-2++ have been proposed [18, 24]. Based on these experiences *TIMED*TTCN-3, [3] a real-time extension for TTCN-3, has been developed and submitted for standardization. *TIMED*TTCN-3 provides the concepts for testing hard real-time requirements.

The test of soft real-time properties, e.g., arrival or loss rates, may be realized in TTCN-3 by executing identical test cases several times or by using external load generators that are controlled by TTCN-3 test components.

In most cases, the testing process is not only based on one, but on several languages: Graphical languages like *Message Sequence Chart* (MSC), the *Specification and Description Language* (SDL) or the *Unified Modelling Language* (UML) are used as a basis for test generation, test specification or test visualisation [1, 10, 15, 20] and data languages like the *Interface Definition Language* (IDL) or the *Abstract Syntax Notation One* (ASN.1) define the information to be exchanged between the *System Under Test* (SUT) and the test system during the test execution.

The usage of MSC [25] in functional black-box testing, including the standardized OSI conformance testing, as a graphical language for test purpose specification and test behaviour visualization has been thoroughly investigated by several authors [1, 2, 12, 15]. MSC is very well suited for the specification of test purposes and allows an automated generation of TTCN-2++ test cases. TTCN-3 is backwards compatible with TTCN-2++ and thus, an automated generation of TTCN-3 test cases is also possible. The visualization of TTCN-2++ and TTCN-3 test cases by means of MSC is more problematic. The reason is that MSC does not provide the right level of abstraction. Test cases are implementations that are written for specific test configurations and include implementation details that cannot be represented easily by MSC diagrams. Furthermore, lots of test information is related to test data and MSC does not support a graphical representation of data. To overcome this problem, the MSC-based *Graphical Presentation Format for TTCN-3* (GFT) [9] focusses on the presentation of test behaviour and extends MSC with special constructs that emphasize test and TTCN-3 specific concepts in behaviour definition, e.g., setting of test verdicts or activation and deactivation of default behaviour. The GFT extensions of MSC are defined as shorthand notations for (in some cases more complex) MSC constructs, i.e., GFT is compatible to MSC.

This paper is about the usage of MSC and GFT for real-time testing. We discuss the specification of real-time test purposes by means of MSC and explain how T_{IMED}TTCN-3 test cases can be generated from MSC real-time test purposes. For the visualization of T_{IMED}TTCN-3 test cases we present a real-time extension for GFT (T_{IMED}GFT), which allows a graphical presentation of T_{IMED}TTCN-3 concepts. For simplicity, all concepts and extensions are explained by using the same simple Inres-based example, which is used in [3] to explain T_{IMED}TTCN-3.

2 Foundations

In this section, we describe the usage of MSC and GFT in the TTCN-3-based testing process, explain the principles of T_{IMED}TTCN-3 and summarize the time concepts of MSC. For the complete understanding of this paper, a basic knowledge of MSC, GFT and the TTCN-3 core notation is required.

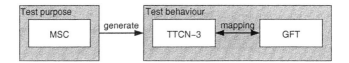

Fig. 1. MSC test purposes, TTCN-3 test behaviour and GFT diagrams

2.1 Test Purpose Definition and Test Behaviour Visualization

The difference between test purposes described in form of MSCs and a test behaviour[1] visualized by GFT diagrams are the different levels of abstraction and points of view. The relations between test purposes and the behaviour of test cases are shown in Fig. 1.

An MSC test purpose is an abstract description of a test. It describes the test from the perspective of the SUT and makes no assumptions about the implementation of the test configuration. By providing information about the test configuration, it is possible to generate TTCN-3 test behaviours. Test behaviour descriptions define tests from the perspective of the test system. They are written for a specific test configuration and include all the activities to coordinate the test components. A TTCN-3 test behaviour description can be visualized in form of a GFT diagram, i.e., there exists a bidirectional mapping between a TTCN-3 test behaviour and a corresponding GFT diagram.

An example may clarify these differences. We want to test an Initiator implementation of the Inres protocol [14]. The SUT can be accessed by using the interfaces ISAP and MSAP. The test purpose is the test of 100 data transfers. For doing this, a connection needs to be established and after the test, the connection has to be released. A formalization of this test purpose in form of an MSC is shown in Fig. 2b[2].

The MSC describes the test purpose from the point of view of the SUT, i.e., only the required information exchange at the ISAP and MSAP is shown. It includes no assumptions about implementation of the test system, e.g., the number of test components and the required synchronization among test components are not specified.

The test case InresRTexample shown in the lines 6-29 of Fig. 3[3] is generated automatically from our test purpose example (Fig. 2b) and the test configuration presented in Fig. 2a. The test configuration only consists of one *Main Test Component* (MTC), which controls both interfaces of the SUT. The test case InresRTexample can be visualized by the GFT diagram shown in Fig. 8[2]. A simple comparison of GFT diagram and textual TTCN-3 description shows that there exists a bi-directional mapping between text and graphics.

---

[1] In the following, the term *test behaviour* refers to TTCN-3 test cases, altsteps, functions and module control.
[2] The meaning and usage of the MSC time constructs, i.e., the dashed lines and arrows with annotations, in Fig. 2b and Fig. 4 will be explained in the following sections.
[3] The *TIMED*TTCN-3 statements, e.g., **now** and **resume**, in Fig. 3 are explained in the following sections.

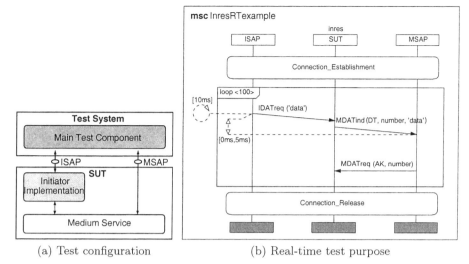

Fig. 2. Test architecture and test purpose for the Inres test case example

2.2 T_{IMED}TTCN-3

T_{IMED}TTCN-3 has been defined in [3]. It introduces the concept of absolute time into TTCN-3, provides a means to specify time synchronized test components, extends the TTCN-3 logging mechanism, supports online and offline evaluation of tests and adds the new test verdict **conf** to the existing TTCN-3 test verdicts.

Absolute time is introduced by means of clocks. Each test component has an associated local clock, which can be read by using a **now** operation (lines 13, 19 and 22 in Fig. 3). In addition, a **resume** statement can be used to wait until a certain point in time (line 16). For example, the statement **resume(self.now+3.0)** defines that the following statement or operation will be performed 3 seconds after the invocation of the **resume** statement. Time is represented as a **float** value that represents the number of seconds counted from a fixed starting point. The starting point is considered to be test system specific and therefore, not defined by T_{IMED}TTCN-3.

T_{IMED}TTCN-3 allows to specify time synchronized test components by means of *timezones*. A timezone is an (optional) attribute, which can be assigned to a test component when the component is created. Test components with the same attribute are considered to be synchronized, i.e., they have the same absolute time. Components without timezone attribute are considered to be not time synchronized with other components.

The TTCN-3 logging mechanism is improved by allowing to put TTCN-3 data structures into the the logfile and by giving access to the logfile in module control after test case termination. This allows to store timestamps of time critical test events as records in the logfile by using **log** statements (lines 19 and 22 in Fig. 3).

T_{IMED}TTCN-3 distinguishes between online and offline evaluation of real-time properties. Online evaluation refers to an evaluation during test execution,

```
(1)  module InresRTexample_module() {
(2)    type record TimestampType {
(3)      float logtime,
(4)      charstring id
(5)    }
(6)    testcase InresRTexample() runs on inres {
(7)      var float sendTime1:=-1.0;
(8)      var integer iterator1:=0;
(9)      var default OtherwiseFailDefault:=activate(OtherwiseFailAltStep);
(10)     Connection_Establishment();
(11)     for (iterator1:=0; iterator1<100; iterator1:=iterator1+1) {
(12)       if (sendTime1==-1.0) {
(13)         sendTime1:=self.now+0.01;
(14)       }
(15)       else {
(16)         resume(sendTime1);
(17)         sendTime1:=sendTime1+0.01;
(18)       }
(19)       log(TimestampType:{self.now,"IDATreq1"});
(20)       ISAP.send(IDATreq:{"data"});
(21)       MSAP.receive(MDATind:{DT,number,"data"});
(22)       log(TimestampType:{self.now,"MDATind2"});
(23)       MSAP.send(MDATreq:{AK,number});
(24)     }
(25)     Connection_Release();
(26)     deactivate(OtherwiseFailDefault);
(27)     setverdict(pass);
(28)     stop;
(29)   }
(30)   control {
(31)     var testrun myTestrun;
(32)     var logfile myLog;
(33)     var verdicttype myVerdict;
(34)     myTestrun:=execute(InresRTexample);
(35)     myVerdict:=myTestrun.getverdict;
(36)     if (myVerdict==pass) {
(37)       myLog:=myTestrun.getlog;
(38)       myVerdict:=evalMultipleDelays("IDATreq1","MDATind2",
                                         0.0,incl,0.005,excl,myLog);
(39)       myTestrun.setverdict(myVerdict);
(40)     }
(41)   }
(42) }
```

Fig. 3. *TIMED*TTCN-3 Inres test case generated from the test purpose in Fig. 2

i.e., the final test verdict is calculated during the test, whereas offline evaluation means that the final evaluation is done after test case termination. Offline evaluation is realized by giving access to the logfile produced by a test case execution.

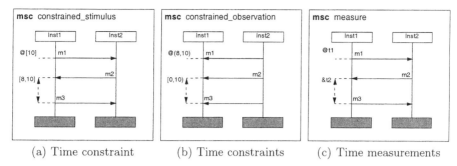

Fig. 4. MSC time constructs

For this, a testrun handle (line 31 in Fig. 3) has been introduced, which gives access to the logfile (line 32) and the final verdict of the functional test execution (line 35). Special functions allow to sort, read and navigate in a logfile. A logfile may be passed into a library function (line 38) that performs the evaluation of the logfile and calculates the final verdict of the T_{IMED}TTCN-3 real-time test case. The final verdict can be assigned by using **setverdict** operation (line 39).

The additional test verdict **conf** has been introduced by T_{IMED}TTCN-3 as an indication that a test case executed correctly with respect to the functional behaviour but failed with respect to a non-functional property. The existing verdict **pass** is used to indicate success with respect to both criteria.

2.3 Time Constructs in MSC

MSC allows to attach time annotations to events. MSC distinguishes between *time constraints* and *time measurements*.

Time constraints are shown in Fig. 4a and 4b. They can be either absolute, i.e., refer to the absolute time of occurrence of one single event, or relative, i.e., constrain the duration between two events. The values of the constraint is specified using intervals. The interval boundaries may be open or closed. An open boundary is indicated by a parenthesis, i.e., '(' or ')', and a closed boundary is defined by a square bracket, i.e., '[' or ']'. An omitted lower bound is treated as zero, an omitted upper bound as ∞. A single value in square brackets represents an interval which contains just that single element. If a time annotation defines no time unit, the default unit *seconds* is used.

Time measurements are presented in Fig. 4c. A measurement may measure the absolute time when a single event occurs or the relative delay between two events. The value of a measurement is stored in a variable which is local to the instance that owns that variable.

In addition to a relative time constraint, Fig. 2 contains a time constraint for a cyclic event (sending message IDATreq("data") every 10ms) inside a loop inline expression. The definition of such periodic events is not supported in the MSC standard. Therefore, we use an extension proposed in [16]. An alternative extension with similar expressive power has been presented in [26].

3 Generating *Timed*TTCN-3 from MSC Test Purposes

Using MSC for test purpose specification and the consecutive test case generation as suggested in this paper is not new and has already been implemented in several academical and industrial tools like, e.g., *Autolink*, *Testcomposer* or *ptk*.

Autolink [20] and *Testcomposer* [15] support the generation of TTCN-2++ test cases either from SDL specifications and MSC test purposes or directly from MSC test purposes. Both tools focus on the generation of tests for testing functional requirements.

The *ptk* tool [1] also generates TTCN-2++ test cases from MSC test purposes. The test generation for real-time constraints, which can be implemented by the TTCN-2++ timer mechanism, is supported by non-standard time annotations in the MSC test purposes. *Autolink* and *ptk* also support the generation of test cases for concurrent test architectures [1, 11].

None of the tools mentioned above is currently able to generate TTCN-3 output. Therefore, we developed an MSC to TTCN-3 generator as an internal case study. While past versions only accepted MSC test purposes containing functional requirements, the latest prototype supports also the real-time constructs described in Section 2.3. Our current prototype does not support concurrent test architectures, i.e., the prototype generates test cases with one main test component (MTC) only. For future versions, we plan to support concurrent test architectures by implementing the approach described in [11].

In the following, we explain how MSC can be used for specifying real-time test purposes and how these test purposes can be automatically transformed into *Timed*TTCN-3 test cases. For simplicity, we restrict ourselves to non concurrent test architectures. For concurrent test architectures, the synchronization among test components has also to be taken into consideration.

3.1 Specification of MSC Real-Time Test Purposes

We use the common practice to specify test purposes using *system level* MSCs. A system level MSC has one designated instance representing the SUT. All other instances correspond to the different interfaces of the SUT. Together with the contained message exchange from and to the interfaces of the SUT, this information can be used to derive corresponding test cases.

In contrast to other approaches (e.g., [1]), we do not generate one test case for each of the traces which is possible due to the interleaving semantics of MSC. Instead, we extract just one single representative *path* from the test purpose MSC, which takes the queue semantics of TTCN-3 into consideration. This will be explained in the next section.

For the specification of MSC real-time test purposes, the MSC real-time constructs described in Section 2.3 are used. We allow to attach MSC time constraints and measurements (see Fig. 2b) to the communication events along the instances that represent the interfaces of the SUT.

While the meaning of allowed MSC time measurements (Fig. 4c) in test purpose descriptions is straightforward (i.e., observation of the point in time when an event occurs and its storage in a T_{IMED}TTCN-3 variable), MSC time constraints can be used with two different aims: They may either describe a timely stimulation of the SUT (*time constrained stimulus*) or a response from the SUT shall arrive within a certain period of time (*time constrained observation*). Both cases look similar, but can be distinguished as follows:

The test system should perform a time constrained stimulus if an absolute time constraint is attached to a send event, or a relative time constraint is attached to a pair of events, where the second event is a send event[4] (cf. Fig. 4a, where Inst2 represents the SUT).

The test system should perform a time constrained observation, if an absolute time constraint is attached to a receive event, or a relative time constraint is attached to a pair of events, where the second event is a receive event[4] (cf. Fig. 4b, where Inst2 represents the SUT).

In our test purpose example (Fig. 2b), both types of relative time constraints are used. The cyclic real-time requirement constraint can be unrolled to a sequence of relative time constraints and requires to send a stimulus every 10ms. The other constraint ([0ms,5ms]), which is attached to the messages IDATreq and MDATind, describes a passive observation of MDATind.

An MSC time measurement and an MSC time constraint used for time constrained observation, result in a similar test behaviour. In both cases the current time is acquired. But, only observing time constraints allow to attach the actual intervals of real-time requirements to the graphical symbols. This identifies them as real-time requirements. In contrast, a real-time requirement cannot be specified by MSC time measurements, because they only allow to specify names for observations, but no concrete time values. As a result of these considerations, MSC time constraints should be the preferred means for specifying real-time test purposes. Nevertheless, MSC time measurements may be useful for gathering time information, which is re-used later as part of other MSC time constraints.

3.2 Transformation of Functional Concepts from MSC to TTCN-3

The transformation of the static aspects of an MSC test purpose into TTCN-3 is very simple. The interfaces of the SUT are described by MSC instances. They are mapped to TTCN-3 ports. For the mapping of MSC events to TTCN-3 statements, only the events along the interface instances are relevant. MSC send events are mapped to TTCN-3 send operations and MSC receive events are mapped onto TTCN-3 receive operations. The MSC message names refer to TTCN-3 data types or signature definitions. MSC message parameters, refer to TTCN-3 templates or define inline templates.

Configuration and data descriptions, like the definition of port types, component types, data types, signatures and templates, cannot be generated automatically from MSC test purposes. They have to be specified manually or imported from other TTCN-3 modules.

---

[4] The type of the first event involved in the relative time constraint is irrelevant, since the time constraint is essentially imposed on the second event.

An MSC test purpose specification may also include references and the inline expressions **alt**, **loop** and **par**. We consider each usage of a reference or an inline expression as one single event, which in case of references or inline expressions may include partially ordered events. This means, these constructs are synchronization points, even though this violates the official MSC semantics [25]. The MSC semantics assumes a weak sequential composition semantics. However, many test case specifiers regard this as counter-intuitive. Hence, the corresponding TTCN-3 constructs suggested in our mapping do not allow that sort of interleaving. Our transformation algorithm maps MSC references to TTCN-3 function calls, MSC **alt** inline expressions to TTCN-3 **alt** statements, MSC **loop** inline expressions to TTCN-3 **for** statements and MSC **par** inline expressions to TTCN-3 **interleave** statements.

Our generation algorithm uses the *Autolink* approach [20] for the calculation of the control flow of TTCN-3 test cases[5]. The algorithm computes a so-called *path* from the partially ordered set of MSC events. Such a path specifies a set of traces, which includes no nondeterminism due to non-receiving events, but considers all interleavings due to receiving events. This path representation takes the port queue semantics of TTCN-3 into consideration. A path can be visualized in form of a tree, where branching is related to alternative receiving events. Our TTCN-3 generation algorithm computes a TTCN-3 test case that allows to test all sequences of events described by the corresponding path.

The T_{IMED}TTCN-3 test case shown in the lines 6-29 of Fig. 3 has been generated automatically from the MSC test purpose shown in Fig. 2b. A comparison of both figures may indicate how functional concepts in MSCs are transformed into TTCN-3.

In this paper, we only have space to present the basic ideas of our TTCN-3 generation procedure. For test purpose specification, we also support HMSCs and allow the usage of timers, actions and conditions in MSC test purposes. The treatment of these constructs and the handling of complex MSC test purposes is investigated in [1, 11, 20]. Our implementation follows these approaches.

3.3 Transforming MSC Real-Time Constructs to T_{IMED}TTCN-3

For deriving test cases, the different usages of real-time constructs in MSC test purposes identified in Section 3.1 have to be handled separately.

Time Constrained Observations.
Time constrained observations in test purposes are translated to T_{IMED}TTCN-3 by generating timestamps for the observed events. For offline evaluation, the timestamps are stored in the logfile produced by the test case. In the online evaluation approach timestamps are compared during the test run, because they are stored in variables. In the following, we focus on offline evaluation, but all

---

[5] Both, TTCN-2++ and TTCN-3 support asynchronous communication over message queues. Therefore, the *Autolink* approach for the generation of TTCN-2++ test cases from MSC test purposes can also be used for the generation of TTCN-3 test cases.

considerations can be generalized to online evaluation, because both approaches are based on the generation of timestamps.

The example test purpose in Fig. 2b contains a time constrained observation. Since the constraint is attached to the messages IDATreq and MDATind, the *TIMED*TTCN-3 statements for generating timestamps (lines 19 and 22 of Fig. 3) are placed directly before and after the associated communication operations (lines 20–21).

Our TTCN-3 generator generates timestamps that contain the value of the local clock when the timestamp is generated and a label of type **charstring**, which is used to identify timestamps afterwards. The value of the label is generated from the message name used in the MSC plus a consecutive number to distinguish between different occurrences of the same message. The type definition for the TimestampType is added automatically at the top of the generated module (lines 2–5). At the bottom, i.e., in the control part, a call to an evaluation function is added (line 38) for each time constrained observation defined in the MSC test purpose.

Evaluation functions may be provided in form of *TIMED*TTCN-3 libraries. For example, the predefined evaluation function evalMultipleDelays is used to retrieve and evaluate a sequence of matching pairs of timestamps from a given logfile. The parameters of the function identify the timestamps to be compared, define the time interval between two timestamps[6] and refer to the logfile, which should be analyzed. If the time difference of each timestamp pair found in the logfile fulfills the given requirement, the verdict **pass** is returned. If the real-time requirement is violated, **conf** is returned and indicates a non-functional failure. This result contributes to the final verdict of the test case (line 39).

Placement of Timestamping Statements.
In a perfect world, no time passes between a send or receive event and the corresponding timestamp generation. Hence, the time stored in the timestamp is the actual time of the event. However, test cases are implemented on real hardware, some time passes between both statements. Thus, we have the choice of putting the **log** statement before or after a time constrained event derived from a test purpose. For **receive** operations, we have to put the **log** statement after the **receive** (lines 21–22 in Fig. 3), because a **receive** operation is blocking. But for a **send** operation, we have the option to put a timestamping **log** statement before or after the **send** operation.

In our example, for the time constraint related to the messages IDATreq and MDATind, the **log** statement associated to the **send** operation in line 20 of Fig. 3 may be inserted before or after this **send** operation. In the first case, the observed duration would be longer than in the second case. The choice of placement depends on whether the time constraint stipulates a minimal or maximal duration.

---

[6] Upper and lower bound of the interval are defined by two **float** values. The parameters incl and excl define whether a boundary is closed or open.

```
(1) float sendTime;
(2) port.receive / port.send
(3) sendTime:=self.now+d;
    ...
(4) resume(sendTime);
(5) port.send
```

Fig. 5. *TIMED*TTCN-3 skeleton for timed stimulus

If the time constraint only has an upper bound (or the lower bound is zero as in e.g., [0ms,5ms)), a maximal duration is specified. In this case, choosing the placement which yields the shorter observed duration might result in a **pass** verdict even though the actual duration violated slightly the real-time constraint. Instead, we place in this case the **log** statement before the **send** operation as shown in lines 19–20.

The opposite considerations hold for testing minimal durations, e.g., (1ms,). We are on the safe side with an observed duration which is shorter than the actual duration. Thus, we insert the **log** statement after the **send** operation.

In the combined case[7], i.e. neither the lower interval bound is omitted or zero nor the upper bound is omitted (e.g. [8ms,10ms]), it is a matter of preference which bound is given the priority. If we assume that encoding for sending and decoding for receiving takes a similar amount of time, we put the **log** statement after the **send** operation because we have to put it also after the **receive** operation. This may lead to a measurement, which is closer to the reality since the extra delay introduced by both operations will eliminate each other.

Time Constrained Stimuli.
A time constrained stimulus in an MSC test purpose description is translated into a *TIMED*TTCN-3 **resume** statement, which is used to schedule the execution of the related **send** operation. A generic *TIMED*TTCN-3 skeleton for a time constrained stimulus is shown in Fig. 5.

If the time constraint consists of a single point in time, e.g., [d], this value can be used as relative offset to the **self.now** expression in line 3 of Fig. 5.

If the time constraint is an interval, e.g., [x,y], any of the values inside the interval is possible as delay of the **send** operation. This may lead to an infinite number of test cases, which is infeasible in practice. Using test data selection heuristics, like domain boundary analysis, an appropriate number of test cases can be selected, e.g., d=x for testing the extreme lower and d=y for testing the extreme upper allowed point in time.

Time Constraints Attached to Inline Expressions and References.
For MSC inline expressions and references several special cases have to be considered. In this paper, we are only able to cover a few of them.

---

[7] Note, that specifying a single element as interval for a time constrained observation is not recommended, because it is very unlikely, that exactly that value is matched.

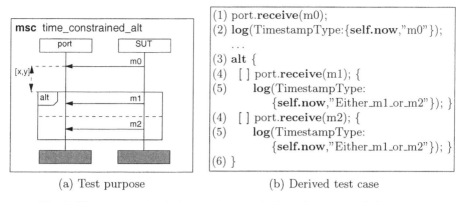

Fig. 6. Time constrained observation attached to first event of alternative

A cyclic time constraint contained in a loop, can be treated nearly like ordinary relative time constraints: In a time constrained cyclic observation, timestamps are not generated for a pair of two communication events, but for a sequence of a single communication event. Hence, a call to a different evaluation function, working on sequences of a single timestamp only, is necessary. For a time constrained cyclic stimulus, a different set of statements than presented in Fig. 5 is required: the first execution of the **send** operation has to be performed immediately, while all subsequent executions need to adhere to the cyclic time constraint. This is achieved by the statements given in lines 7, 12–18 of Fig. 3.

For MSC inline expressions and references, MSC allows to impose time constraints on the first or last occurring MSC event which is contained in the inline expression or reference, respectively. This means, that for every possible first or last event, a timestamp has to be generated. Since a magnitude of different permutations is possible, we present just the example in Fig. 6.

4 TIMEDGFT

GFT [9] is one of the standardized presentation formats of TTCN-3. It provides an exact way of displaying TTCN-3 behaviour descriptions, i.e., test cases, altsteps, functions and module control, graphically. An one-to-one mapping from TTCN-3 behaviour descriptions to GFT diagrams and vice versa can be found in the appendices C and D of [9].

GFT is based on the MSC standard. It uses a subset of MSC and extends this subset with test specific symbols and keywords. For all TTCN-3 statements, there exists an appropriate GFT symbol. Thus, only the real-time extensions of TIMEDTTCN-3 have to be considered in order to define our real-time extension of GFT, which is called TIMEDGFT. The TIMEDGFT presentation of the special TIMEDTTCN-3 concepts and statements is summarized in Fig. 7.

4.1 Timezones

For the specification of clock synchronized test components, TIMEDTTCN-3 introduced the concept of timezones. A timezone is an attribute, which is assigned

Concept	TimedTTCN3 Realization	TimedGFT Presentation
Timezones	new parameter of **create** statement	MyTC:=CType.**create** (Berlin)
	new parameter of **execute** statement	MyTestCase(Berlin)
	timezone operation	no special symbol
Absolute time	**now** operation	no special symbol
	resume statement	@[t+3.0]
Logging	extension of **log** statement	MyTemplate
Logfile handling	**first, next, previous** and **retrieve** operations	no special symbols
Testrun handling	**getlog** operation	no special symbol
	overwriting of verdicts in control part	myTestrun. **setverdict(fail)**
Non-functional verdict	**conf** verdict	conf

Fig. 7. Real-time constructs of $\mathcal{T}_{IMED}$GFT

to a test component during its creation. In $\mathcal{T}_{IMED}$GFT, the creation of a test component is related to create and execute symbols. The assignment of a timezone is an additional parameter in a create or execute symbol (Fig. 7).

A test component may read its timezone attribute by means of a **timezone** operation. $\mathcal{T}_{IMED}$GFT provides no special symbol for the **timezone** operation. Depending on the usage, the **timezone** operation may appear in several symbols. For example, a **timezone** operation will appear in an action box, if it is used in an assignment, or a **timezone** operation will be presented within a reference symbol, if it defines the actual parameter of an altstep or function call.

4.2 Absolute Time

$\mathcal{T}_{IMED}$TTCN-3 supports the usage of absolute time by providing the **now** operation and the **resume** statement.

The *now* Operation.
The **now** operation is used for the retrieval of the current local time of a test

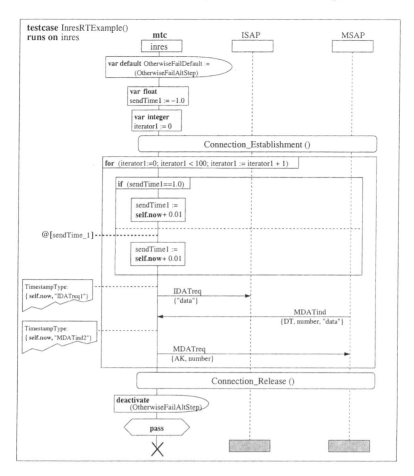

Fig. 8. T_{IMED}GFT presentation of the Inres test case in Fig. 3

component. The local character of the **now** operation is reflected by its application to the **self** handle. T_{IMED}GFT provides no special symbol for the **now** operation. Depending on its usage, the **now** operation appears in different symbols. In Fig. 8, the **now** operation is used in assignments and in **log** statements. Therefore, they appear as inscriptions in action boxes and log symbols.

The *resume* Statement.
The **resume** statement provides the ability to delay the execution of a test component until a specified absolute time is reached. T_{IMED}GFT has adopted the absolute time constraint symbol of MSC to present the **resume** statement graphically. In Fig. 8, the visualization of **resume**(sendTime1) can be found. Contrary to MSC, the dashed time line is attached to an instance and not to an event. The reason is that the **resume** statement is a statement of its own and can not be related to other events. For the mapping of T_{IMED}GFT to MSC, the

dashed time line may be attached to the event, which is deferred by the **resume** statement[8].

4.3 Logging Mechanism

T_{IMED}TTCN-3 refines the TTCN-3 **log** statement by allowing to write structured T_{IMED}TTCN-3 timestamp data values into logfiles. In GFT, **log** statements are presented in action boxes. T_{IMED}GFT introduces a new log symbol. The reason for this new symbol is that we want to emphasize places in the test behaviour, where time and other information are collected in the logfile. Figure 8 presents the new symbol. It is used for logging the two timestamps {**self**.**now**, "IDATreq1"} and {**self**.**now**, "MDATind2"} of type TimestampType.

4.4 Testrun and Logfile Handling

The result of the execution of a T_{IMED}TTCN-3 test case is a test verdict and a log file. In the module control part, T_{IMED}TTCN-3 gives access to both by a testrun handle, which is returned by the **execute** statement.

For the graphical presentation of a module control part, GFT provides a control diagram, which includes one control instance only. Fig. 9 shows the control diagram of the example Inres T_{IMED}TTCN-3 module.

The **getlog** operation can be applied to a testrun handle in order to access the log file, e.g. for offline evaluation. The context of the **getlog** operation determines the symbol in which it is presented. For example, in Fig. 9, the **getlog** operation is used in an assignment and therefore, presented in an action box.

The actual evaluation function evalMultipleDelays, which might be imported from a library, is not shown. Such functions typically operate on logfiles. In order to sort and navigate in logfiles, T_{IMED}TTCN-3 provides the functions **first**, **next**, **previous** and **retrieve**. They have no special T_{IMED}GFT presentation. In the same manner as for the **getlog** operation, their presentation depends on the context in which they are used.

4.5 Verdict Handling

In addition to the existing verdicts, T_{IMED}TTCN-3 extends the verdict handling of TTCN-3 by introducing the additional verdict **conf** and by allowing to access and overwrite a test verdict in the control part of a module.

The handling of verdicts in T_{IMED}GFT is almost identical to their handling in GFT. The **setverdict** operation is always presented in a condition symbol (bottom of Fig. 9). There exists no special symbol to emphasize the **getverdict** operation. The context determines the symbol in which it is presented.

The only difference of T_{IMED}GFT to GFT with respect to verdict handling is the setting of verdicts in the module control part or in functions called by

---

[8] The mapping of GFT to MSC is defined in Annex F of [9]. A similar mapping of the T_{IMED}GFT to MSC will be defined when T_{IMED}GFT becomes part of GFT.

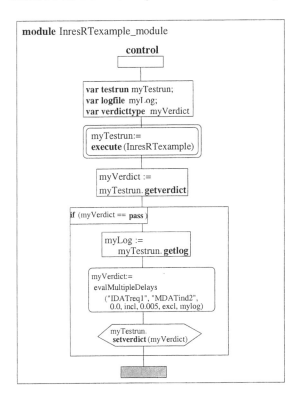

Fig. 9. *TIMED*GFT control diagram for Fig. 3

the module control part. In these cases, the complete *TIMED*TTCN-3 statement including the testrun handle has to be written into the condition symbol, because without the testrun handle, the relation between verdict and testrun would not be clear. The retrieval and the setting of a verdict by means of a testrun handle inside a control diagram is shown in Fig. 9.

4.6 Evaluation Functions

On- and offline evaluation involves the application of different pre-defined functions and operators which operate on the gathered timestamps. *TIMED*GFT provides no special symbols for the call of evaluation functions. They may be presented in action or reference symbols.

5 Outlook

In this paper, we presented the MSC-based specification of real-time test purposes and the generation of *TIMED*TTCN-3 test cases from MSC test purposes. Furthermore, we introduced *TIMED*GFT, a real-time extension of GFT, which allows to present TIMEDTTCN-3 graphically. This paper can be seen as the continuation of our work on TIMEDTTCN-3 [3].

At present, we develop tools, which support the generation of TIMEDTTCN-3 test cases from MSC test purposes. A first prototype of our tool is already available and our experiments with more complex test purposes are promising. Furthermore, we work on the prototype of a TIMEDGFT tool.

We also started to define the semantics of TIMEDTTCN-3 by enriching the semantics of TTCN-3 with real-time concepts. We plan to submit this semantics extension together with TIMEDGFT to the ETSI standardization process in order to have a full integration of TIMEDTTCN-3 into the TTCN-3 standards series.

References

1. P. Baker, P. Bristow, C. Jervis, D. King, and B. Mitchell. Automatic Generation of Conformance Tests From Message Sequence Charts. In *Proceedings of the 3rd SAM (SDL and MSC) Workshop*, 2002.
2. P. Baker, E. Rudolph, and I. Schieferdecker. Graphical Test Specification – The Graphical Format of TTCN-3. In R. Reed and J. Reed, editors, *SDL2001 – Meeting UML*. Springer, 2001.
3. Z.R. Dai, J. Grabowski, and H. Neukirchen. Timed TTCN-3 – A Real-Time Extension for TTCN-3. In I. Schieferdecker, H. König, and A. Wolisz, editors, *Testing of Communicating Systems*, volume 14, Berlin, March 2002. Kluwer.
4. Danet TTCN Toolbox – TTCN-3.
   http://www.bss.danet.de/solution/ttcn/ttcn_toolbox_ttcn-3_uk.htm, 2002.
5. Da Vinci Communications Terzo tools product information.
   http://www.davinci-communications.com/products_ttcn3.html, 2002.
6. M. Ebner, A. Yin, and M. Li. Definition and Utilisation of OMG IDL to TTCN-3 Mappings. In I. Schieferdecker, H. König, and A. Wolisz, editors, *Testing of Communicating Systems - Application to Internet Technologies and Services*, volume 14. Kluwer, 2002.
7. ETSI Technical Report (TR) 101 666 (1999-05): Information technology - Open Systems Interconnection Conformance testing methodology and framework; The Tree and Tabular Combined Notation (TTCN) (Ed. 2++). European Telecommunications Standards Institute (ETSI), Sophia-Antipolis (France), 1999.
8. ETSI European Standard (ES) 201 873-1 V2.2.1 (2002-08): The Testing and Test Control Notation version 3; Part 1: TTCN-3 Core Language. European Telecommunications Standards Institute (ETSI), Sophia-Antipolis (France), 2002.
9. ETSI European Standard (ES) 201 873-3 V2.2.1 (2002-09): The Testing and Test Control Notation version 3; Part 3: Graphical Presentation Format for TTCN-3 (GFT). European Telecommunications Standards Institute (ETSI), Sophia-Antipolis (France), 2002.
10. N. Goga. Comparing TorX, Autolink, TGV and UIO Test Algorithms. In R. Reed and J. Reed, editors, *SDL2001 – Meeting UML*. Springer, 2001.
11. J. Grabowski, B. Koch, M. Schmitt, and D. Hogrefe. SDL and MSC Based Test Generation for Distributed Test Architectures. In R. Dssouli, G. von Bochmann, and Y. Lahav, editors, *SDL'99 - The next Millenium*. Elsevier Science Publishers B.V., 1999.

12. J. Grabowski and T. Walter. Visualisation of TTCN test cases by MSCs. In Y. Lahav, A. Wolisz, J. Fischer, and E. Holz, editors, *Proceedings of the 1st Workshop of the SDL Forum Society on SDL and MSC - SAM'98*, 1998.
13. J. Grabowski, A. Wiles, C. Willcock, and D. Hogrefe. On the Design of the New Testing Language TTCN-3. In H. Ural, R.L. Probert, and G. von Bochmann, editors, *Testing of Communicating Systems*, volume 13. Kluwer, 2000.
14. D. Hogrefe. Report on the Validation of the Inres System. Technical Report IAM-95-007, Universität Bern, November 1995.
15. A. Kerbrat, T. Jéron, and R. Groz. Automated test generation from SDL specifications. In R. Dssouli, G. von Bochmann, and Y. Lahav, editors, *SDL'99 - The next Millenium*. Elsevier Science Publishers B.V., 1999.
16. H. Neukirchen. Corrections and extensions to Z.120, November 2000. Delayed Contribution No. 9 to ITU-T Study Group 10, Question 9.
17. I. Schieferdecker, S. Pietsch, and T. Vassiliou-Gioles. Systematic Testing of Internet Protocols - First Experiences in Using TTCN-3 for SIP. In *Proceedings of the 5th IFIP Africom Conference on Communication Systems, Cape Town (South Africa)*, May 2001.
18. I. Schieferdecker, B. Stepien, and A. Rennoch. PerfTTCN, a TTCN Language Extension for Performace Testing. In M. Kim, S. Kang, and K. Hong, editors, *Testing of Communicating Systems*, volume 10. Chapman & Hall, 1997.
19. I. Schieferdecker and B. Stepien. Automated Testing of XML/SOAP based Web Services. In *Proceedings of the 13th. Fachkonferenz der Gesellschaft für Informatik (GI) Fachgruppe "Kommunikation in verteilten Systemen" (KiVS), Leipzig (Germany)*, Feb. 26.–28. 2003.
20. M. Schmitt, A. Ek, J. Grabowski, D. Hogrefe, and B. Koch. Autolink – Putting SDL-based test generation into practice. In A. Petrenko and N. Yevtuschenko, editors, *Testing of Communicating Systems*, volume 11. Kluwer, 1998.
21. J.Z. Szabó. Experiences of TTCN-3 Test Executor Development. In I. Schieferdecker, H. König, and A. Wolisz, editors, *Testing of Communicating Systems - Application to Internet Technologies and Services*, volume 14. Kluwer, 2002.
22. Telelogic Tau/Tester product information. http://www.tautester.com/, 2002.
23. Testing Technologies TT Tool Series product information. http://www.testingtech.de/products/TTToolSeries.html, 2002.
24. T. Walter and J. Grabowski. A Framework for the Specification of Test Cases for Real Time Distributed Systems. *Information and Software Technology*, 41:781–798, 1999.
25. Recommendation Z.120: Message Sequence Charts (MSC). International Telecommunication Union (ITU-T), Geneve, 1999.
26. T. Zheng and F. Khendek. An extension to MSC-2000 and its application. In *Proceedings of the 3rd SAM (SDL and MSC) Workshop*, 2002.

Interoperabolity Events Complementing Conformance Testing Activities

Philippe Cousin

ETSI Interoperability Service Manager
Philippe.Cousin@etsi.fr

Abstract. Euclid of Alexandria, the most prominent mathematician of antiquity, is also the leading mathematics teacher of all times due to his treatise on mathematics, 'The Elements'. When once asked whether there wasn't an easier way to study geometry than 'The Elements', Euclid was said to respond: 'There is no royal road to geometry'.
Interoperability is what most standards are about. It is the only thing that these standards are about. So is there a royal road to interoperability? Unfortunately, interoperability does not do any better than geometry – there is no royal road to interoperability. Interoperability is extremely difficult to achieve.
Two ways of validating a standard and an implementation are 'conformance testing' and 'interoperability testing'. In conformance testing, manufacturers run their implementation against a set of 'golden' test suites; in 'interoperability testing', they run their implementations against each other. Frequently, 'interoperability testing' is done at 'bake-offs', i.e. open events where developers get together to debug a standard and their implementations.
Views on which method is best are at times strongly divergent. Theoretically, if a standard is tight and has only a limited number of options, implementations that pass conformance tests will automatically be interoperable. In practice, however, a 'pass' in conformance testing does not necessarily guarantee that two implementations are interoperable. Likewise, if two implementations are interoperable, they may not necessarily conform to the standard. In reality, things probably lie somewhere in the middle, and both conformance testing and interoperability testing pave the road towards interoperability.

This presentation then will introduce the ETSI Interoperability service, which was set up at the end of 1999 as a professional service specializing in the organization of interoperability events. In 2001 it was established on a permanent basis with dedicated support, and adopted the PLUGTESTS name.

End of 2002, the ETSI interoperability Service organised 26 interoperability events which pooled altogether 2000 engineers from about 600 companies. This new activity within ETSI and in Europe is getting a strong success also thanks to the efficient cooperation with the ETSI PTCC (Protocol and Testing Competence Centre) bringing expertise in methodologies and conformance testing.

Testing Transition Systems with Input and Output Testers

Alexandre Petrenko[1], Nina Yevtushenko[2], and Jia Le Huo[3]

[1] CRIM, Centre de recherche informatique de Montréal
550 Sherbrooke West, Suite 100, Montreal, Quebec, H3A 1B9, Canada
Petrenko@crim.ca
[2] Tomsk State University, 36 Lenin Street, Tomsk, 634050, Russia
Yevtushenko@elefot.tsu.ru
[3] Department of Electrical and Computer Engineering, McGill University
3480 University Street, Room 633, Montreal, Quebec, H3A 2A7, Canada
Jiale@macs.ece.mcgill.ca

Abstract. The paper studies testing based on input/output transition systems, also known as input/output automata. It is assumed that a tester can never prevent an implementation under test (IUT) from producing outputs, while the IUT does not block inputs from the tester, either. Thus, input from the tester and output from the IUT may occur simultaneously and should be queued in finite buffers between the tester and the IUT. A framework for so-called queued-quiescence testing is developed, based on the idea that the tester should consist of two test processes, one applying inputs via a queue to an IUT and the other reading outputs from a queue until it detects no more outputs of the IUT, i.e., the tester detects quiescence of the IUT. The testing framework is then extended with so-called queued-suspension testing by considering a tester that has several pairs of input and output processes. Test derivation procedures are elaborated with a fault model in mind.

Keywords: conformance testing, test generation, input/output transition system, fault model

1 Introduction

The problem of deriving tests from state-oriented models that distinguish between input and output actions is usually addressed with one of the two basic assumptions about the relationships between inputs and outputs. Assuming that a pair of input and output constitutes an atomic action of a system, in other words, that the system cannot accept the next input before producing output as a reaction to a previous input, one relies on the input/output Finite State Machine (FSM) model. There is a large body of work on test generation from FSM with various fault models and test architectures, for references see, e.g., [6] and [1]. A system, where the next input can arrive even before an output is produced in response to a previous input, is usually modeled by the input/output automaton model [5], also known as the input/output transition system (IOTS) model (the difference between them is marginal, at least from the testing perspective). Compared to the FSM model, this model has received a far less attention in the testing community, see, e.g., [2], [9], [10]. In this paper, we consider the IOTS model and take a close look on some basic assumptions underlying the existing IOTS testing frameworks.

An important publication on test generation from labeled transition systems (LTS) with inputs and outputs is [12]. In this paper, it is assumed that a tester interacting with an implementation under test (IUT) is an LTS. The LTS composition operator used to formalize this interaction does not distinguish between inputs and outputs, and the tester is not input-enabled. Due to the synchronous nature of the LTS composition, the tester preempts output of the IUT any time it decides to send input to the IUT. Although this allows the tester to avoid choosing between inputs and outputs, the tester overrides the principle that "output actions can never be blocked by the environment" [12, p.106]. An IOTS "generates output and internal actions autonomously" [5], so such an IUT can be synchronously composed only with a tester that is receptive to the IUT's output.

Another assumption about the tester is taken by Tan and Petrenko [11]. In this work, it is recognized that the tester cannot block the IUT's outputs. It is only assumed that the tester can detect the situation when it offers input to the IUT, but the latter, instead of consuming it, issues an output (a so-called "exception"). An exception halts a current test run (as the tester has lost control over the test execution) and results in the verdict **inconclusive**. Notice that the tester of [12] has only two verdicts, **pass** and **fail**.

Either approach relies on an assumption that is not always justified in a real testing environment. As an example, consider the situation when the tester cannot directly interact with an IUT because of a context, such as queues or interfaces, between them. As pointed out in [15], to apply the test derivation algorithm of [12], one has to take into account the presence of a queue context. It also states "the assumption that we can synthesize every stimulus and analyze every observation is strong", so that some problems in observing quiescence occur.

The case when IOTS is tested via infinite queues is investigated by Verhaard et al [14]. The proposed approach relies on a specification of a given IOTS explicitly combined with a queue context, so it is not clear how this approach could be implemented in practice. This context is also considered in [4], where a stamping mechanism is proposed to order the outputs with respect to the inputs, while quiescence is ignored. A stamping process has to be synchronously composed with an IUT as the tester in [12].

We also notice that we are aware of the only work [11] that uses fault models in test derivation from IOTS. In [12] and [14], a test case is derived from a trace provided by the user.

The above discussion indicates a need for another approach that does not rely on such strong assumptions about the testing environment and incorporates a fault model to derive tests that can be characterized in terms of fault detection. In this paper, we report on our findings in attempts to elaborate such an approach. In particular, we introduce a framework for testing IOTS, assuming that a tester can never prevent an IUT from producing outputs, while the IUT does not block inputs from the tester either, and thus, input and output actions may occur simultaneously and should be queued in finite buffers between the tester and the IUT.

The paper is organized as follows. In Section 2, we introduce some basic definitions and define a composition operator for IOTS based on a refined notion of compatibility of IOTS first defined in [5]. Section 3 presents our framework for so-called queued-quiescence testing, based on the idea that the tester should consist of two test processes: one process applies inputs to an IUT via a finite input queue and the other reads outputs that the IUT puts into a finite output queue until the second process detects no more outputs from the IUT, i.e., the tester detects quiescence of the IUT. We elaborate such a tester and formulate several implementation relations that can be tested with a queued-quiescence tester. In Section 4, we discuss how queued-quiescence tests can be derived for a given specification and fault model that comprises a finite set of implementations. In Section 5, we extend our testing framework with so-called queued-suspension testing by allowing a tester to have several pairs of input and output processes and demonstrate that a queued-suspension tester can check finer implementation relations than a queued-quiescence tester. We conclude by comparing our contributions with the previous work and discussing further work. An earlier version of this paper is published in an INRIA preprint [7].

2 Preliminaries

A *labeled transition system* (LTS) is a 4-tuple $L = <S, \Sigma, \lambda, S_0>$, where S is a finite set of states with a non-empty set of initial states $S_0 \subseteq S$; Σ is a finite set of actions; $\lambda \subseteq S \times (\Sigma \cup \{\tau\}) \times S$ is a transition relation. The special symbol $\tau \notin \Sigma$ represents the internal action. We call an LTS *deterministic* if it contains no internal action, has a single initial state, and for transitions $(s, a, s'), (s, a, s'') \in \lambda, s' = s''$. (As opposed to the preprint [7], this paper considers LTS that might be non-deterministic.) After [12], we only consider strongly converging LTS, i.e., the LTS that contain no loop of internal actions.

Let $L_1 = <S, \Sigma_1, \lambda_1, S_0>$ and $L_2 = <T, \Sigma_2, \lambda_2, T_0>$, the *parallel composition* $L_1 \parallel L_2$ is the LTS $<R, \Sigma_1 \cup \Sigma_2, \lambda, R_0>$, where $R_0 = S_0 \times T_0$ is the set of initial states; the set of states $R \subseteq S \times T$ and the transition relation λ are the smallest sets obtained by application of the following inference rules:

- if $a \in \Sigma_1 \cap \Sigma_2, (s, a, s') \in \lambda_1$, and $(t, a, t') \in \lambda_2$ then $(st, a, s't') \in \lambda$;
- if $a \in \{\tau\} \cup \Sigma_1 \backslash \Sigma_2, (s, a, s') \in \lambda_1$, then $(st, a, s't) \in \lambda$;
- if $a \in \{\tau\} \cup \Sigma_2 \backslash \Sigma_1, (t, a, t') \in \lambda_2$, then $(st, a, st') \in \lambda$.

We use the LTS model to define a transition system with inputs and outputs. The difference between these two types of actions is that no system can deny an input action from its environment, while it is completely up to the system when to produce an output, so that the environment cannot block the output. Formally, an *input/output transition system* (IOTS) L is an LTS in which the set of actions Σ is partitioned into two sets, the set of input actions I and the set of output actions O. We use $<S, I, O, \lambda, S_0>$ to represent an IOTS $<S, I \cup O, \lambda, S_0>$ with $I \cap O = \emptyset$. Further, we use $IOTS(I, O)$ to denote the set of all possible IOTS over the input set I and output set O.

Given state s of L, we further denote $init(s)$ the set of actions defined at s, i.e., $init(s) = \{a \in (\Sigma \cup \{\tau\}) \mid \exists s' \in S \text{ s.t. } ((s, a, s') \in \lambda)\}$. The IOTS is (strongly) *input-enabled* if each input action is enabled at any state, i.e., $I \subseteq init(s)$ for each s. In this paper, we consider only input-enabled IOTS specifications, while an implementation IOTS (that models an IUT) is always assumed to be input-enabled. We notice that

IOTS here corresponds to IOLTS in [4], and input-enabled IOTS to IOA in [5]. State s of the IOTS is called *unstable* if $init(s) \cap (O \cup \{\tau\}) \neq \emptyset$. Otherwise, the state is *stable*. A non-empty sequence $\alpha \in \Sigma^*$ is called a *trace* of L in state s if there exist actions $a_1, \ldots, a_k$ in $\Sigma \cup \{\tau\}$ and states $s_1, \ldots, s_{k+1}$ such that $(s_i, a_i, s_{i+1}) \in \lambda$ for all $i = 1, \ldots, k$; $s_1 = s$; and the projection of $a_1 \ldots a_k$ onto the action set Σ is the sequence α. We use $traces(s)$ to denote the set of traces of L in state s, and $traces(P)$ to denote the union of traces of L in the states in P, where P is a set of states of *Spec*. Sometimes, we use L to refer to the set of initial states of the IOTS L, e.g., $traces(L)$ denotes the union of traces of L in its initial states. We call an IOTS L *oscillating* if there exist a state s reachable from an initial state and a sequence $o_1 o_2 \ldots o_k \in O^*$ such that $(o_1 o_2 \ldots o_k)^* \subseteq traces(s)$. Following [13] and [12], we refer to a trace that takes the IOTS from a given state to a stable state as a *quiescent* trace. We use $qtraces(P)$ to denote the set of quiescent traces of *Spec* in P.

When we compose two IOTS using the parallel composition of LTS, an output action enabled in one IOTS is blocked from happening by the other IOTS if the action is not enabled in the second IOTS. Such a situation, however, cannot be justified by our assumption about the IOTS model, i.e., outputs from an IOTS are under the control of the IOTS itself. On the other hand, the composition operator for IOA defined in [5], which does not have this problem, is only applicable to input-enabled IOTS. This discussion suggests that we need to define a composition operator for IOTS that are not necessarily input-enabled. To this end, we first state compatibility conditions that define when two IOTS can be composed by relaxing the original conditions of [5]. We use $L_1 \parallel L_2$ for IOTS L_1 and L_2 to denote the parallel composition of the LTS L_1 and L_2 when the difference between their inputs and outputs is neglected.

Definition 1. Let two IOTS $L_1 = \langle S, I_1, O_1, \lambda_1, S_0 \rangle$ and $L_2 = \langle T, I_2, O_2, \lambda_2, T_0 \rangle$ be such that the set $O_1 \cap O_2 = \emptyset$. Let st be a state of the composition $L_1 \parallel L_2$. The IOTS L_1 and L_2 are *compatible in state st* if

- $a \in init(s)$ implies $a \in init(t)$ for any $a \in I_2 \cap O_1$ and
- $a \in init(t)$ implies $a \in init(s)$ for any $a \in I_1 \cap O_2$.

L_1 and L_2 are said to be *compatible* if they are compatible in each initial state in $S_0 \times T_0$. L_1 and L_2 are *fully compatible* if they are compatible in all the states of $L_1 \parallel L_2$.

Clearly, two input-enabled IOTS with $I_1 = O_2$ and $I_2 = O_1$ are fully compatible, but the converse is not true. Based on the notion of compatibility we define what we mean by a parallel composition of two IOTS. We notice that the parallel composition $\parallel$ of any two IOTS that are not fully compatible violates the assumption that outputs of an IOTS cannot be blocked. Therefore, we define a parallel composition of IOTS only for fully compatible ones.

Definition 2. The *parallel composition*][of two fully compatible IOTS $L_1 \in IOTS(I_1, O_1)$, and $L_2 \in IOTS(I_2, O_2)$, where the sets $I_1 \cap I_2$ and $O_1 \cap O_2$ are empty, is an IOTS defined as $L_1][L_2 = L_1 \parallel L_2$, with inputs $(I_1 \cup I_2) \setminus (O_1 \cup O_2)$ and outputs $O_1 \cup O_2$.

For fully compatible IOTS, the results of both operators, $\parallel$ and][, coincide. For the IOTS that are not fully compatible, the composition][is not defined.

3 Framework for Queued-Quiescence Testing

In defining a framework for testing systems modeled by IOTS, we first assume that testers are modeled by IOTS. We then require that any tester possess the following properties in addition to the usual soundness requirement. First, due to our assumption about the IOTS model, a tester should not preempt output of any IOTS. Second, a tester should always reach a verdict in finite steps, and once a verdict is reached, the tester should not change it later in the same test run. Third, a tester should be deterministic, meaning that it should have no internal actions and at most a single output action is enabled in any state. Finally, a tester should not make choice between inputs and outputs.

In a typical testing framework, it is usually assumed that a tester is a single process applying inputs to an IUT and observing outputs from the IUT. The two systems, the tester and the IUT, form a closed system. This means that if L_1 is an IOTS modeling a tester, while L_2 is an IOTS modeling the IUT, then $I_1 = O_2$, and $I_2 = O_1$. To be fully compatible with all the IOTS in $IOTS(I_2, O_2)$ the tester should be input-enabled. However, input-enabledness of testers, while making them meet the first requirement, may cause violation of the remaining ones.

An input-enabled tester may yield an infinite test run because the IOTS modeling the tester includes cycles. The test execution may never terminate when the tester interacts with an IUT with proper cycles. This, however, could simply be resolved by defining a tester whose only cycles are self-loops in the states labeled with verdicts. An IUT may continuously interact with such a tester, but the tester still reaches a verdict in a finite number of steps and remains in a state with the reached verdict. However, an arbitrary IUT may produce a wrong output after the tester has reached the verdict **pass**, which cannot be reversed because of the self-loops. To solve this problem, we require that states with the verdict **pass** only be reached when the quiescence of an IUT is detected. This feature of the tester immediately excludes oscillating specifications from further consideration, but still leaves us a wide class of specifications. Thus, we will define testers with the above stated features.

Another problem of input-enabled testers is that such a tester needs choosing between inputs and outputs. In fact, in any state where the tester has to produce an output to an IUT, all the inputs are enabled as well. So the tester has to choose between doing input or output, violating the last requirement.

It turns out that a tester processing inputs separately from outputs may resolve the problem. It is sufficient to decompose the tester into two processes, one for inputs and another for outputs. Intuitively, this could be done as follows. The input test process only sends to the IUT via input buffer a given (finite) number of consecutive test stimuli. In response to the submitted input sequence, the IUT produces outputs that are stored in another (output) buffer. The output test process, that is simply an observer, only accepts outputs of the IUT by reading the output buffer. All the output sequences that the specification can produce in response to the submitted input sequence should take the output test process into a state labeled with the verdict **pass**, while any other output sequence produced by an IUT should take the output test process to a state labeled with the verdict **fail**. Since the notion of a tester is based on the definition of a set of output sequences that the specification IOTS can produce in response to a submitted input sequence, we formalize both notions as follows.

Let $pref(\alpha)$ denote the set of all the prefixes of a sequence $\alpha \in \Sigma^*$ over the set Σ. The set $pref(\alpha)$ has the empty sequence ε. Also given a set $\Gamma \subseteq \Sigma^*$, let $\{\beta \in pref(\gamma) \mid \gamma \in \Gamma\} = pref(\Gamma)$.

Definition 3. Given an input word $\alpha \in I^*$, the *input test process* with α for $L \in$ IOTS(I, O) is an (deterministic) IOTS $\alpha = <pref(\alpha), \varnothing, I, \lambda_\alpha, \{\varepsilon\}>$, where the state set is $pref(\alpha)$ with ε as the only initial state, the set of inputs is empty, the set of outputs is I, and the transition relation $\lambda_\alpha = \{(\beta, a, \beta a) \mid \beta a \in pref(\alpha)\}$.

We slightly abuse α to denote both the input sequence and the input test process that executes this sequence. It is easy to see that each input test process is fully compatible with any IOTS in IOTS(I, O) that is input-enabled.

To define an output test process that complements the input test process α, we have first to determine all the output sequences, valid and invalid, the output test process has to expect from the IUT. The number of valid output sequences is finite, as the specification does not oscillate by our assumption. Thus, in response to α, the IOTS $Spec \in$ IOTS(I, O) can execute any trace that is a completed trace [3] of the IOTS α][$Spec$ leading into a terminal state, i.e., into state g, where $init(g) = \varnothing$. Let $ctraces(\alpha$][$Spec)$ be the set of all such traces. It turns out that the set $ctraces(\alpha$][$Spec)$ is closely related to the set of quiescent traces of the specification $qtraces(Spec)$, viz. it includes each quiescent trace β whose input projection, denoted $\beta_{\downarrow_I}$, is the sequence α.

Proposition 4. $ctraces(\alpha$][$Spec) = \{\beta \in qtraces(Spec) \mid \beta_{\downarrow_I} = \alpha\}$.

Thus, the set $ctraces(\alpha$][$Spec)_{\downarrow_O} = \{\beta_{\downarrow_O} \mid \beta \in qtraces(Spec) \ \& \ \beta_{\downarrow_I} = \alpha\}$ contains all the output sequences that can be produced by $Spec$ in response to the input sequence α.

Given a quiescent trace $\beta \in qtraces(P)$, where P is a set of states of $Spec$, the sequence $\beta_{\downarrow_I}\beta_{\downarrow_O}\delta$ is said to be a *queued-quiescent trace* of $Spec$ in P, where $\delta \notin \Sigma$ is a designated symbol indicating that no more outputs follows, in other words, that $Spec$ becomes quiescent as it has reached a stable state. We use $Qqtraces(P)$ to denote the set of queued-quiescent traces of P $\{(\beta_{\downarrow_I}\beta_{\downarrow_O}\delta) \mid \beta \in qtraces(P)\}$ and $Qqtraces_o(P, \alpha)$ to denote the set $\{\beta_{\downarrow_O}\delta \mid \beta \in qtraces(P) \ \& \ \beta_{\downarrow_I} = \alpha\}$. Next, we define the output test process and the test case.

Given the input test process α and the set $Qqtraces_o(Spec, \alpha)$, we define a set of output sequences $out(\alpha)$ that the output test process can receive from an IUT. It is sufficient to consider all the shortest invalid output sequences along with all the valid ones. Any valid sequence should not be followed by any further output action, as the specification becomes quiescent, while any premature quiescence indicates that the observed sequence is not valid. The set $out(\alpha)$ is defined as follows. For each $\beta \in pref(Qqtraces_o(Spec, \alpha))$ the sequence $\beta \in out(\alpha)$ if $\beta \in Qqtraces_o(Spec, \alpha)$, otherwise $\beta a \in out(\alpha)$ for all $a \in O \cup \{\delta\}$ such that $\beta a \notin pref(Qqtraces_o(Spec, \alpha))$.

Definition 5. The *output test process* for the IOTS $Spec$ and the input test process α is an (deterministic) IOTS $<pref(out(\alpha)), O \cup \{\delta\}, \varnothing, \lambda_{out(\alpha)}, \{\varepsilon\}>$, where certain states are labeled with verdicts **pass** or **fail**. and the state set is $pref(out(\alpha))$ with ε as the only initial state, the input set is $O \cup \{\delta\}$, and the output set is empty. State $\beta \in$

$pref(out(\alpha))$ is labeled with the verdict **pass** if $\beta \in Qqtraces_o(Spec, \alpha)$ or with the verdict **fail** if $\beta \in out(\alpha)\backslash Qqtraces_o(Spec, \alpha)$. The transition relation $\lambda_{out(\alpha)} = \{(\beta, a, \beta a) \mid \beta a \in pref(out(\alpha))\} \cup \{(\beta, \delta, \beta) \mid \beta \text{ is labeled } \textbf{pass}\} \cup \{(\beta, a, \beta) \mid a \in O \cup \{\delta\} \text{ \& } \beta \text{ is labeled } \textbf{fail}\}$.

For a given input test process α, where $\alpha \in I^*$, we reuse $out(\alpha)$ to denote the output test process that complements the input test process α. The pair $(\alpha, out(\alpha))$ is called a *queued-quiescence tester* or simply a *test case* for the IOTS *Spec*.

The self-looping transitions at the states labeled **pass** and **fail** are added to make the output test process fully compatible with any IUT in the set IOTS(I, O). These self-loops are the only cycles of the output test process, so verdicts **pass** or **fail** can be reached in finite steps. Once verdicts are reached, they are not changed. Therefore, the states with verdicts indicate the end of the test execution. We assume that once the output test process detects the quiescence, the IUT cannot produce any visible output later, which justifies why the **pass** states have only a self-loop on quiescence.

To describe the execution of a queued-quiescence test case, we define a new operator $\delta \Downarrow O$. For IOTS L, $L_{\delta \Downarrow_o}$ is an IOTS obtained by first augmenting all the stable states of L by self-looping transitions labeled with δ, then projecting the augmented automaton onto the alphabet $O \cup \delta$, and finally determinizing the obtained automaton. The execution of a queued-quiescence test case $(\alpha, out(\alpha))$ against an IOTS $Imp \in IOTS(I, O)$ is described by the IOTS $(\alpha\ ][\ Imp)_{\delta \Downarrow_o}\ ][\ out(\alpha)$. Each trace leading this IOTS into a state, where the output test process is in a state labeled with **pass** or **fail**, is a *test run*. Notice that we treat the symbol δ as an input of the output test process, assuming that the tester executing δ just detects the fact that its buffer has no more symbols to read. Since the outputs of an IUT are stored in a finite queue, any implementation that, in response to the input sequence α, can produce an output sequence longer than the queue length may overflow the queue. To solve this problem, we should determine a lower bound of the output queue length so that the buffer is not overflowing until the tester reaches a verdict. The bound depends on the input sequence α and *Spec*, and it is finite because *Spec* does not oscillate.

The queued-quiescence tester $(\alpha, out(\alpha))$ meets all the requirements stated above. We use the term *verdict* state to refer to a state of the IOTS $(\alpha\ ][\ Imp)_{\delta \Downarrow_o}\ ][\ out(\alpha)$ such that the IOTS $out(\alpha)$ is in a state with a verdict.

Proposition 6. For a queued-quiescence test case $(\alpha, out(\alpha))$ of *Spec* and any IOTS $Imp \in IOTS(I, O)$

- the IOTS $(\alpha\ ][\ Imp)_{\delta \Downarrow_o}$ and $out(\alpha)$ are fully compatible;
- at least one verdict state is reachable from every state in the IOTS $(\alpha\ ][\ Imp)_{\delta \Downarrow_o}\ ][\ out(\alpha)$ and every cycle in the IOTS involves only verdict states, in other words, the tester always reaches a verdict in finite steps;
- both α and $out(\alpha)$ are deterministic;
- there is no state in α or $out(\alpha)$ where both inputs and outputs are enabled;
- if a verdict state is reached in the IOTS $(\alpha\ ][\ Spec)_{\delta \Downarrow_o}\ ][\ out(\alpha)$, the output tester $out(\alpha)$ is in a state with the verdict **pass**, i.e., the test case $(\alpha, out(\alpha))$ is sound.

The composition $(\alpha\,][\,Imp)_{\delta\Downarrow_o}\,][\,out(\alpha)$ has one or several verdict states. In a particular test run, one of these states with the verdict **pass** or **fail** is reached. Considering the distribution of verdicts in the verdict states of the composition, three cases are possible:

Case 1. All the states have **fail**.
Case 2. States have **pass** as well as **fail**.
Case 3. All the states have **pass**.

These cases lead us to various relations between an implementation and the specification that can be established by the queued-quiescence testing.

In the first case, the implementation is distinguished from the specification in a single test run.

Definition 7. Given IOTS *Spec* and *Imp*, *Imp* is *queued-quiescence separable* from *Spec*, if there exists a test case $(\alpha, out(\alpha))$ for *Spec* such that all the verdict states of the IOTS $(\alpha\,][\,Imp)_{\delta\Downarrow_o}\,][\,out(\alpha)$ are labeled with the verdict **fail**.

In the second case, the implementation can also be distinguished from the specification if a proper run is taken by the implementation during the test execution.

Definition 8. Given IOTS *Spec* and *Imp*, *Imp* is *queued-quiescence distinguishable* from *Spec*, if there exists a test case $(\alpha, out(\alpha))$ for *Spec* such that at least one verdict state of $(\alpha\,][\,Imp)_{\delta\Downarrow_o}\,][\,out(\alpha)$ is labeled with the verdict **fail**.

Clearly, *Imp* queued-quiescence separable from *Spec* is also queued-quiescence distinguishable from it. Consider now case 3, when for a given test case $(\alpha, out(\alpha))$ all the states have **pass**. In this case, the implementation does nothing illegal when the test case is executed, as it produces only valid output sequences. Two situations can yet be distinguished here. Either there exists a **pass** state of the output test process that is not included in any verdict state of $(\alpha\,][\,Imp)_{\delta\Downarrow_o}\,][\,out(\alpha)$ or there is no such a state. The difference is that with the given test case in the former situation, the implementation could still be distinguished from its specification, while in the latter, it could not. This motivates the following definition.

Definition 9. Given IOTS *Spec* and *Imp*,

- *Imp* is *queued-quiescence weakly-distinguishable* from *Spec* if there exists a test case $(\alpha, out(\alpha))$ for *Spec* such that the verdict states of $(\alpha\,][\,Imp)_{\delta\Downarrow_o}\,][\,out(\alpha)$ does not include all the **pass** states of $out(\alpha)$.
- *Imp* is *queued-quiescent trace-included* in *Spec* if for all $\alpha \in I^*$ all the verdict state of the IOTS $(\alpha\,][\,Imp)_{\delta\Downarrow_o}\,][\,out(\alpha)$ are labeled with the verdict **pass**.
- *Imp* and *Spec* are *queued-quiescent trace-equivalent* if for all $\alpha \in I^*$ all the verdict states of the IOTS $(\alpha\,][\,Imp)_{\delta\Downarrow_o}\,][\,out(\alpha)$ are labeled with the verdict **pass** and include all the **pass** states of $out(\alpha)$.

By definition, *Imp* is queued-quiescence weakly-distinguishable from *Spec* if *Imp* is queued-quiescence distinguishable from *Spec*, or if *Imp* is queued-quiescent trace-included in but not queued-quiescent trace-equivalent to *Spec*. We characterize the above relations in terms of traces and queued-quiescent traces.

Proposition 10. Given IOTS *Spec* and *Imp*,

1. *Imp* is queued-quiescence separable from *Spec* iff there exists an input sequence α such that $Qqtraces_o(Imp, \alpha) \cap Qqtraces_o(Spec, \alpha) = \emptyset$.
2. *Imp* is queued-quiescence distinguishable from it iff there exists an input sequence α such that $traces(\alpha)[\![Imp]\!]_{\delta\Downarrow_O} \not\subseteq traces(\alpha)[\![Spec]\!]_{\delta\Downarrow_O}$.
3. *Imp* is queued-quiescence weakly-distinguishable from it iff there exists an input sequence α such that $traces(\alpha)[\![Imp]\!]_{\delta\Downarrow_O} \neq traces(\alpha)[\![Spec]\!]_{\delta\Downarrow_O}$.
4. *Imp* is queued-quiescent trace-included into *Spec*, iff *Imp* does not oscillate and $Qqtraces(Imp) \subseteq Qqtraces(Spec)$.
5. *Imp* and *Spec* are queued-quiescent trace-equivalent iff *Imp* does not oscillate and $Qqtraces(Imp) = Qqtraces(Spec)$.

$traces(\alpha)[\![Imp]\!]_{\delta\Downarrow_O}$ is the set of traces of $(\alpha)[\![Imp]\!]_{\delta\Downarrow_O}$ in the initial states of the IOTS. $traces(\alpha)[\![Imp]\!]_{\delta\Downarrow_O} = traces(\alpha)[\![Imp]\!]_{\Downarrow_O} \cup Qqtraces_o(Imp, \alpha)\delta^*$. The relation between $traces(\alpha)[\![Imp]\!]_{\Downarrow_O}$ and $traces(\alpha)[\![Spec]\!]_{\Downarrow_O}$ is used to deal correctly with oscillating implementations [10]. Fig. 1 provides an example of the systems that are not quiescent trace equivalent, but are queued-quiescent trace-equivalent. Indeed, the quiescent trace $aa1\delta$ of the IOTS L_2 is not a trace of the IOTS L_1. In both, the input sequence a yields the queued-quiescent trace $a1\delta$, aa yields the queued-quiescent traces $aa1\delta$ and $aa2\delta$, any longer input sequence results in the same output sequences as aa.

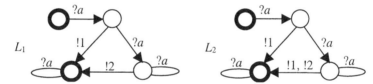

Fig. 1. Two IOTS that have different sets of quiescent traces, but are queued-quiescent trace-equivalent. Inputs are decorated with "?", outputs with "!"; and stable states are drawn in bold

The IOTS L_1 and L_2 are considered indistinguishable in our framework, while according to the **ioco** relation [12], they are distinguishable. The IOTS L_2 has the quiescent trace $aa1$ that is not a trace of L_1, therefore, to distinguish the two systems, the tester has to apply two consecutive inputs a. The output 1 appearing after the second input a indicates that the system being tested is, in fact, L_2 and not L_1. However, to reach such a conclusion, the tester should be able to prevent the appearance of the output 1 after the first input a. This is not possible under our assumption that the tester cannot block outputs of the IUT. The tester interacts with the IUT via queues and has no way of knowing in which state the output is actually produced. The presence of a testing context, which is a pair of finite queues in our case, makes implementation relations that could be tested via the context coarser, as is usually the case [8].

4 Deriving Queued-Quiescence Test Cases

Proposition 10 indicates the way test derivation could be performed for the IOTS *Spec* and an explicit fault model that includes a finite set of implementations. Namely, for each *Imp* in the fault model, we may first attempt to determine an input sequence α such that $Qqtraces_o(Imp, \alpha) \cap Qqtraces_o(Spec, \alpha) = \emptyset$. If fail we could next try to find α such that $traces(\alpha)[\![Imp]\!]_{\aleph_o} \not\subseteq traces(\alpha)[\![Spec]\!]_{\aleph_o}$. If $traces(\alpha)[\![Imp]\!]_{\aleph_o} \subseteq traces(\alpha)[\![Spec]\!]_{\aleph_o}$ for each α the question is about an input sequence α such that $traces(\alpha)[\![Imp]\!]_{\aleph_o} \neq traces(\alpha)[\![Spec]\!]_{\aleph_o}$, thus $traces(\alpha)[\![Imp]\!]_{\aleph_o} \subset traces(\alpha)[\![Spec]\!]_{\aleph_o}$. Based on the found input sequence, a queued-quiescence test case for *Imp* at hand can be constructed, as explained in the previous section. If no input sequence with this property can be determined we conclude that the IOTS *Spec* and *Imp* are queued-quiescent trace-equivalent, they cannot be distinguished by the queued-quiescence testing.

Search for an appropriate distinguishing input sequence could be performed in a straightforward way by considering input sequences of increasing length. To do so, we just parameterize Definitions 7, 8 and 9 and accordingly Proposition 10 with the length of input sequences. Given a length of input sequences k, then, e.g., *Imp* and *Spec* are queued-quiescent k-trace-equivalent iff $traces(\alpha)[\![Imp]\!]_{\aleph_o} = traces(\alpha)[\![Spec]\!]_{\aleph_o}$ for all $\alpha \in I^{\leq k}$, where $I^{\leq k}$ denotes the set of all input sequences of length equal or less than k. If the length of α such that $traces(\alpha)[\![Imp]\!]_{\aleph_o} \not\subseteq traces(\alpha)[\![Spec]\!]_{\aleph_o}$ is k then *Imp* is said to be queued-quiescence k-distinguishable from *Spec*. With these parameterized definitions, we examine all the input sequences starting from an empty sequence. The procedure terminates when the two IOTS are distinguished or when the value of k reaches a predefined maximum defined by the input buffer of the IUT available for queued testing.

Consider the example in Fig. 2. *Imp* is not queued-quiescence 1-distinguishable from *Spec*, for both produce the output 1 in response to the input a. However, *Imp* is queued-quiescence 2-distinguishable from *Spec*. Indeed, in response to the sequence aa the *Spec* can produce the output 1 or 12. While *Imp* - 2 or 12.

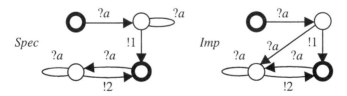

Fig. 2. The IOTS that are queued-quiescence 2-distinguishable, but not queued-quiescence 1-distinguishable

The search for a distinguishing input sequence relies on a procedure that verifies whether a given input sequence α satisfies $traces(\alpha)[\![Imp]\!]_{\aleph_o} \subseteq traces(\alpha)[\![Spec]\!]_{\aleph_o}$. Instead of elaborating this procedure, we give a more general procedure that accepts a regular language defined over the input set. Let E denote such a language, $E \subseteq I^*$, following the definition of the input test process, we also use E to denote the (deterministic) IOTS whose trace set is $pref(E)$. Since E is regular, such an IOTS exists. In the following proposition, we generalize item 2 in Proposition 10 by considering the inclusion relation between $traces(E)[[Imp)_{\delta\Downarrow_o}$ and $traces(E)[[Spec)_{\delta\Downarrow_o}$.

Proposition 11. Given two IOTS *Spec* and *Imp*, *Imp* is queued-quiescence distinguishable from *Spec* iff there exists a regular language $E \subseteq I^*$ such that $traces(E)[[Imp)_{\delta\Downarrow_o} \nsubseteq traces(E)[[Spec)_{\delta\Downarrow_o}$. Moreover, any trace $\beta \in traces(E)[[Imp)$ such that $\beta_{\Downarrow_o}\delta^* \cap (traces(E)[[Imp)_{\delta\Downarrow_o} \setminus traces(E)[[Spec)_{\delta\Downarrow_o}) \neq \emptyset$ yields a queued-quiescence test case $(\beta_{\Downarrow_I}, out(\beta_{\Downarrow_I}))$ that, when executed against *Imp*, produces the verdict **fail**.

Proof: (If) If there exists a regular language $E \subseteq I^*$ such that $traces(E)[[Imp)_{\delta\Downarrow_o} \nsubseteq traces(E)[[Spec)_{\delta\Downarrow_o}$, the part after "moreover" in the proposition indicates how a corresponding test case can be derived.

(Only if) If *Imp* is queued-quiescence distinguishable from *Spec*, according to the definition of the queued-quiescence distinguishability, there exists an input sequence α, which is a regular language with a single word, such that $traces(\alpha)[[Imp)_{\delta\Downarrow_o} \nsubseteq traces(\alpha)[[Spec)_{\delta\Downarrow_o}$. QED.

We call a language E satisfying the properties in Proposition 11 a *distinguishing input set* of *Spec* and *Imp*. The proposition suggests a test case derivation procedure.

Procedure 12. For deriving a test case of *Spec* that *Imp* fails if a given language E is a distinguishing input set.
Input: IOTS *Spec* and *Imp*, and a regular language E.
Output: E is not a distinguishing input set or a test case $(\alpha, out(\alpha))$.
  Step 1. Construct the deterministic automata that accept $traces(E)[[Imp)_{\delta\Downarrow_o}$ and $traces(E)[[Spec)_{\delta\Downarrow_o}$, respectively.
  Step 2. Using the direct products of the obtained automata, determine a sequence $\rho \in traces(E)[[Imp)_{\delta\Downarrow_o} \setminus traces(E)[[Spec)_{\delta\Downarrow_o}$. If such a sequence exists, go to Step 3; otherwise, return the result that E is not a distinguishing input set.
  Step 3. Construct a deterministic automaton by composing *Imp* with the LTS $<pref(\rho_{\Downarrow_o}), O, \lambda_\rho, \{\varepsilon\}>$, where $\lambda_\rho = \{(\beta, a, \beta a) \mid \beta a \in pref(\rho_{\Downarrow_o})\}$. Determine a trace γ of the obtained LTS with $\gamma_{\Downarrow_I} \in E$ and $\gamma_{\Downarrow_o} = \rho_{\Downarrow_o}$ and the queued-quiescence test case $(\gamma_{\Downarrow_I}, out(\gamma_{\Downarrow_I}))$.

Proposition 13. Given two IOTS *Spec* and *Imp*, and a distinguishing input set E, let $(\alpha, out(\alpha))$ be the queued-quiescence test case derived by the above procedure. Then the queued-quiescence test case executed against *Imp* produces the verdict **fail**.

If we consider every $E \in \{\{\alpha\} \mid \alpha \in I^{\leq k}\}$, the test cases that queued-quiescence k-distinguish *Imp* from *Spec* are derived. We notice that if the set E is I^*, Procedure 12 reduces to the test case derivation procedure reported earlier [7].

It is interesting to know that the notion of k-distinguishability applied to the IOTS and FSM models exhibits different properties. In particular, two k-distinguishable FSM are also $k+1$-distinguishable. This does not always hold for IOTS. The system *Imp* in Fig. 3 is queued-quiescence 1-distinguished from *Spec*; however, it is not queued-quiescence k-distinguished from *Spec* for any $k > 1$. This indicates that a spe-

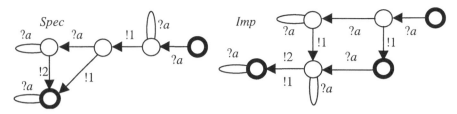

Fig. 3. The IOTS that are queued-quiescence 1-distinguishable, but not queued-quiescence k-distinguishable for $k > 1$

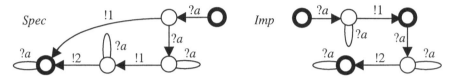

Fig. 4. Two queued-quiescent trace equivalent IOTS

cial care has to be taken when one attempts to adapt FSM-based methods to the queued testing of IOTS.

5 Queued-Suspension Testing

In the previous sections, we explored the possibilities for distinguishing IOTS based on their traces and queued-quiescent traces. The latter are pairs of input and output projections of quiescent traces. If two non-oscillating systems with different quiescent traces have the same sets of queued-quiescent traces, queued-quiescence testing may not differentiate them. However, sometimes such IOTS can still be distinguished by a queued testing, as we demonstrate below.

Consider the example in Fig. 4. Here the two IOTS have different sets of quiescent traces, however, they have the same set of queued-quiescent traces {$a1\delta$, $aa1\delta$, $aa12\delta$, $aaa1\delta$, $aaa12\delta$, ... }. In the testing framework presented in Section 3, they are not distinguishable. Indeed, we cannot tell them apart when a single input is applied to their initial states. Moreover, in response to the input sequence aa and to any longer sequence, they produce the same output sequence 12. The difference is that IOTS *Imp*, while producing the output sequence 12, becomes quiescent just before the output 2 and the IOTS *Spec* does not. The problem is that this quiescence is not visible through the output queue by the output test process that expects either 1 or 12 in response to aa. The queued-quiescence tester can detect the quiescence after reading the output sequence 12 as an empty queue, but it cannot detect an "intermediate" quiescence of the system. It has no way of knowing whether the system becomes quiescent before a subsequent input is applied. Both inputs are in the input buffer and it is completely up to the system when to read the second input.

Further decomposing the tester for *Spec* into two input and two output test processes could solve the problem. In this case, testing is performed as follows. The first input test process issues the input a. The first output test process expects the output 1 followed by a quiescence δ, when the quiescence is detected, the control is transferred to the second input test process that does the final a. Then the second output test process expects quiescence. If, instead, it detects the output 2 it produces the verdict **fail** which indicates that the IUT is *Imp* and not *Spec*. As opposed to a queued-quiescence tester, such a tester can detect intermediate quiescence of the IUT.

The example motivates the definition of a new type of testers. Such a tester is defined for a given sequence of input words $\alpha_1 \ldots \alpha_p$, in which $\alpha_i \neq \varepsilon$ for $i = 2, \ldots, p$. The tester is a finite tree with queued-quiescence test cases as nodes connected by transfer of control. The root node $(\alpha_1, out(\alpha_1))$ is a queued-quiescence test case of *Spec*, and is executed first. If the IUT passes the queued-quiescence test case, one of the node's children is selected based on the output of the IUT $\beta_1\delta \in Qqtraces_o(Spec, \alpha_1)$ and control is transferred to this node; otherwise, the IUT fails the tester and the test execution is terminated. We use *Spec*-**after**-(α, β) to denote the set of stable states that are reached by *Spec* when it executes all possible quiescent traces with the input projection α and output projection β. If we also use *Spec*-**after**-(α, β) to denote the IOTS obtained from *Spec* by initializing it in these states, the selected child node is a queued-quiescence test case of *Spec*-**after**-(α_1, β_1). The input test process of the child node executes α_2. The process continues until the IUT fails or a verdict of a leaf node is reached.

We define a sequence of output words $\beta_1 \ldots \beta_i$ to be *consistent* with the corresponding sequence of input words $\alpha_1 \ldots \alpha_i$ if $\beta_1 \in Qqtraces_o(Spec, \alpha_1)$ and $\beta_j \in Qqtraces_o(Spec\text{-}\mathbf{after}\text{-}(\alpha_1 \ldots \alpha_{j-1}, \beta_1 \ldots \beta_{j-1}), \alpha_j)$ for each $j = 2, \ldots, i$. Every node in the tree is identified by a consistent output sequence that leads the tester to the node. Given α_i, we use $(\alpha_i, out(\alpha_i, \beta_1 \ldots \beta_{i-1}))$ to denote the queued-quiescence test case of *Spec*-**after**-$(\alpha_1 \ldots \alpha_{i-1}, \beta_1 \ldots \beta_{i-1})$. We have the definition of the tester based on the discussions above.

Definition 14. Given a finite sequence of input words $\alpha_1 \ldots \alpha_p$, a *queued-suspension tester* or a *queued-suspension test case* $(\alpha_1 \ldots \alpha_p, Out(\alpha_1 \ldots \alpha_p))$ is a tree $(N, P, (\alpha_1, out(\alpha_1)))$, in which

- N is the set of nodes, $N = \{(\alpha_1, out(\alpha_1))\} \cup \{(\alpha_i, out(\alpha_i, \beta_1 \ldots \beta_{i-1})) \mid \beta_1 \ldots \beta_{i-1}$ is an output sequence consistent with $\alpha_1 \ldots \alpha_{i-1}, i = 2, \ldots, p\}$;
- P is the transition relation, $P = \{(\alpha_1, out(\alpha_1)) \rightarrow (\alpha_2, out(\alpha_2, \beta_1)) \mid \beta_1$ is an output sequence consistent with $\alpha_1\} \cup \{(\alpha_i, out(\alpha_i, \beta_1 \ldots \beta_{i-1})) \rightarrow (\alpha_{i+1}, out(\alpha_{i+1}, \beta_1 \ldots \beta_i)) \mid \beta_1 \ldots \beta_i$ is an output sequence consistent with $\alpha_1 \ldots \alpha_i, i = 2, \ldots, p\text{-}1\}$;
- $(\alpha_1, out(\alpha_1))$ is the root node.

It is clear that for a single input word, a queued-suspension tester reduces to a queued-quiescence tester. The queued-suspension testing is more discriminative than queued-quiescence testing, as Fig. 4 illustrates. In fact, consider a queued-quiescence tester derived from a single sequence $\alpha_1 \ldots \alpha_p$ and a queued-suspension tester derived from the sequence of p words $\alpha_1, \ldots, \alpha_p$, the former uses just the output projection of quiescent traces that have the input projection $\alpha_1 \ldots \alpha_p$ while the latter additionally partitions the quiescent traces into p quiescent sub-traces. Then the two systems that cannot be distinguished by the queued-suspension testing have to produce the same output projection, moreover, the output projections have to coincide up to the partition defined by the partition of the input sequence. This leads us to the notion of queued-suspension traces.

Given a finite sequence of finite input words $\alpha_1\ldots\alpha_p$, a sequence $(\alpha_1\beta_1\delta)\ldots(\alpha_p\beta_p\delta)$ is called a *queued-suspension trace* of *Spec* if $\beta_1\ldots\beta_p$ is an output sequence consistent with $\alpha_1\ldots\alpha_p$. We use *Qstraces(Spec)* to denote the set of queued-suspension traces of *Spec* in the initial states.

We define the relations that can be established by queued-suspension testing similar to Definitions 7, 8, and 9.

Definition 15. Given IOTS *Spec* and *Imp*,

- *Imp* is *queued-suspension separable* from *Spec*, if there exists a test case $(\alpha_1\ldots\alpha_p, Out(\alpha_1\ldots\alpha_p))$ for *Spec* such that for any consistent output sequence $\beta_1\ldots\beta_{p-1}$ all the verdict states of the IOTS $(\alpha_p$][*Imp*-**after**-$(\alpha_1\ldots\alpha_{p-1}, \beta_1\ldots\beta_{p-1}))\delta\Downarrow_O$][$out(\alpha_p, \beta_1\ldots\beta_{p-1})$ are labeled with the verdict **fail**.
- *Imp* is *queued-suspension distinguishable* from *Spec*, if there exist a test case $(\alpha_1\ldots\alpha_p, Out(\alpha_1\ldots\alpha_p))$ for *Spec* and a consistent output sequence $\beta_1\ldots\beta_{p-1}$ such that at least one verdict state of the IOTS $(\alpha_p$][*Imp*-**after**-$(\alpha_1\ldots\alpha_{p-1}, \beta_1\ldots\beta_{p-1}))\delta\Downarrow_O$][$out(\alpha_p, \beta_1\ldots\beta_{p-1})$ is labeled with the verdict **fail**.
- *Imp* is *queued-suspension weakly-distinguishable* from *Spec* if there exist a test case $(\alpha_1\ldots\alpha_p, Out(\alpha_1\ldots\alpha_p))$ for *Spec* and a consistent output sequence $\beta_1\ldots\beta_{p-1}$ such that the verdict states of the IOTS $(\alpha_p$][*Imp*-**after**-$(\alpha_1\ldots\alpha_{p-1}, \beta_1\ldots\beta_{p-1}))\delta\Downarrow_O$][$out(\alpha_p, \beta_1\ldots\beta_{p-1})$ does not include all the **pass** states of $out(\alpha_p, \beta_1\ldots\beta_{p-1})$.
- *Imp* is said to be *queued-suspension trace-included* in *Spec* if for all $\alpha \in I^*$ and all possible partitions of α into words $\alpha_1, \ldots, \alpha_p$, all the verdict states of IOTS $(\alpha_p$][*Imp*-**after**-$(\alpha_1\ldots\alpha_{p-1}, \beta_1\ldots\beta_{p-1}))\delta\Downarrow_O$][$out(\alpha_p, \beta_1\ldots\beta_{p-1})$ are labeled with the verdict **pass**.
- *Imp* and *Spec* are *queued-suspension trace-equivalent* if for all $\alpha \in I^*$, all possible partitions of α into words $\alpha_1, \ldots, \alpha_p$, and all consistent output sequence $\beta_1\ldots\beta_{p-1}$, all the verdict states of the IOTS $(\alpha_p$][*Imp*-**after**-$(\alpha_1\ldots\alpha_{p-1}, \beta_1\ldots\beta_{p-1}))\delta\Downarrow_O$] [$out(\alpha_p, \beta_1\ldots\beta_{p-1})$ are labeled with the verdict **pass** and include all the **pass** states of $out(\alpha_p, \beta_1\ldots\beta_{p-1})$.

Accordingly, the following is a generalization of Proposition 10.

Proposition 16. Given IOTS *Spec* and *Imp*,

- *Imp* is queued-suspension separable from *Spec* iff there exists a finite sequence of input words $\alpha_1\ldots\alpha_i$ such that $Qqtraces_o(Imp\text{-}\mathbf{after}\text{-}(\alpha_1\ldots\alpha_{i-1}, \gamma_1\ldots\gamma_{i-1}), \alpha_i) \cap Qqtraces_o(Spec\text{-}\mathbf{after}\text{-}(\alpha_1\ldots\alpha_{i-1}, \gamma_1\ldots\gamma_{i-1}), \alpha_i) = \varnothing$ for any consistent $\gamma_1\ldots\gamma_{i-1}$.
- *Imp* is queued-suspension distinguishable from it iff there exist a finite sequence of input words $\alpha_1\ldots\alpha_i$ and consistent $\gamma_1\ldots\gamma_{i-1}$ such that $traces(\alpha_i$][*Imp*-**after**-$(\alpha_1\ldots\alpha_{i-1}, \gamma_1\ldots\gamma_{i-1}))\delta\Downarrow_O \nsubseteq traces(\alpha_i$][*Spec*-**after**-$(\alpha_1\ldots\alpha_{i-1}, \gamma_1\ldots\gamma_{i-1}))\delta\Downarrow_O$.
- *Imp* is queued-suspension weakly-distinguishable from it iff there exist a finite sequence of input words $\alpha_1\ldots\alpha_i$ and consistent $\gamma_1\ldots\gamma_{i-1}$ such that $traces(\alpha_i$][*Imp*-**after**-$(\alpha_1\ldots\alpha_{i-1}, \gamma_1\ldots\gamma_{i-1}))\delta\Downarrow_O \neq traces(\alpha_i$][*Spec*-**after**-$(\alpha_1\ldots\alpha_{i-1}, \gamma_1\ldots\gamma_{i-1}))\delta\Downarrow_O$.

- *Imp* is queued-suspension trace-included into *Spec*, iff *Imp* does not oscillate and $Qstraces(Imp) \subseteq Qstraces(Spec)$.
- *Imp* and *Spec* are queued-suspension trace-equivalent iff *Imp* does not oscillate and $Qstraces(Imp) = Qstraces(Spec)$.

The queued-suspension testing needs input and output buffers as the queued-quiescence testing. The size of the input buffer is defined by the longest input word in a chosen test case $(\alpha_1...\alpha_p, Out(\alpha_1...\alpha_p))$, while that of the output buffer by the longest output sequence produced in response to any input word. We assume the size of the input buffer k is given and use it to define queued-suspension k-traces and accordingly, to parameterize Definition 15 obtaining appropriate notions of k-distinguishability. In particular, a queued-suspension trace of *Spec* $\alpha_1\beta_1\delta...\alpha_p\beta_p\delta \in Qstraces(Spec)$ is called a queued-suspension k-trace of *Spec* if $|\alpha_i| \leq k$ for all $i = 1, ..., p$. The set of all these traces $Qstraces^{\leq k}(Spec)$ has a finite representation.

Definition 17. Let S_{stable} be the set of all stable states of an IOTS *Spec* = $<S, I, O, \lambda, S_0>$. A *queued-suspension k-machine* for *Spec* is a tuple $<R, I^{\leq k}O^*\delta, \lambda^k_{stable}, r_0>$, denoted $Spec^k_{susp}$, where the starting state $r_0 = \{S_0\}$ and the set of states $R \subseteq P(S_{stable}) \cup \{S_0\}$ ($P(S_{stable})$ is a powerset of S_{stable}), and the transition relation λ^k_{stable} are the smallest sets obtained by application of the following rules:
- $(r, \alpha\beta\delta, r') \in \lambda^k_{stable}$ if $\alpha \in I^{\leq k}, \beta \in O^*$ and r' is the set of states r-**after**-(α, β) in *Spec*.
- In case that some initial state of *Spec* is unstable $(r_0, \varepsilon\beta\delta, r') \in \lambda^k_{stable}$ if $\beta \in O^*$ and $\beta \neq \varepsilon$, and $r' = S_0$-**after**-(ε, β).

Notice that each system that does not oscillate has at least one stable state.

Proposition 18. The set of traces of $Spec^k_{susp}$ coincides with the set of queued-suspension k-traces of *Spec*.

Corollary 19. A non-oscillating IOTS *Imp* is queued-suspension k-distinguishable from *Spec* iff Imp^k_{stable} has a trace that is not a trace of $Spec^k_{susp}$.

Fig. 1 depicts the IOTS that are queued-suspension trace equivalent, recall that they are also queued-quiescent trace-equivalent, but not quiescent trace equivalent.

We notice that a queued-suspension k-machine can be viewed as an FSM with the input set $I^{\leq k}$ and output set $O^{\leq m}$ for an appropriate integer m, so that FSM-based methods could be adapted to derive queued-suspension test cases.

6 Conclusion

We addressed the problem of testing from transition systems with inputs and outputs and elaborated a testing framework based on the idea of decomposing a tester into input and output processes. Input test process applies inputs to an IUT via a finite input queue and output test process reads outputs that the IUT puts into a finite output queue until it detects no more outputs from the IUT, i.e., the tester detects quiescence of the IUT. In such a testing architecture, input from the tester and output from the IUT may occur simultaneously. We call such a testing scenario queued testing. We elaborated two types of queued testers, the first consisting of single input and single output test processes, a so-called queued-quiescence tester, and the second consisting of several such pairs of processes, a so-called queued-suspension tester. We defined implementation relations that can be checked by the queued testing with both types of testers and proposed test derivation procedures.

Our work differs from the previous work in several important aspects. First of all, we make a liberal assumption on the way the tester interacts with an IUT, namely that the IUT can issue output at any time and the tester cannot determine exactly the stimulus that causes the output. We believe that this assumption is less restrictive than any other assumption known in the testing literature [2], [6]. Testing with this assumption requires buffers between the IUT and tester. To make our approach practical, these buffers are considered finite, opposed to the case of infinite queues considered earlier [14]. We demonstrated that the implementation relations that can be verified by the queued testing are coarser than those previously considered. The test derivation procedures were elaborated with a fault model in mind. The resulting test suite becomes finite and related to the assumptions about potential faults, as opposed to the approach of [12], where the number of test cases is, in fact, uncontrollable and not driven by any assumption about faults. The finiteness of test cases allows us, in addition, to check equivalence relations and not only preorder relations as in, e.g., [12].

Concerning future work, we believe that this paper may trigger research in various directions. It is interesting, for example, to see to which extent one could adapt FSM-based test derivation methods driven by fault models, as is done in [11] with a more restrictive assumption about a tester in mind.

Acknowledgment

The first author acknowledges fruitful discussions with Andreas Ulrich about testing IOTS. This work was in part supported by the NSERC grant OGP0194381. The second author acknowledges a partial support of the program "Russian Universities". The third author acknowledges a partial support of ReSMiQ.

References

1. Bochmann, G. v., Petrenko, A.: Protocol Testing: Review of Methods and Relevance for Software Testing. In: The Proceedings of the ACM International Symposium on Software Testing and Analysis, ISSTA'94. USA (1994)
2. Brinksma, E., Tretmans, J.: Testing Transition Systems: An Annotated Bibliography. In: Cassez, F., Jard, C., Rozoy, B., Ryan, M. (eds.): Modeling and Verification of Parallel Processes. Lecture Notes in Computer Science, Vol. 2067. Springer-Verlag, Berlin Heidelberg New York (2001)
3. van Glabbeek, R. J.: The Linear Time-Branching Time Spectrum. In: The Proceedings of CONCUR'90. Lecture Notes In Computer Science, Vol. 458. Springer-Verlag, Berlin Heidelberg New York (1990)
4. Jard, C., Jéron, T., Tanguy, L., Viho, C.: Remote Testing Can Be as Powerful as Local Testing. In: The Proceedings of the IFIP Joint International Conference, Methods for Protocol Engineering and Distributed Systems, FORTE XII/PSTV XIX. China (1999)

5. Lynch, N., Tuttle, M. R.: An Introduction to Input/Output Automata. In: CWI Quarterly, Vol. 2, Issue 3 (1989)
6. Petrenko, A.: Fault Model-Driven Test Derivation from Finite State Models: Annotated Bibliography. In: Cassez, F., Jard, C., Rozoy, B., Ryan, M. (eds.): Modeling and Verification of Parallel Processes. Lecture Notes in Computer Science, Vol. 2067. Springer-Verlag, Berlin Heidelberg New York (2001)
7. Petrenko, A., Yevtushenko, N.: Queued Testing of Transition Systems with Inputs and Outputs. In: Hierons, R., Jeron, T. (eds.): INRIA preprint, the Proceedings of the Workshop on Formal Approaches to Testing of Software, FATES'02, A Satellite Workshop of CONCUR'02. Czech Republic (2002)
8. Petrenko, A., Yevtushenko, N., Bochmann, G. v., Dssouli, R.: Testing in Context: Framework and Test Derivation. In: Computer Communications, Vol. 19 (1996)
9. Phalippou, M.: Executable Testers. In: The Proceedings of the IFIP Sixth International Workshop on Protocol Test Systems, IWPTS'93. France (1993)
10. Segala, R.: Quiescence, Fairness, Testing and the Notion of Implementation. In: The Proceedings of CONCUR'93. Lecture Notes in Computer Science, Vol. 715. Springer-Verlag, Berlin Heidelberg New York (1993)
11. Tan, Q. M., Petrenko, A.: Test Generation for Specifications Modeled by Input/Output Automata. In: The Proceedings of the IFIP 11th International Workshop on Testing of Communicating Systems, IWTCS'98. Russia (1998)
12. Tretmans, J.: Test Generation with Inputs, Outputs and Repetitive Quiescence. In: Software-Concepts and Tools, Vol. 17, Issue 3 (1996)
13. Vaandrager, F.: On the Relationship between Process Algebra and Input/Output Automata. In: The Proceedings of Sixth Annual IEEE Symposium on Logic in Computer Science (1991)
14. Verhaard, L., Tretmans, J., Kim, P., Brinksma, E.: On Asynchronous Testing. In: The Proceedings of the IFIP 5th International Workshop on Protocol Test Systems, IWPTS'92. Canada (1992)
15. de Vries, R. G., Belinfante, A., Feenstra, J.: Automated Testing in Practice: The Highway Tolling System. In: The Proceedings of the IFIP 14th International Conference on Testing of Communicating Systems, TestCom'2002. Berlin, Germany (2002)

Generating Checking Sequences for a Distributed Test Architecture

Hasan Ural and Craig Williams

School of Information Technology and Engineering
University of Ottawa, Ottawa, Ontario, K1N 6N5, Canada
{ural,cwilliam}@site.uottawa.ca

Abstract. The objective of testing is to determine whether an implementation under test conforms to its specification. In distributed test architectures involving multiple testers, this objective can be complicated by the fact that testers may encounter problems relating to controllability and observability during the application of tests. The controllability problem manifests itself when a tester is required to send the current input and because it did not send the previous input nor did it receive the previous output it cannot determine when to send the input. The observability problem manifests itself when a tester is expecting an output in response to either the previous input or the current input and because it is not the sender of the current input, it cannot determine when to start and stop waiting for the output. Based on a distinguishing sequence, a checking sequence construction method is proposed to yield a sequence that is free from controllability and observability problems.

1 Introduction

Determining, under certain assumptions, whether a given "black box" implementation N of a Finite State Machine (*FSM*) M is functioning correctly is referred to as a *fault detection* (*checking*) *experiment*. Foundations of fault detection experiments can be found in sequential circuit testing literature [4, 7]. This experiment is based on an input sequence called a *checking sequence*, constructed from a given deterministic and minimal FSM M with a designated initial state, that determines whether a given FSM N is a correct or faulty implementation of M. The construction of a checking sequence must deal with the "black box" nature of the given implementation N, which allows only limited controllability and observability of N. The limited controllability refers to not being able to directly transfer N to a designated state and the limited observability refers to not being able to directly recognize the current state of N. To overcome the restrictions imposed by the limited controllability and observability, special input sequences must be utilized in the construction of a checking sequence such that the output sequences produced by N in response to these input sequences provide sufficient information to deduce that every state transition of M is implemented correctly by N.

In order to verify the implementation of a transition from state a to b under input x, 1) N must be transferred to the state recognized as a, 2) the input x is applied and the output produced in response by N must be as specified in M, and 3) the state reached after the application of x must be recognized as state b. Hence, a crucial part of testing the correct implementation of each transition is recognizing the starting and terminating states of the transition. The recognition of a state of an FSM M can be achieved by a distinguishing sequence (DS) [7], a characterization set [7] or a unique input-output (UIO) sequence [14]. It is known that a distinguishing sequence may not exist for every minimal FSM [11], and that determining the existence of a distinguishing sequence for an FSM is PSPACE-complete [12]. Nevertheless, based on distinguishing sequences, various methods have been proposed in the literature to test FSMs [5, 7, 18, among others].

Testing an implementation N of an FSM M can be carried out as a fault detection experiment in some specific test architectures. One such architecture is the distributed test architecture shown in Figure 1 [9] where the lower interface and the upper interface of the implementation N may be controlled and observed indirectly by the lower tester (L) and directly by the upper tester (U), respectively. A similar architecture is given in [10].

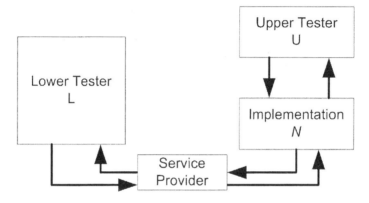

Fig. 1. A Distributed Test Architecture

In this architecture, U and L are two remote testers that are required to coordinate the application of a preset checking sequence. However, this requirement may lead to controllability and observability problems, in addition to those that stem from the black box nature of N. The *controllability (synchronization) problem* manifests itself when L (or U) is expected to send an input to N after N responds to an input from U (or L) with an output to U (or L), but L (or U) is unable to determine whether N sent that output. It is therefore important to construct a synchronizable checking sequence that causes no controllability problems during its application in the distributed test architecture. For some FSMs, a checking sequence can be constructed such that no two consecutive inputs will cause a controllability problem, and hence the coordination among testers is achieved indirectly through their interactions with N [6]. However, for some other FSMs, there may not exist a checking sequence in which the testers can coordinate solely via their interactions with N [6]. In this case it is necessary for testers to communicate directly by exchanging external coordination messages over a dedicated channel during the application of the checking sequence. An external coordination message exchange relating to controllability is denoted $<-C_{L(U)}, +C_{U(L)}>$, where "$-C_{L(U)}$" denotes the sending of an external coordination message to tester L(or U) from tester U(or L), and "$+C_{U(L)}$" denotes the receipt of an external coordination message from tester U(or L) by tester L(or U) [2].

During the application of even a synchronizable checking sequence in a distributed test architecture, the observability problem manifests itself when L (or U) is expected to receive an output from N in response to either the previous input or the current input and because L (or U) is not the one to send the current input, L (or U) is unable to determine when to start and stop waiting. Such observability problems hamper the detectability of output shift faults in N i.e., an output associated with the current input is generated by N in response to either the previous input or the next input. To ensure the detectability of output shift faults in N the checking sequence needs to be augmented either by additional input subsequences selected from FSM M or by external coordination message exchanges between testers such that during the application of the checking sequence testers can determine whether the output observed is received in response to the correct input as specified in M. An external coordination message exchange relating to observability is denoted <-$O_{L(U)}$, +$O_{U(L)}$>, where "-$O_{L(U)}$" denotes the sending of an external coordination message to tester L(or U) from tester U(or L), and "+$O_{U(L)}$" denotes the receipt of an external coordination message from tester U(or L) by tester L(or U) [2].

This paper proposes a method for constructing a checking sequence that does not pose controllability and observability problems during its application in a distributed test architecture. Earlier work on the controllability problem [1, 3, 15, 16, 17] and the observability problem [2, 13, 19, 20] consider the construction of a test sequence rather than a checking sequence, except [6, 8] which use UIO sequences for the construction of a checking sequence. It is well known that the complete fault coverage of a checking sequence cannot always be achieved by a test sequence where transition verification is not necessarily based on state verification.

The rest of the paper is organized as follows: Related terminology is reviewed in Section 2. In Section 3, the proposed method is presented along with an illustrative example. Section 4 concludes with a summary and future research directions.

2 Preliminaries

2.1 FSM and Its Graphical Representation

For ease of presentation and readability, the proposed method will be presented using a 2-port FSM. The generalization of the method to n-port ($n \geq 2$) FSM is simply a matter of adapting a different notation as the one given in [19]. A *2-port Finite State Machine* (2p-FSM) $M = (S, \Sigma, \Gamma, \delta, \lambda, s_1)$ where
- S is a finite set of states of M,
- $s_1 \in S$ is the initial state of M,
- $\Sigma = \Sigma_U \cup \Sigma_L$, where $\Sigma_{U(L)}$ is the input alphabet of port U (L), and $\Sigma_U \cap \Sigma_L = \emptyset$. Let $I = \Sigma_U \cup \Sigma_L \cup \{-\}$, where - means *null* input,
- $\Gamma = \Gamma_U \cup \Gamma_L$, where $\Gamma_{U(L)}$ is the output alphabet of port U (L), and $\Gamma_U \cap \Gamma_L = \emptyset$. Let $O = \{<a_U, a_L> \mid \exists a_{U(L)} \in \Gamma_{U(L)} \cup \{-\} \}$, where - means *null* output,
- δ is the transition function that maps $S \times I$ to S, i.e., $\delta: S \times I \to S$, and
- λ is the output function that maps $S \times I$ to O, i.e., $\lambda: S \times I \to O$.

Henceforth, a 2p-FSM M or N will be called simply FSM M or N, respectively.

An FSM M is *deterministic* if, for each input $x \in I$, there is at most one transition defined at each state of M. An FSM M is said to be *minimal* if none of its states are equivalent (i.e., $\forall\ s_i, s_j \in S, s_i \neq s_j, \exists$ an input sequence $X \in I^*$ such that $\lambda(s_i, X) \neq \lambda(s_j, X)$). An FSM M is said to be *completely specified* if, for each input $x \in I$, there is a transition defined at each state of M.

An FSM M can be represented by a directed graph $G = (V, E)$ where a set of vertices V represents the set S of states of M, and a set of directed edges E represents all specified transitions of M. A *transition* of an FSM M is a triple $t_{jk} = (s_j, s_k; x/y)$, where $s_j, s_k \in S, x \in I$, and $y \in O$ such that $\delta(s_j, x) = s_k$, $\lambda(s_j, x) = y$, and x/y is known as an *input/output pair*. Each edge $e_{jk} = (v_j, v_k; x/y) \in E$ represents a state transition from state s_j to state s_k with input x and output y where the input/output pair x/y is the *label* of e_{jk}, denoted by $label(e_{jk})$, v_j is called the *head* of e_{jk}, denoted by $head(e_{jk})$, and v_k is called the *tail* of e_{jk}, denoted by $tail(e_{jk})$.

A *path* $P = (v_1, v_2; x_1/y_1)(v_2, v_3; x_2/y_2)\ldots(v_{k-1}, v_k; x_{k-1}/y_{k-1})$, $k>1$, in $G = (V, E)$ is a finite sequence of adjacent (but not necessarily distinct) edges in G, where v_1 and v_k are $head(P)$ and $tail(P)$, and $x_1/y_1, x_2/y_2, \ldots, x_{k-1}/y_{k-1}$ is the *label* of P, denoted $label(P)$. A path P is represented by $(v_1, v_k; X/Y)$ where $label(P) = X/Y$ is the *input/output sequence* $(x_1/y_1)(x_2/y_2)\ldots(x_{k-1}/y_{k-1})$, input sequence $X = (x_1 x_2 \ldots x_{k-1})$ is the *input portion* of X/Y, and output sequence $Y = (y_1 y_2 \ldots y_{k-1})$ is the *output portion* of X/Y. The *cost* or *length* of each edge of G is equal to the number of input/output pairs in its label. The cost of a path (or length of a path) P in G is the sum of the costs (or lengths) of edges included in P and is denoted $cost(P)$. The first transition $(v_1, v_2; x_1/y_1)$ of path P is denoted $first(P)$ and the last transition $last(P)$. The *concatenation* of a path A and a path B is denoted $A @ B$.

A sequence $(i_1 i_2 \ldots i_k)$ is a *subsequence* of $(x_1 x_2 \ldots x_m)$ if there exists a Δ, $0 \leq \Delta \leq m-k$, such that for all j, $1 \leq j \leq k$, $i_j = x_{j+\Delta}$. A sequence $(i_1 i_2 \ldots i_k)$ is a *prefix* of $(x_1 x_2 \ldots x_m)$ if $\forall\ j$, $1 \leq j \leq k$, $i_j = x_j$. An FSM M has a *reset* function if there exists an input $r \in I$ which takes M from any state s_i to the initial state s_1 with a single transition $(s_i, s_1; r/<-,->)$.

A digraph $G = (V, E)$ is *strongly connected* if, for any pair of vertices v_j and v_k, there exists a path from v_j to v_k. It is *weakly connected* if its underlying undirected graph is connected. A *tour* of G is a path in G that starts and ends at the same vertex of G. An *Euler tour* of G is a tour that contains every edge of E exactly once. A *postman tour* (PT) of G is a tour that contains every edge in E at least once. A *rural postman tour* (RPT) of G over a set $E_C \subseteq E$ is a tour traversing every edge in E_C at least once. A *Chinese postman tour* (CPT) is a minimum-cost PT. A *rural Chinese postman tour* (RCPT) of G over a set $E_C \subseteq E$ is a minimum-cost RPT over E_C. A *rural postman path* (RPP) from v_i to v_j over $E_C \subseteq E$ is a path from v_i to v_j that includes every edge in E_C. A *rural Chinese postman path* (RCPP) from v_i to v_j over $E_C \subseteq E$ is a minimum-cost RPP.

Given a vertex $v \in V$, *in-degree*(v), is defined as $|\{(u, v; x/y): (u, v; x/y) \in E\}|$ and *out-degree*(v), is defined as $|\{(v, w; x/y): (v, w; x/y) \in E\}|$.

Given an FSM M, an input sequence X is a *distinguishing sequence* (D) if the output sequence Y produced by M in response to X is different for each state. $DS(s_i)$ denotes the transition sequence induced by the application of D at state s_i. A *test seg-*

ment for a transition $t_{ij} = (s_i, s_j; x/y)$ is the transition sequence induced by the application of xD at state s_i.

Given an FSM M, let $\Phi(M)$ be the set of FSMs each of which has at most $|S|$ states and the same input and output sets as M. Let N be an FSM of $\Phi(M)$. N is *isomorphic* to M if there is a one-to-one and onto function f on the state sets of M and N such that for any state transition $(s_i, s_j; x/y)$ of M, $(f(s_i), f(s_j); x/y)$ is a transition of N. A *checking sequence* of M is an input sequence starting at a specific state of M that distinguishes M from any N of $\Phi(M)$ that is not isomorphic to M. In the context of testing, this means that in response to this input sequence, any faulty implementation N will produce an output sequence different than the expected output, thereby indicating the presence of a fault(s).

2.2 Controllability (Synchronization) Problem

Given an FSM M and a global input/output sequence $\omega = x_1/y_1\, x_2/y_2 \ldots x_m/y_m$ of M, where $x_i \in I$ and $y_i \in O$, $1 \le i \le m$, a *controllability (synchronization) problem* occurs when, in the labels x_j/y_j and x_{j+1}/y_{j+1} of any two consecutive transitions, there exists a tester k that sends x_{j+1} that is neither the one sending x_j nor one of those receiving an output belonging to y_j, $1 \le j \le m-1$.

Given an FSM M and a global input/output sequence $\omega = x_1/y_1\, x_2/y_2 \ldots x_m/y_m$ of M, where $x_i \in I$ and $y_i \in O$, $1 \le i \le m$, any two consecutive transitions t_{ij} and t_{jk} whose labels are x_j/y_j and x_{j+1}/y_{j+1} form a *synchronizable pair* of transitions if t_{jk} can follow t_{ij} without causing a synchronization problem. For a transition $t_{ij} = (v_i, v_j; x/y_j)$, each transition $t_{jk} = (v_j, v_k; x_k/y_k)$ that forms a synchronizable pair of transitions with t_{ij} is called an *synchronizable successor* of t_{ij}. Any (sub)sequence of transitions in which every pair of transitions is synchronizable is called a *synchronizable transition (sub)sequence*. A global input/output sequence is said to be *synchronizable* if it is the label of a synchronizable transition sequence.

2.3 Observability Problem

Given an FSM M and a global input/output sequence $\omega = x_1/y_1\, x_2/y_2 \ldots x_m/y_m$ of M, where $x_i \in I$ and $y_i \in O$, $1 \le i \le m$, a *1-shift output fault* in an implementation N of M exists when, in the labels x_j/y_j and x_{j+1}/y_{j+1} of any two consecutive transitions, there exists one $a_{L(U)} \in \Gamma_{L(U)}$ in y_j of M which occurs in y_{j+1} in N (and not in y_j in N) or there exists one $a_{L(U)} \in \Gamma_{L(U)}$ in y_{j+1} of M which occurs in y_j in N (and not in y_{j+1} in N), $1 \le j \le m-1$. An instance of the observability problem manifests itself as an *undetectable 1-shift output fault* if there is a 1-shift output fault related to $a_{L(U)} \in \Gamma_{L(U)}$ in any two consecutive transitions whose labels are x_j/y_j and x_{j+1}/y_{j+1}, such that tester $L(U)$ satisfies the condition ($a_{L(U)}$ is in y_j XOR $a_{L(U)}$ is in y_{j+1}) AND $x_{j+1} \notin \Sigma_{L(U)}$. In this case, we say that tester $L(U)$ is *involved* in the shift, and would not be able to detect it.

3 The Proposed Method

Let $M = (S, \Sigma, \Gamma, \delta, \lambda, s_1)$ hereafter stand for a minimal, (in)completely specified FSM which is represented by a strongly connected digraph $G = (V, E)$ and has a distinguishing sequence D. Let $|S|$ be n and $s_1 \in S$ be the initial state of M. It is assumed that any implementation N of M correctly implements a reset function which takes M from any state s_i to the initial state s_1 with a single transition $(s_i, s_1; r/<-,->)$. The construction of a synchronizable global checking sequence of M is based on the construction of two sets which represent all potential controllability and observability problems for M. A state cover and transition cover are generated, and an auxiliary graph G'' representing these sequences is constructed. A rural Chinese postman path on G'' yields a checking sequence which can be applied in a distributed test architecture without encountering controllability or observability problems.

3.1 Identifying Controllability and Observability Problems

In the first phase of the proposed method, the set of all controllability problems, T_C, and the set of all observability problems, T_O, are generated from the digraph $G = (V, E)$ of the given FSM M. The set T_C is constructed as follows:

for each vertex $v_j \in V$ do
  for each edge e_{ij} (say t_{ij}) = $(v_i, v_j; x_j/y_j)$ entering vertex v_j do
    for each edge e_{jk} (say t_{jk})= $(v_j, v_k; x_{j+1}/y_{j+1})$ leaving vertex v_j do
      if $x_j \in \Sigma_U$ AND $x_{j+1} \in \Sigma_L$ AND $a_L = -$ in y_j
        then add $(t_{ij}, t_{jk}; <-C_L, +C_U>)$ to T_C
      else if $x_j \in \Sigma_L$ AND $x_{j+1} \in \Sigma_U$ AND $a_U = -$ in y_j
        then add $(t_{ij}, t_{jk}; <-C_U, +C_L>)$ to T_C

Each transition pair added to T_C forms a controllability problem as the sender of x_{j+1} is not the sender of x_j and does not receive an output in y_j. Given a path P on G representing a sequence of transitions, the sequence can be made synchronizable as follows: For each pair of consecutive transitions t_{mn}, t_{no} in P, if $(t_{mn}, t_{no}; <-C_{U(L)}, +C_{L(U)}>) \in T_C$ then insert the external coordination message exchange $<-C_{U(L)}, +C_{L(U)}>$ relating to controllability between t_{mn} and t_{no} in the label of P.

Consider the FSM $M1$ shown in Figure 2. Applying the above procedure results in a set consisting of 1 non-synchronizable transition pair, i.e., $T_C = \{(t3, t10, <-C_U, +C_L>)\}$.

The set T_O of all triples corresponding to transition pairs with a potential undetectable 1-shift output fault is generated next. The set T_O is constructed from $G = (V, E)$ as follows:

for each vertex $v_j \in V$ do
  for each edge e_{ij} (say t_{ij}) = $(v_i, v_j; x_j/y_j)$ entering vertex v_j do
    for each edge e_{jk} (say t_{jk})= $(v_j, v_k; x_{j+1}/y_{j+1})$ leaving vertex v_j do
      if for output $a_{L(U)} \in \Gamma_{L(U)}$, $a_{L(U)}$ is in y_j XOR $a_{L(U)}$ is in y_{j+1} AND $x_{j+1} \notin \Sigma_{L(U)}$
      AND $(t_{ij}, t_{jk}; <-C_{U(L)}, +C_{L(U)}>) \notin T_C$
        then add $(t_{ij}, t_{jk}, <-O_{L(U)}, +O_{U(L)}>)$ to T_O,

where $U(L)$ is the tester sending the input x_{j+1} in t_{jk} and $L(U)$ is the tester involved in the shift.

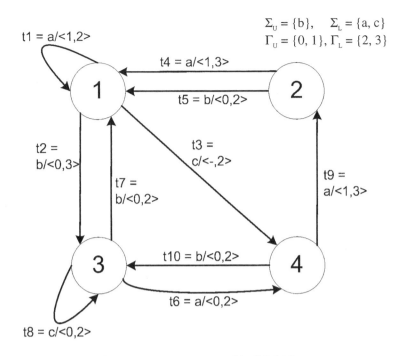

Fig. 2. Digraph $G = (V, E)$ of 2p-FSM $M1$

The set T_o identifies *only the necessary subset* of all potential undetectable 1-shift output faults in G as defined in Section 2.3. Specifically, the *if-statement* in the algorithm limits T_o to only those transition pairs that form a synchronizable pair of transitions in G. Given the input and output alphabets shown in Figure 2, consider a pair of consecutive transitions $t_{ij} = (s_i, s_j; a/<-,2>)$ and $t_{jk} = (s_j, s_k; b/<0,->)$. This transition pair forms a potential undetectable forward shift fault of the output '2', i.e. L cannot determine whether '2' is output by a correctly implemented t_{ij}, or by faulty implementations of both t_{ij} and t_{jk}, i.e., $t_{ij} = (s_i, s_j; a/<-,->)$ and $t_{jk} = (s_j, s_k; b/<0,2>)$. However, note that any instance of t_{ij} followed by t_{jk} would not form a synchronizable pair of transitions and hence would require the insertion of an external coordination message exchange $<-C_U, +C_L>$ relating to controllability. If we justifiably assume that L waits to receive the '2' from t_{ij} before sending the external coordination message $-C_U$ relating to controllability to U, the observability problem is resolved; i.e. if L waits and does not receive '2' before it sends $-C_U$, we conclude that the implementation of t_{ij} is faulty. A similar argument and intuitive treatment does not apply in any case of backward shifts if the output $<-,->$ is not allowed for any transition, other than the reset transitions.

Applying the above procedure to FSM $M1$ produces a set of 5 potential undetectable 1-shift output faults, i.e.: $T_o = \{(t1, t3, <-O_U, +O_L>), (t3, t9, <-O_U, +O_L>), (t4, t3, <-O_U, +O_L>), (t5, t3, <-O_U, +O_L>), (t7, t3, <-O_U, +O_L>)\}$.

3.2 Construction of Test Segments

The second phase of the proposed method generates a test segment $test(t_{ij}) = t_{ij}@DS(s_j)$ for each transition t_{ij}, where $DS(s_i)$ is the transition sequence induced by D on G at v_i. Observability and controllability problems within $DS(s_i)$ and between t_{ij} and $first(DS(s_j))$ are resolved by inserting the corresponding external coordination message from T_O or T_C, respectively. Formally, this phase consists of two steps:

Step 1. For each state s_i, find the transition sequence $DS(s_i)$ induced by D on G at v_i. For each pair of consecutive transitions t_{mn}, t_{no} in $DS(s_i)$:

- If $(t_{mn}, t_{no}, <-O_{U(L)}, +O_{L(U)}>) \in T_O$ then insert the external coordination message exchange $<-O_{U(L)}, +O_{L(U)}>$ between t_{mn} and t_{no} in $label(DS(s_i))$ and remove $(t_{mn}, t_{no}, <-O_{U(L)}, +O_{L(U)}>)$ from T_O.
- If $(t_{mn}, t_{no}, <-C_{U(L)}, +C_{L(U)}>) \in T_C$ then insert the external coordination message exchange $<-C_{U(L)}, +C_{L(U)}>$ between t_{mn} and t_{no} in $label(DS(s_i))$.

Step 2. For each transition $t_{ij} = (v_i, v_j; x/y_j)$:

- Construct $test(t_{ij}) = t_{ij}@DS(s_j)$.
- If there is a triple $(t_{mn}, t_{no}, <-O_{U(L)}, +O_{L(U)}>) \in T_O$ where $t_{mn} = t_{ij}$ and $t_{no} = first(DS(s_j))$ then insert the external coordination message exchange $<-O_{U(L)}, +O_{L(U)}>$ between t_{ij} and $DS(s_j)$ in $label(test(t_{ij}))$ and remove $(t_{mn}, t_{no}, <-O_{U(L)}, +O_{L(U)}>)$ from T_O.
- If there is a triple $(t_{mn}, t_{no}, <-C_{U(L)}, +C_{L(U)}>) \in T_C$ where $t_{mn} = t_{ij}$ and $t_{no} = first(DS(s_j))$ then insert the external coordination message exchange $<-C_{U(L)}, +C_{L(U)}>$ between t_{ij} and $DS(s_j)$ in $label(test(t_{ij}))$.

The distinguishing sequence D for FSM $M1$ is ab. The transition sequences induced by this D at each state are shown in Table 1, and the test segments generated in step 2 are shown in Table 2. Note that Step 2 removes triple $(t3, t9, <-O_1, +O_2>)$ from T_O as each potential 1-shift output fault needs only be handled once. Removing triples ensures that the remaining phases of the proposed method do not unnecessarily avoid transition pairs whose potential 1-shift output fault is already handled in some $test(t_{ij})$.

Table 1. $label(DS(s_i))$ for 2p-FSM $M1$

State s_i	$label(DS(s_i))$
s_1	$t1\ t2$
s_2	$t4\ t2$
s_3	$t6\ t10$
s_4	$t9\ t5$

3.3 Selection of Preambles

This phase of the proposed method generates a preamble for each state s_i, which is a transition sequence that transfers M from the initial state s_1 to state s_i. In choosing a preamble for state s_i, denoted $preamble(s_i)$, three goals must be considered:

Table 2. $label(test(t_{ij}))$ for 2p-FSM $M1$

Transition t_{ij}	$label(test(t_{ij}))$
t1	t1 t1 t2
t2	t2 t6 t10
t3	t3 <-O_U, +O_L> t9 t5
t4	t4 t1 t2
t5	t5 t1 t2
t6	t6 t9 t5
t7	t7 t1 t2
t8	t8 t6 t10
t9	t9 t4 t2
t10	t10 t6 t10

1) Minimize observability and controllability problems in $preamble(s_i)$
2) Minimizing observability and controllability problems between the last transition of $preamble(s_i)$ and transitions starting at s_i.
3) Minimize the length of $preamble(s_i)$

These goals may conflict; the preamble for s_i that requires the fewest external coordination message exchanges may end with a transition which causes significant problems when followed by the transitions starting at s_i. As our goal is to minimize the number of external coordination message exchanges introduced by the use of $preamble(s_j)$, goals 1 and 2 are given precedence and both these goals are considered in choosing $preamble(s_j)$. This is accomplished by first calculating a cost for a transition t_{ij} based on the number of controllability and observability problems that will be introduced if a transition sequence ending with transition t_{ij} is chosen as $preamble(s_j)$. Using these costs, preambles for each state $s_j \neq s_1$ are found by first constructing a graph $G' = (V', E')$. Edges in E' are assigned a cost based on the controllability and observability problems caused by transition pairs represented by adjacent edges. For each state s_j, a graph G_j is created from G'. An RCPT over a selected edge in G_j selects the preamble for state s_j that causes the fewest controllability and observability problems in the resulting state cover and transition cover sequences. Formally, this phase proceeds as follows:

Step 1. The sum of the external coordination message exchange costs for each transition t_{ij} (that may be the last transition in a preamble for state s_j) followed by any of its adjacent transitions is calculated as follows:
for each vertex $v_j \in V$, $v_j \neq v_1$, do
   for each edge e_{ij} (say $t_{ij}) = (v_i, v_j; x/y_j)$ entering vertex v_j where $v_i \neq v_j$ do
   let sum_cost(t_{ij}) = 0
   for each edge e_{jk} (say $t_{jk}) = (v_j, v_k; x_{j+1}/y_{j+1})$ leaving vertex v_j do
   if $(t_{ij}, t_{jk}, <-C_{U(L)}, +C_{L(U)}>) \in T_c$ then
     if $t_{jk} = first(DS(s_j))$

then $sum\_cost(t_{ij}) = sum\_cost(t_{ij}) + 2w$ (as the pair $t_{ij}t_{jk}$ will occur twice, once in verifying s_j and once to verify t_{jk})
else $sum\_cost(t_{ij}) = sum\_cost(t_{ij}) + w$
if $(t_{ij}, t_{jk}, <-O_{U(L)}, +O_{L(U)}>) \in T_o$
then $sum\_cost(t_{ij}) = sum\_cost(t_{ij}) + w$

Step 2. Construct the graph $G' = (V', E')$ from $G = (V, E)$ by the following steps:
create a vertex v_1 in V'
for each edge e_{1k} (say $t_{1k}) = (v_1, v_k; x_k/y_k)$ leaving vertex $v_1 \in V$ where $v_k \ne v_1$ do
create a vertex labelled "v_1-t_{1k}-v_k" in V'
add an edge from v_1 to "v_1-t_{1k}-v_k" labelled t_{1k}, i.e., $(v_1, v_1$-t_{1k}-$v_k; t_{1k})$
let $cost(t_{1k}) = 1$
for each vertex $v_j \in V$, $v_j \ne v_1$
for each edge e_{ij} (say $t_{ij}) = (v_i, v_j; x/y_j)$ entering vertex $v_j \in V$, $head(e_{ij}) \ne v_j$:
for each edge e_{jk} (say $t_{jk}) = (v_j, v_k; x_k/y_k)$ leaving vertex $v_j \in V$, $tail(e_{jk}) \ne v_j$, $tail(e_{jk}) \ne v_1$:
create a vertex labelled "v_i-t_{ij}-v_j" in V' if it doesn't exist already
create a vertex labelled "v_j-t_{jk}-v_k" in V' if it doesn't exist already
add an edge $e'_{jk} = (v_i$-t_{ij}-v_j, v_j-t_{jk}-$v_k; t_{jk})$ in E'
if $\exists$ a triple $(t_{ij}, t_{jk}, <-O_{U(L)}, +O_{L(U)}>) \in T_o$ then $cost(e'_{jk}) = 1 + w$
if $\exists$ a triple $(t_{ij}, t_{jk}, <-C_{U(L)}, +C_{L(U)}>) \in T_c$
then $cost(e'_{jk}) = 1 + w*(m + 1)$
else $cost(e'_{jk}) = 1$

Step 3. For each state $s_j \ne s_1$, create a graph $G_j = (V_j, E_j)$ from $G' = (V', E')$ as follows:
Initially, $G_j = G'$
create a vertex v' in V_j and add an edge $Z = (v', v_1; \text{"Z"})$ to E_j
for each vertex labelled "v_i-t_{ij}-v_j" in V'
add a dashed edge $e'_{ij} = (v_i$-t_{ij}-$v_j, v'; -)$ to E_j
let $cost(e'_{ij}) = sum\_cost(t_{ij})$ (as calculated in step 1)

Step 4. For each state $s_j \ne s_1$: Find an RCPT P of G_j, starting at v_1, over the single edge Z. Let $preamble(s_j) = label(P')$ where P' is the subpath of P from v_1 to v_i-t_{ij}-v_j.
For each pair of consecutive transitions t_{mn}, t_{no} in $preamble(s_j)$:
- if $(t_{mn}, t_{no}, <-C_{U(L)}, +C_{L(U)}>) \in T_c$ then insert the external coordination message exchange $<-C_{U(L)}, +C_{L(U)}>$ between t_{mn} and t_{no} in $label(preamble(s_j))$.
- if $(t_{mn}, t_{no}, <-O_{U(L)}, +O_{L(U)}>) \in T_o$ then insert the external coordination message exchange $<-O_{U(L)}, +O_{L(U)}>$ between t_{mn} and t_{no} in $label(preamble(s_j))$.

In Step 2, the cost of each edge leaving v_1 is 1. The cost of every remaining edge e'_{jk} in E' depends on whether the transition pair $(v_i, v_j; x/y_j) (v_j, v_k; x_k/y_k)$ appears in T_o or T_c. The variable w represents a high cost to be associated with external coordination message exchanges. The cost of $w*(m + 1)$ for a controllability problem is based on the following: The purpose of G' is to aid in choosing preambles which minimize the number of external coordination message exchanges required in the resulting sequences forming the state cover and transition cover. Each preamble will be used once to form a state cover sequence for state s_j, and m times to verify each of the m transitions starting at s_j. Therefore if a transition pair in a preamble contains a controllability problem, the resulting number of external coordination message exchanges

Table 3. $sum\_cost(t_{ij})$ for transition t_{ij}

Transition t_{ij}	$sum\_cost(t_{ij})$
t2	0
t6	0
t10	0
t3	w
t9	0

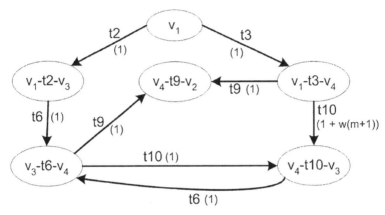

Fig. 3. Digraph $G' = (V', E')$ of 2p-FSM M1

introduced is $(m + 1)$, so the cost assigned in E' is $w*(m + 1)$. When constructing the graph G_j, the outdegree of node v_j in G is then substituted for m. In contrast, observability problems result in a cost of w as each potential 1-shift output fault need only be checked once. If the pair $(v_i, v_j; x/y_j)$ $(v_i, v_k; x/y_k)$ does not appear in a triple in T_O or T_C, the cost of edge e'_{jk} is 1.

For FSM M1, Step 1 calculates the costs shown in Table 3. Note that the transitions not included are those entering state s_1, as s_1 does not require a preamble, and those that start and end at the same state.

Applying Step 2 to FSM M1 generates the graph $G' = (V', E')$ shown in Figure 3, with the cost of each edge shown in parentheses. Based on graphs G_2, G_3, and G_4, preambles selected for states s_2, s_3, s_4, of FSM M1 are t3 t9, t2, and t2 t6, respectively. As an example, the Digraph $G_3 = (V_3, E_3)$ created for s_3 is shown in Figure 4.

3.4 Generating the Checking Sequence

The final phase of the proposed method first generates sequences that form the state cover and transition cover, using the preambles found in the previous phase. Following prefix elimination, an RCPP on an auxiliary graph G'' yields a synchronizable ordering of the remaining sequences. The input portion of the transition sequence

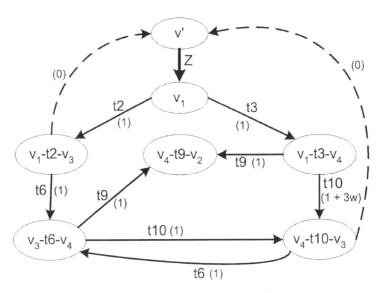

Fig. 4. Digraph $G_3 = (V_3, E_3)$ of 2p-FSM $M1$

represented by the label of this path is a synchronizable checking sequence with no potential undetectable 1-shift output faults for FSM M. This phase proceeds as follows:

Step 1. For each state s_j:
- Let $state\_cover(s_j) = preamble(s_j)@DS(s_j)$
- If $\exists$ a triple $(t_{mn}, t_{no}, <-C_{U(L)}, +C_{L(U)}>) \in T_c$ where $t_{mn} = last(preamble(s_j))$ and $t_{no} = first(DS(s_j))$ then insert the external coordination message exchange $<-C_{U(L)}, +C_{L(U)}>$ between t_{mn} and t_{no} in $label(state\_cover(s_j))$.
- If $\exists$ a triple $(t_{mn}, t_{no}, <-O_{U(L)}, +O_{L(U)}>) \in T_O$ where $t_{mn} = last(preamble(s_j))$ and $t_{no} = first(DS(s_j))$ then insert the external coordination message exchange $<-O_{U(L)}, +O_{L(U)}>$ between t_{mn} and t_{no} in $label(state\_cover(s_j))$.

Step 2. For each transition $t_{jk} = (s_j, s_k; x/y)$:
- Let $trans\_cover(t_{jk}) = preamble(s_j)@test(t_{jk})$
- If $\exists$ a triple $(t_{mn}, t_{no}, <-C_{U(L)}, +C_{L(U)}>) \in T_c$ where $t_{mn} = last(preamble(s_j))$ and $t_{no} = t_{jk}$ then insert the external coordination message exchange $<-C_{U(L)}, +C_{L(U)}>$ between t_{mn} and t_{no} in $label(trans\_cover(t_{jk}))$.
- If $\exists$ a triple $(t_{mn}, t_{no}, <-O_{U(L)}, +O_{L(U)}>) \in T_O$ where $t_{mn} = last(preamble(s_j))$ and $t_{no} = t_{jk}$ then insert the external coordination message exchange $<-O_{U(L)}, +O_{L(U)}>$ between t_{mn} and t_{no} in $label(trans\_cover(t_{jk}))$.

Step 3. Let C be the set of all transition sequences in the state and transition covers. For every transition sequence $c_p \in C$, if c_p is a prefix of some c_q, $c_q \in C$, $c_p \neq c_q$, then for any external coordination message exchange $<-O_{U(L)}, +O_{L(U)}>$ relating to observability between transitions t_{mn} and t_{no} in $label(c_p)$, insert this observability message between t_{mn} and t_{no} in $label(c_q)$. Then eliminate c_p.

Step 4. For the m sequences $c_1, \ldots, c_m$ remaining in C, let $l_i = label(c_i)$, $1 \leq i \leq m$.
For every subsequence $label(t_{mn})$ $<$-$O_{U(L)}$, $+O_{L(U)}>$ $label(t_{no})$ in any l_i, remove $<$-$O_{U(L)}$, $+O_{L(U)}>$ from any subsequent occurrence of this subsequence in any l_j.

In the following steps, h is the tester sending the *last* input of D, and h' is the other tester.

Step 5. Create four vertices in V'' labelled 1^U, 1^L, T^B, and T^h.

Step 6. For each sequence $c_i \in C$, a solid edge is added to E'' as follows:
- $(1^U, T^h; l_i)$, if sender of x in $first(c_i)$ is U, and $last(c_i)$ sends output only to h,
- $(1^U, T^B; l_i)$, if sender of x in $first(c_i)$ is U, and $last(c_i)$ sends output to h',
- $(1^L, T^h; l_i)$, if sender of x in $first(c_i)$ is L, and $last(c_i)$ sends output only to h,
- $(1^L, T^B; l_i)$, if sender of x in $first(c_i)$ is L, and $last(c_i)$ sends output to h'.

Step 7. Add the following dashed edges to E'' representing reset transitions:
- $(T^B, 1^U;$ "rfU/-") and $(T^B, 1^L;$ "rfL/-")
- $(T^h, 1^h;$ "rf$h$/-")

Step 8. Add a dashed edge $(1^h, 1^{h'}; <$-$C_{h'}, +C_h>)$ to E''.
After this step is complete, the resulting digraph will be known as $G'' = (V'', E'')$.

Step 9. Beginning at $1^{h'}$, find a rural Chinese postman path (RCPP) over the solid edges in E''. The input portion of the label of this path represents a checking sequence for FSM M.

Step 10. Eliminate any external coordination message exchanges in the resulting checking sequence that relate to potential 1-shift output faults that can be rendered detectable by some subsequence in the checking sequence, as in [13].

The first two steps of this phase generate the state and transition covers, respectively. In the prefix elimination in Step 3, a transition sequence c_p is eliminated if it is a prefix (without considering external coordination message exchanges) of some other sequence c_q. However, if $label(c_p)$ contains external coordination messages relating to observability, these messages are first copied into $label(c_q)$ to ensure that all potential undetectable output shift faults remain detectable. Redundant external coordination message exchanges relating to observability are removed in Step 4.

In Step 5, vertices 1^L and 1^U are created as all sequences in C begin at the initial state with input from U or L. Step 6 adds solid edges representing sequences $c_1, \ldots, c_m$. Edges that terminate at T^B can be followed by a reset input from either tester. If the last transition of c_i sends output only to h, the edge terminates at T^h and may only be followed by a reset input from h. In Step 7, dashed edges representing these reset inputs from U and L are added to E'' as 'rfU/-' and 'rfL/-' respectively. The dashed edge $(1^h, 1^{h'}; <$-$C_{h'}, +C_h>)$ added in Step 8 represents an external coordination message exchange relating to controllability that may be used during the construction of the rural Chinese postman path over the set of solid edges in E''.

Applying steps 1 and 2 to the example FSM $M1$ yield the state and transition covers shown in Table 4 and Table 5, respectively.

Table 4. State cover for 2p-FSM M1

State s_i	label(state_cover(s_i))
s_1	t1 t2
s_2	t3 t9 t4 t2
s_3	t2 t6 t10
s_4	t2 t6 t9 t5

Table 5. Transition cover for 2p-FSM M1

Transition t_{ij}	label(trans_cover(t_{ij}))
t1	t1 t1 t2
t2	t2 t6 t10
t3	t3 <-O_U, +O_L> t9 t5
t4	t3 t9 t4 t1 t2
t5	t3 t9 t5 t1 t2
t6	t2 t6 t9 t5
t7	t2 t7 t1 t2
t8	t2 t8 t6 t10
t9	t2 t6 t9 t4 t2
t10	t2 t6 t10 t6 t10

Table 6. Sequences eliminated in Step 3 for 2p-FSM M1

Sequence Eliminated	Prefix of
state_cover(s_3)	trans_cover(t10)
state_cover(s_4)	trans_cover(t6)
trans_cover(t2)	trans_cover(t10)
trans_cover(t3)	trans_cover(t5)

In Step 3, 4 sequences are eliminated as shown in Table 6. The set of 10 remaining sequences is shown in Table 7; note that the observability message in the label of the eliminated sequence trans_cover(t3) has been copied into l_5 in Step 4.

Figure 5 shows the digraph G" for M1 obtained by Step 5 to Step 8. A rural Chinese postman path over the solid edges obtained in Step 9 yields a synchronizable checking sequence, with no potential undetectable 1-shift output faults, represented on G" by the input portion of the path represented by the label sequence:

rfL/- l_1 rfL/- l_2 rfL/- l_3 rfL/- l_4 rfL/- l_5 rfU/- l_6 rfU/- l_7 rfU/- l_8 rfU/- l_9 rfU/- l_{10}

The input portion of the path represented by this label sequence on G" corresponds to a checking sequence composed of 10 reset inputs, 41 non-reset inputs, 1 external coordination message exchange relating to observability, and no external coordination message exchanges relating to controllability.

Table 7. Checking sequence subsequences for 2p-FSM M1

Label	Corresponding Sequence	Verifies
l_1	t1 t2	s_1
l_2	t3 t9 t4 t2	s_2
l_3	t1 t1 t2	t1
l_4	t3 t9 t4 t1 t2	t4
l_5	t3 $<-O_U, +O_L>$ t9 t5 t1 t2	t3, t5
l_6	t2 t6 t9 t5	s_4, t6
l_7	t2 t7 t1 t2	t7
l_8	t2 t8 t6 t10	t8
l_9	t2 t6 t9 t4 t2	t9
l_{10}	t2 t6 t10 t6 t10	s_3, t2, t10

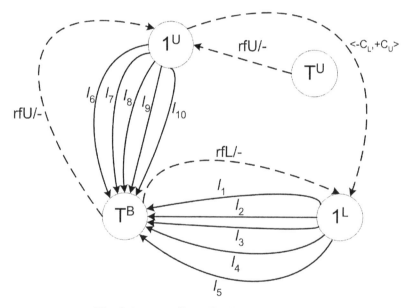

Fig. 5. Digraph $G'' = (V'', E'')$ for 2p-FSM M1

In Step 10, the subsequence t6 t9 t5 in the checking sequence above is found to be sufficient to detect the potential backward shift of '1' in t3 t9, thus rendering the external coordination message exchange $<-O_U, +O_L>$ between t3 and t9 unnecessary. As a result, the checking sequence generated by our method for FSM M1 requires no external coordination message exchanges relating to controllability and observability.

4 Concluding Remarks

A method for constructing a checking sequence of a given $2p$-FSM M using a distinguishing sequence has been proposed. The resulting checking sequence does not pose controllability and observability problems during its application in a distributed test architecture. The method can easily be generalized to np-FSMs, $n \geq 2$, by adapting a different notation as the one given in [19].

One alternative approach is to first generate a checking sequence using the D-method [18] and then identify controllability and observability problems and insert external coordination message exchanges relating to controllability and observability. In the D-method, the shortest path from s_1 to s_j is chosen as $preamble(s_j)$. However, this may introduce controllability and/or observability problems that are avoided by our proposed method. For example, applying the D-method to the example FSM $M1$ yields a checking sequence composed of 10 reset inputs, 38 non-reset inputs, 1 external coordination message exchange relating to observability, and 1 external coordination message exchange relating to controllability. As a result of the controllability problem, this checking sequence requires a test architecture which supports direct communication among testers, while the checking sequence generated by our method for this same FSM does not. In general, for a given FSM and a given distinguishing sequence, if there exists a set of checking sequences that do not require any external coordination message exchanges relating to controllability or observability, then one such checking sequence will be selected by the proposed method.

Our method would tie the D-method in terms of the number of non-reset inputs for FSM $M1$ if triples in T_o were not removed during the construction of test segments in Section 3.2. In that case, the preamble $t2$ $t6$ $t9$ would be chosen for s_2 instead of $t3$ $t9$; while this is a longer preamble, prefix elimination results in a checking sequence of only 38 non-reset inputs and no external coordination message exchanges relating to controllability and observability. Unfortunately, considering the effect of prefix elimination during the selection of preambles would be computationally expensive. The same remark applies to the D-method as there may be numerous shortest paths to some states. Another computationally expensive issue stems from the possible existence of more than one distinguishing sequence for a given FSM. It is therefore recognized that our heuristic approach may not yield the shortest synchronizable checking sequence for a given FSM.

A given implementation N may not implement the reset function correctly. A method for generating checking sequences for a distributed test architecture that does not assume the correct implementation of the reset function is part of our current research.

Acknowledgments

This work is supported in part by the Natural Sciences and Engineering Research Council of Canada under grant OGP00000976 and Post Graduate Studies Scholarship Program.

References

1. S. Boyd and H. Ural, "The synchronization problem in protocol testing and its complexity," *Information Processing Letters*, vol. 40, pp. 131-136, 1991.
2. L. Cacciari and O. Rafiq, "Controllability and observability in distributed testing," *Information and Software Technology*, vol. 41, pp. 767-780, 1999.
3. W. Chen and H. Ural, "Synchronizable checking sequences based on multiple UIO sequences," *IEEE/ACM Transactions on Networking*, vol 3, pp. 152-157, 1995.
4. A. Gill, *Intro. to the Theory of Finite-State Machines*, New York: McGraw-Hill, 1962.
5. G. Gonenc, "A method for the design of fault detection experiments", *IEEE Trans. on Computers*, vol. 19, pp. 551-558, 1970.
6. S. Guyot and H. Ural, "Synchronizable checking sequences based on UIO sequences," Proc. *IFIP IWPTS'95*, Evry, France, 395-407, Sept. 1995.
7. F.C. Hennie, "Fault detecting experiments for sequential circuits", *Proc. Fifth Ann. Symp. Switching Circuit Theory and Logical Design*, pp. 95-110, Princeton, N.J., 1964.
8. R.M. Hierons and H. Ural, "UIO sequence based checking sequences for distributed test architectures", Accepted for publication in *JIST*.
9. ISO/IEC Information technology – Opens Systems Interconnection – Conformance testing methodology and framework, 9646-1, Part 1: General Concepts, 1995.
10. ISO/IEC Open Distributed Processing, Reference Model, 10748, Parts 1-4, 1995.
11. Z. Kohavi, *Switching and Finite Automata Theory*, McGraw-Hill, Inc.: New York, N.Y.
12. D. Lee and M. Yannakakis, "Testing finite state machines: State identification and verification," *IEEE Transactions on Computers*, vol. 43, pp. 306-320, 1994.
13. G. Luo, R. Dssouli, G. v. Bochmann, P. Venkataram and A. Ghedamsi, "Test generation with respect to distributed interfaces," *Computer Standards and Interfaces*, vol. 16, pp. 119-132, 1994.
14. K.K. Sabnani and A.T. Dahbura, "A protocol test generation procedure," *Computer Networks*, vol. 15, pp. 285-297, 1988.
15. B. Sarikaya and G. v. Bochmann, "Synchronization and specification issues in protocol testing," *IEEE Transactions on Communications*, vol. 32, pp. 389-395, Apr. 1984.
16. K.C. Tai and Y.C. Young, "Synchronizable test sequences of finite state machines," *Computer Networks*, vol. 13, pp. 1111-1134, 1998.
17. H. Ural and Z. Wang, "Synchronizable test sequence generation using UIO sequences," *Computer Communications*, vol.16, pp. 653-661, 1993.
18. H.Ural, X. Wu and F. Zhang, "On minimizing the lengths of checking sequences," *IEEE Transactions on Computers*, vol. 46, pp. 93-99, 1997.
19. D. Whittier, "Solutions to Controllability and Observability Problems in Distributed Testing," Master's thesis, University of Ottawa, Canada, 2001.
20. Y.C. Young and K.C. Tai, "Observation inaccuracy in conformance testing with multiple testers," Proc. *IEEE WASET*, 80-85, 1998.

Conformance of Distributed Systems

Maximilian Frey[1] and Bernd-Holger Schlingloff[2,3]

[1] O$_2$ (Germany) GmbH & Co. OHG
Georg-Brauchle-Ring 23-25, 80992 Munich, Germany
Maximilian.Frey@o2.com
[2] Humboldt-Universität zu Berlin, Institut für Informatik
Rudower Chaussee 25, 12489 Berlin
[3] Fraunhofer Institut für Rechnerarchitektur und Softwaretechnik FIRST
Kekuléstr. 7, 12489 Berlin
Holger.Schlingloff@FIRST.FhG.DE

Abstract. This paper introduces a new conformance relation between a specification and an implementation of a distributed system. It is based on a local view which allows to avoid or reduce the state explosion problem. The conformance relation is defined via Petri nets and shows not only equivalence between transitions but also equivalence between local states. This equivalence depends on the structural properties of the Petri net and is independent of any specific initial marking. We compare our notion of conformance to classical ones and give model checking and test case generation algorithms for it.

1 Introduction

When testing telecommunication systems the testing devices, implementations and specifications usually are distributed systems. Testing of telecommunication systems means testing at different layers. Layer 1 or layer 2 according to the ISO reference model can be tested with purely sequential models, since there is only one end node. When testing higher layers (for example layer 3 or the network layer), at least the testing architecture is distributed.

Here is a short example in mobile communication to demonstrate this point. One specific interface within mobile GSM networks is the A-interface [GSM03.02]. This interface is positioned between the access network (also called BSS) and the core network. In order to test the correct implementation of an MSC (core network switch), the A-interface is of utmost importance, since all calls of the mobile customers are handled by that interface. A typical scenario is depicted in Figure 1 on the following page. Layer 2 of the A-interface is implemented by the SCCP (see reference [SCCP]) which is an SS7 protocol. The MSC is an end node for the SSCP; therefore, distributed testing is not necessary. It is sufficient to use a single point of control and observation (PCO). Layer 3 of the A-interface contains the call control protocol [GSM04.08].

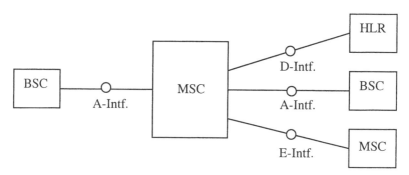

Fig. 1. Interfaces of the MSC

For testing the Call Control (CC) at least two different parties, the calling and the called party, are necessary. If the interfaces for the parties are different we have a protocol interworking scenario. For example interworking occurs between mobility management (MM) and MAP [MAP] at the D-interface or between CC and the ISUP [ISUP] on the E-interface. In that case the implementation is usually assumed to be a black box.

Extending the interworking from one node of the network to subnetworks like the access network or the core network, interworking between different nodes and a distributed implementation occurs. If new nodes or new versions of software are introduced in the network, integration tests in a distributed implementation must be performed. In that case the implementation can not be modelled as a black box, since the interfaces between the different nodes are traceable.

Distributed testing environments for telecommunication systems have been introduced some time ago. In TTCN-2 [TTCN-2] distributed testing components can be statically generated. Additionally, in TTCN-3 [TTCN-3] dynamic generation of distributed testing components is introduced. In [1999_GSBWL] a concept for a distributed testing environment based on TTCN and CORBA has been proposed.

Often, the specification of a telecommunication system refers to distributed parts. For example, the specifications of the Mobile Application Part Protocol [MAP], the DSS1 Protocol [DSS1] or the ISUP Protocol [ISUP2] contain SDL-specifications. A specification in SDL [SDL] consists of processes communicating via unbounded queues. The protocol specifications are divided into one protocol machine for incoming and one for outgoing calls. Furthermore, there is a supervision process which specifies how the different protocol machines are interacting. Each protocol machine has an external queue and at least one queue shared with the supervision process. Most of the layer-3 protocol specifications such as DSS1 or GSM/UMTS CC contain messages which have only local meaning as well as messages which also have global meaning. For example, a *call proceeding* message has only local meaning to the protocol machine, while a *setup* message has global meaning and is transferred to the supervision process.

Thus, there are causal independent events like the sending of a *call processing* message as local answer to a *setup* message and the global sending of the *setup* message through the network.

Especially in testing of telecommunication applications it is necessary to deal with distributed systems. Most existing methods for generating tests for communicating systems use finite state machines (see for example [Yao Petrenko Bochmann 1993], [Anido Cavalli 95] [Lee Yannakakis 1996]). This modelling formalism has the disadvantage that it does not differentiate between nondeterminism and parallelism.

Definition 1: A finite state machine (FSM) M is a quintuple M=(I,O,S,δ,λ) where I,O, and S are finite and nonempty sets of *input symbols*, *output symbols*, and *states*, resp. δ: S × I → S is the *transition function* and λ: S × I → O is the *output function*.

In validation methods based on FSMs, distribution and parallelism is handled by building the cartesian product, i.e., by generating global states. This leads to a state explosion problem. To define some kind of test coverage, FSM based methods often define a conformance relation between FSMs. Applying these methods to distributed systems, often a huge amount of similar test cases is generated, where many individual test cases represent just different interleavings of the same distributed testing scenario. The cause for these unnecessary interleavings is in the state explosion which occurs by generating global from sets of local states. The generation of global states leads to two sorts of high complexity: Firstly, by the generation of global states from sets of local states, and secondly, by the derivation of distributed test scenario from the various generated sequences (see [Petrenko Ulrich Chapenko 1998]).

Some authors propose to use distributed test case generation. Mostly, the proposed methods define a distributed transition tour through the system (see, e.g., [Ulrich König 1999] [Jard 2001] [Kim Shin Janson Kang 1999], and others). Compared to the methods for FSMs, transition tour methods are not able to test the conformance of FSMs, because they do not test the equivalence of states (see [Ramalingam Das Thulasiraman 95].

To define methods which are effective for distributed systems and test more than transition tours, a specification of conformance for distributed models is necessary. A well known model for distribution and parallelism are Petri nets. To model communication, we introduce an extension of Petri nets to include input and output symbols. A a new notion of correctness, we define a conformance relation between these nets. This relation can be used for model checking and testing purposes. Since our conformance relation is entirely based on local states, we avoid both of the drawbacks mentioned above.

Our paper is organized as follows. In Section 2, we give some basic definitions. Then, in Section 3 we define executions, simulations and conformance on these extended models. In Section 4 we compare the conformance relation on extended Petri nets to the conformance relation on FSMs. In Section 5 we sketch how conformance can be verified by model checking, and how tests can be automatically generated from Petri net specifications. Finally we give some conclusions and hints for further work.

2 Modelling of Distributed Systems

In order to model a subset of a telecommunication network, we have to model different nodes. Thus, the system under test (SUT) is the network. For the observation of this SUT, ports are used. They allow to observe the message flow on the external interfaces between the nodes.

For testing purposes, we distinguish between two kinds of ports.

- Ports which are points of control and observation (PCOs). The testing environment is applying input symbols to the SUT and receiving output symbols from the SUT.
- Ports which are points of observation (POs). These ports are in between two components of the SUT for internal communication. The input symbols of one component are the output symbols of the other component. The testing environment can only observe these input/output symbols.

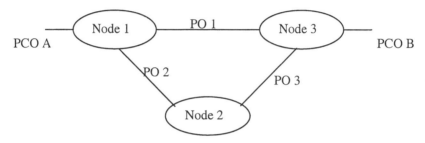

Fig. 2. Distributed system with PCOs and POs

Figure 2 shows an example for a distributed system with PCOs and POs. The system described contains three nodes. In a GSM system, PCO A and PCO B are PCOs on the CC protocol of an A-interface, Node1 and Node3 are MSCs and Node 2 is an HLR. PO 2 and PO 3 are POs on the MAP-Protocol of a C-interface.

Subsequently, we model distributes systems as the one above by Petri nets (cf. e.g. [Reisig 88]). The testing environment can send input symbols to the SUT only at PCOs. Therefore, Petri nets are extended with a set of input symbols and a set of output symbols. Similar to a transition in an FSM an event of a Petri Net receives an input symbol from the environment and generates an output symbol to the environment. In order to specify the input and output symbol for an event, two functions are added to the usual definition of a Petri Net.

Furthermode, PCOs and POs are modelled by sets of input and output symbols which occur at the specified port. Thus, the Petri net formalism additionally is extended by a set of PCOs and a set of POs. As a special case, the set of POs can be empty if the SUT consists of only one node.

Communication between different nodes is asynchronous, as within most telecommunication protocols. For any event e which has an output symbol of an PO there must exists a succecessor event e' such that the output symbol of e is the input symbol of e'. An internal communication via a PO has taken place. Between both events is exactly one condition. It is not possible that a further event is between both, because that event would be an internal event which is not visible to any PO and PCO. In the same way each output and input event of one PO are connected.

Local (and also global) states of a extended Petri Net are *stable*, which means that they can only be left by an input from the environment. Similar to the FSM model, we require that all of our Petri nets are deterministic.

Definition 2: A extended Petri net P is a tuple P=(B,E,R,I,O,PCO,PO,ι,o), where
- B is a finite and nonempty set of *conditions*,
- E is a finite and nonempty set of *events*, (B∩E=∅)
- R ⊆ (B×E) ∪ (E×B) is the *flow relation*, where $x^\bullet = \{y \mid (x,y) \in R\}$ and $^\bullet x = \{y \mid (y,x) \in R\}$

- I is a finite set of *input symbols*,
- O is a nonempty, finite set of *output symbols*,
- PCO is a set of *ports* which are *points of control and observation*, i.e., a set of tuples (I_i, O_i) such that $I_i \subseteq I \setminus O$ and $O_i \subseteq O \setminus I$. PO is a set of ports which are *points of observation*, i.e., a set of sets (O_j) such that $O_j \subseteq I \cap O$.
- ι: E→ I is the *input function*,
- o: E→ O is the *output function*,

- For each $i \in I \setminus O$ there is exactly one PCO (I_i, O_i) such that $i \in I_i$. For each $o \in O \setminus I$ there is exactly one PCO (I_i, O_i) such that $o \in O_i$. For each $o \in I \cap O$ there is exactly one PO (O_j) such that $o \in O_j$.
- For any event e such that $o(e) \in I \cap O$ there exists a condition b and an event e' such that $^\bullet b = e$, $b^\bullet = e'$, and ι(e')= o(e).
  For any event e such that $ι(e) \in I \cap O$ there exists a condition b and an event e' such that $b^\bullet = e$, $^\bullet b = e'$, and ι(e)= o(e').
- P is loop-free: $\forall x \in B \cup E\ (x^\bullet \cap {^\bullet x} = \emptyset)$
- P is deterministic: $\forall b \in B\ \forall e,e' \in b^\bullet\ (e \neq e' \rightarrow ι(e) \neq ι(e'))$

The distinction between PCOs and POs is for application purposes only; conceptually, the subsequent theory could also be based on one sort of ports.

Definition 3:
A *marking* of an extended Petri net P=(B,E,R,I,O,PCO,PO,ι,o) is any subset of B. Marking c' is *obtained* from c *by firing* event e (c[e⟩c') if ($^\bullet e \subseteq c \land c' = (c - {^\bullet e}) \cup e^\bullet$). Marking c' is *reachable* from marking c (c[+⟩c'), if $\exists e_1, e_2, \ldots, e_n \in E\ \exists c_1, \ldots, c_{n-1} \subseteq B$ ($c[e_1⟩c_1 \land c_1[e_2⟩c_2 \land \ldots \land c_{n-1}[e_n⟩c'$). We write c[*⟩c' if c=c' ∨ c[+⟩c'. A net is *one-safe from* c iff $\forall c' \subseteq B\ \forall e \in E\ (c[*⟩c' \land {^\bullet e} \subseteq c' \rightarrow e^\bullet \cap c' = \emptyset)$.

Definition 4: An extended Petri net P=(B,E,R,I,O,PCO,PO,ι,o) is *strongly connected* if $\forall x,x' \in E \cup B: (x,x') \in R^\bullet$.

Thus, an extended Petri net is strongly connected if all conditions and events are mutually reachable. We use nets to model reactive systems which are cyclic. Moreover, the communication at POs between nodes is in both directions. Therefore, all of our specifications are strongly connected.

The property of being strongly connected is not part of our definition of an extended Petri net. This is because we model the execution of Petri Nets by an unfolding of this net, which again is a net. Obviously an unfolding is not cyclic.

Also, the notion of *initial marking* is not part of our definition of a net. This is because we are aiming at a structural morphism between nets independent of the actual starting state. Each subnet is supposed to model a separate process, executing continuously and synchronizing by message passing.

Extended Petri nets can be used to model much more general systems than the telecommunication networks described above. To be closer to these, Petri nets would have to be built from components. For each PCO and each side of an PO a sequential, strongly connected component net would have to exist. However, such a restriction is not necessary for our subsequent considerations.

3 Conformance of Extended Petri Nets

To define a conformance relation on extended Petri Nets, we reconsider the conformance relation on FSMs. The execution of an FSM is a sequence of transitions observable by an input and output symbol each. Since our FSMs are deterministic, for any initial state and sequence of input signals exactly one execution of the FSM can be observed by the according sequence of output signals. The usual conformance relation between FSMs is a relation between states which means that not only the same transitions and transition paths must exist in two conforming FSMs but also the corresponding states are equivalent. To check whether a state is the inital state of an FSM is a state verification problem. If the implementation is considered as a black box, this can not be determined by testing [Lee Yannakakis 1996].

Definition 5: Let M'=(I,O,S',δ',λ') and M=(I,O,S,δ,λ) be two FSMs.
- For an initial state s_1 an input sequence i_1, i_2, ... , i_n takes the FSM to states s_{i+1}= $\delta(s_i, i_i)$ and generates a output sequence $\lambda(s_1, i_1)$, $\lambda(s_2, i_2)$, ... , $\lambda(s_n, i_n)$.
- A state s∈ S is equivalent to a state s'∈ S if all input sequences M starting at s and M' starting at s' generate the same output sequences.
- Assume that M models an implementation and M' a specification. M conforms to M', if
  (1) For each state s in M there is a state s' in M' such that s is equivalent to s',
  (2) For each state s' in M' there is a state s in M such that s is equivalent to s'

Often, only part (1) or part (2) of the third condition is used, yielding a refinement relation. For FSMs, requiring (1) and (2) amounts to isomorphism of the corresponding minimal machines. This does not hold for more general computational models. We do not model executions of extended Petri Nets by sequences, but by causal nets which preserve the partial ordering of causality in the original net. Here, conformance is more differentiating.

Definition 6: An extended causal net is an extended Petri net $K=(B_K, E_K, R_K, I_K, O_K, PCO_K, PO_K, \iota_K, o_K)$ where $\forall b \in B_K((|b^\bullet| \leq 1) \land (b,b) \notin R_K^+)$.

When generating a transition system from a net, each reachable marking represents a state. For a net with n conditions, there might be up to 2^n reachable markings. This is the so-called state explosion problem. We want to avoid the exponential blow-up by generating executions which start with a single condition of the net. However, executions of a Petri net should not be deadlocking. Therefore, we let each execution start at a marking which contains the chosen condition, plus some extra conditions which are necessary to avoid deadlock. An example can be found in figure 3. Assume that we want to generate an execution starting with condition b_1. Then, we also have to add condition b_6 to the initial marking, because otherwise event e_6 and all successor events of it would never be enabled, and the net would deadlock at $\{b_3, b_4\}$. For each condition in the net, there might be several markings containing it from which the net is live. As a representative we choose the minimal marking with respect to execution of events. The respective marking for the condition b_6 in figure 3 is indicated by a dashed line.

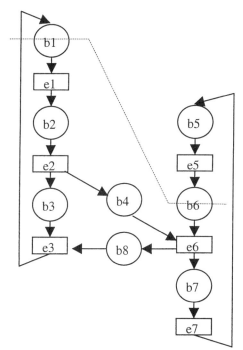

Fig. 3. Example of a strongly connected Petri net

Now, we define the execution of a net on a given input. In FSMs, an execution is determined by a sequence of input symbols supplied to the system. In extended Petri nets, there are several PCOs, where input symbols can be supplied independently and in parallel. In order to model a distributed input for extended Petri nets, the following alternatives exist:

- A single sequence of input symbols is used. Symbols of different PCOs are totally ordered within this sequence.
- A partial order of input symbols is used, where input symbols of one PCO are totally ordered. Input symbols of different PCOs can be ordered or independent.
- There is no order on input symbols of different PCOs. The distributed sequence is represented by a set of sequences which contain a sequence of input symbols for each PCO.

We use the last of these alternatives, because it implies no additional causal relation on the execution of the Petri net. The other choices would lead to additional causal dependencies which are not covered in the specification net.

Definition 6: We say that SEQ={$seq_1, seq_2, ..., seq_n$} is an *admissible input* for the net P=(B,E,R, I,O,PCO,PO,ι,o), where PCO contains n ports (I_i, O_i), if each seq_i is a non-empty and finite sequence of input symbols from I_i.

Let P be an extended Petri net P=(B,E,R, I,O,PCO,PO,ι,o). Each execution starts from a certain initial condition for one process. All nondeterministic choices for this process should be represented in the initial marking.

Definition 7: An initial marking of P for the condition b, denoted by c(b), is a set of conditions such that all events having b as precondition are enabled in c(b); i.e., $\forall e \in E\ (e \in b^{\bullet} \rightarrow {}^{\bullet}e \subseteq c(b))$.

Since we are aiming at a partial order model of program runs, we require that each execution of a net is a (contact-free) causal net.

Definition 8: Let P be an extended Petri net P=(B,E,R, I,O,PCO,PO,ι,o), let $b_i \in B$ be some condition from P, and let SEQ be an admissible input for P. Furthermore, let K be an extended causal net K=(B_K, E_K, R_K, I,O,PCO,PO,ι_K, o_K) with the same input and output alphabets and PCOs and POs as P. We say that K is an *execution of P starting at b_i according to SEQ* (denoted by $K \in [P, b_i, SEQ\rangle$) if a homomorphism h: K→ P and an initial marking $c(b_i)$ exist such that
- $h(B_K) = B$
- $h(E_K) = E$
- $\forall e \in E_K\ ((h(e^{\bullet}) = h(e)^{\bullet}) \wedge (h({}^{\bullet}e) = {}^{\bullet}h(e)))$
- $\forall b, b' \in B_K\ (((b',b) \notin R_K^* \wedge (b,b') \notin R_K^*) \rightarrow h(b) \neq h(b'))$
- $\forall b \in B_K - h(c(b_i))\ (|{}^{\bullet}b| = 1)$
- $\forall b \in h(c(b_i))\ (|{}^{\bullet}b| = 0)$.
- $\forall e \in E_K\ (h(\iota_K(e)) = \iota(h(e)))$
- $\forall e \in E_K\ (h(o_K(e)) = o(h(e)))$
- ($\exists e_1, e_2, ... \in E\ \exists c_1, c_2, ... \subseteq B\ \exists i_1, i_2, ... \in I\ ({}^{\bullet}e_1 \subseteq c(b) \wedge c_i = (c(b) - {}^{\bullet}e_i) \cup e_i^{\bullet} \wedge \iota(e_i) = i_1 \wedge {}^{\bullet}e_i \subseteq c_{i-1} \wedge c_i = (c_{i-1} - {}^{\bullet}e_i) \cup e_i^{\bullet} \wedge \iota(e_i) = i_i \wedge \forall seq \in SEQ\ (\exists (I',O') \in PCO\ (\forall x \in seq\ (x \in I') \wedge i_1, i_2, ...|_{I'} = seq))$

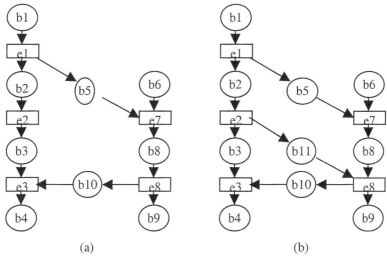

Fig. 4. Run of the specification (a) and the implementation (b) for the same input sequences

Thus, the execution of an extended Petri net is a causal net which describes the causal dependencies between conditions and events together with the input symbols which have been consumed by the events and the output symbols which have been generated by the events. An algorithm to construct executions from nets is given in Section 5 below.

In the definition of conformance for FSMs the equivalence of states is based on the equality of generated output sequences. For extended Petri nets, we model executions by causal nets; thus, conformance will be based on a relation between causal nets. In the following Definition 9, we propose two different alternatives. In both cases a homomorphism between the execution of the specification and the execution of the implementation is used. Also the input and output function is preserved by the homomorphism.

Definition 9: Let $Ki=(B_{Ki},E_{Ki},R_{Ki},I,O,PCO,PO,\iota_{Ki},o_{Ki})$ and $Ks=(B_{Ks},E_{Ks},R_{Ks},c_{Ks}, I,O,PCO,PO,\iota_{Ks},o_{Ks})$ be two extended causal nets with the same input and output symbols and the same ports. Intuitively, Ki is an execution of some implementation I, and Ks is an execution of some specification S.

- Ki is *weakly simulating* Ks if a mapping h: Ks → Ki exists such that
  $\forall\ x,x'\in E_{Ks}\cup B_{Ks}\ ((x, x') \in R_{Ks}\ \to (h(x), h(x')) \in R_{Ki}\ ) \wedge$
  $\forall\ e\in E_{Ks}\ (o_{Ks}(e) = o_{Ki}(h(e)) \wedge \iota_{Ks}(e) = \iota_{Ki}(h(e)))$
- Ki is *strongly simulating* Ks if a mapping h: $E_{Ks}\to E_{Ki}$ exists such that
  $\forall\ e,e'\in E_{Ks}\cup B_{Ks}\ ((x, x') \in R_{Ks}\ \leftrightarrow (h(x), h(x')) \in R_{Ki}\ ) \wedge$
  $\forall\ e\in E_{Ks}\ (o_{Ks}(e) = o_{Ki}(h(e)) \wedge \iota_{Ks}(e) = \iota_{Ki}(h(e)))$

The difference between the weak and strong simulation relation is that the strong simulation relation preserves the causal relation of the specification. The weak simulation relation allows that the implementation contains additional causal dependencies, but no additional events. An example is given in figure 4, where the execution (b) is weakly, but not strongly simulating (a).

Conformance between implementation and specification is defined similar as for FSMs. The conformance relation is based on (weak) simulation between conditions.

Definition 10:
Let $I=(B_I, E_I, R_I, I, O, PCO, PO, \iota_I, o_I)$ and $S=(B_S, E_S, R_S, I, O, PCO, PO, \iota_S, o_S)$ be strongly connected extended Petri nets (the implementation and specification, respectively).
- We say that condition $b_I \in B_I$ is *(weakly) simulating* condition $b_S \in B_S$ if for all admissible inputs SEQ and execution K_I and K_S such that $K_I \in [I, b_I, SEQ\rangle$, $K_S \in [S, b_S, SEQ\rangle$ it holds that $K_I \in$ is (weakly) simulating K_S.
- I *(weakly) conforms to* S if
  1. $\forall \ b_S \in B_S (\exists \ b_I \in B_I (b_I \text{ is (weakly) simulating } b_S))$, and
  2. $\forall \ b_I \in B_I (\exists \ b_S \in B_S (b_I \text{ is (weakly) simulating } b_S))$

For functional correctness, it is necessary that an implementation has the same causal dependencies as the specification. Otherwise the implementation could contain a race condition, which is not allowed by the specification. This is an error which occurs frequently in practical applications and is very difficult to detect. On the other side, if an implementation execution has more dependencies than the corresponding specification execution, this indicates that the degree of parallelism is reduced in the implementation. This might cause a performance problem, but does not influence the functional correctness. Whereas the specification and verification of performance is beyond the scope of the present paper, additional dependencies in the implementation leading to deadlock or similar errors can be found with our weak simulation relation. In this case the I/O-behavior of the implementation would be different to the behavior of the specification for at least one state and one set of input sequence. Thus, (weak) simulation is sufficient for testing of correct functionality and is used in the above definition.

4 Conformance in Distributed and Sequential Models

A conformance relation based on a distributed and parallel model has been introduced above. In this section this relation is compared with the conformance relation on FSMs in two ways. We ask the following questions:

1. Are both confomance relations equivalent for sequential systems?
2. Is the conformance relation on extended Petri nets equivalent to the conformance relation on the global FSM of the Petri net?

For the first question we define a subclass of extended Petri nets, sequential Petri nets which can describe sequential systems only. This class will be equivalent to FSMs. Therefore the number of PCOs is one and no parallelism is generated.

Definition 11: An extended Petri Net $P=(B,E,R,I,O,PCO,PO,\iota,o)$ is *sequential*, if $\forall$ $e \in E$ ($|{}^\bullet e| \leq 1$ and $|e^\bullet| \leq 1$) and $\forall \ b \in B$ ($|c(b)| \leq 1$) and $|PCO|=1$ and $|PO|=0$.

Theorem 1 shows that the class of sequential Petri nets is equivalent to FSMs.

Conformance of Distributed Systems   173

Theorem 1: Strongly connected sequential Petri Nets and strongly connected FSM are equivalent.

Proof: Since $|PCO|=1$ and $|PO|=0$, we have $PCO=\{(I,O)\}$. Since $\forall\ e \in E$ ($|{}^\bullet e|\leq 1$ and $|e^\bullet|\leq 1$) and $\forall\ b \in B$ ($|c(b)| \leq 1$), all events can be used to define relations λ and δ: $\forall\ e \in E\ \forall\ b,b' \in B\ (\ ((b,e) \in R$ and $(e,b') \in R) \rightarrow (\delta(b,\iota(e))=b' \wedge\ \lambda(b,\iota(e))=o(e)\ ))$. This way we have defined an FSM with (I,O,B,λ,δ).

Vice versa, from an FSM a Petri net can be generated in a similar way.

In order to show that the conformance relation on FSMs is equivalent to the conformance relation on sequential Petri nets, we first have to show that executions of sequential Petri nets can be represented by sequences of output symbols. The second step of the proof is to show that the weak simulation relation is equivalent to the equality on output sequences.

Lemma 1: Each execution of strongly connected sequential Petri Nets P starting at a condition b and according to a set SEQ of sequences of input symbols $[P,b,h,SEQ\rangle$ represents a sequence of output symbols.

Proof:

$K=(B_K,E_K,R_K,I,O,PCO,PO,\iota_K,o_K) = [P,b,h,SEQ\rangle$ is an execution of a sequential Petri Net $P=(B,E,R,I,O,PCO,PO,\iota,o)$. Since $|PCO|=1$, SEQ contains exacly one sequence. Since P is sequential and $\forall\ e \in E$ ($|{}^\bullet e|\leq 1$ and $|e^\bullet|\leq 1$), it is valid that $\forall\ e \in E_K$ ($|{}^\bullet e|\leq 1$ and $|e^\bullet|\leq 1$).Because P is strongly connected, it is valid that $\forall\ e \in E_K$ ($|{}^\bullet e|=1$ and $|e^\bullet|=1$) Because K is $[P,b,h,SEQ\rangle$, it is valid, that $\forall\ s \in ((B_K - h(c(b)))\cup E_K)$ ($|{}^\bullet s|=1$ and $|s^\bullet|=1$ and $(s,s) \notin R_K^+$). $\forall\ s \in h(c(b))$ ($|{}^\bullet s|=0$ and $|s^\bullet|=1$ and $(s,s) \notin R_K^+$). K describes a sequence $h(c(b))$, e_1, b_1, e_2, b_2, e_3, b_3, e_4, ... The output sequence is $o(e_1), o(e_2), o(e_3), o(e_4), ...$

Lemma 2: For two strongly connected sequential Petri Nets $I=(B_I,E_I,R_I, I,O,PCO,PO,\iota_I,o_I)$ and $S=(B_S,E_S,R_S,\ I,O,PCO,PO,\iota_S,o_S)$ the condition $b_I \in B_I$ is weakly simulating the condition $b_S \in B_S$ if and only if for all SEQ, $[I,b_I,h_I,SEQ\rangle$ and $[S,b_S,h_S,SEQ\rangle$ are representing the same output sequence.

Proof: " $\rightarrow$ "

From Lemma1: $[I,b_I,h_I,SEQ\rangle$ is represented by $h_I(c(b_I))$, e_{I1}, b_{I1}, e_{I2}, b_{I2}, e_{I3}, ..., and $[S,b_S,h_S,SEQ\rangle$ is represented by $h_S(c(b_S))$, e_{S1}, b_{S1}, e_{S2}, b_{S2}, e_{S3}, ... Since b_I is weakly simulating b_S for all SEQ $[I,b_I,h_I,SEQ\rangle$ is weakly simulating $[S,b_S,h_S,SEQ\rangle$. Because S and I are sequential, all possible SEQ contain only one sequence of input symbols. $[I,b_I,h_I,SEQ\rangle=(B_{Ki},E_{Ki},R_{Ki},I,O,PCO,PO,\iota_{Ki},o_{Ki})$ and $[S,b_S,h_S,SEQ\rangle =(B_{Ks},\ E_{Ks},\ R_{Ks},\ c_{Ks}$, I, O, PCO, PO, ι_{Ks}, o_{Ks}) and it exists an homomorphism $h':[S,b_S,h_S,SEQ\rangle \rightarrow [I,b_I,h_I,SEQ\rangle$ such that $h'(E_{Ks})= E_{Ki}$ and $\forall\ x,x' \in E_{Ks} \cup B_{Ks}\ ((x, x') \in R_{Ks} \rightarrow (h(x), h(x')) \in R_{Ki}\ )$ and $\forall\ e \in E_{Ks}\ (o_{Ks}(e)= o_{Ki}(h(e)) \wedge\ \iota_{Ks}(e)= \iota_{Ki}(h(e))\ )$. $h([S,b_S,SEQ\rangle)$ is represented by $h'(h_S(c(b_S)))$, $h'(e_{S1})$, $h'(b_{S1})$, $h'(e_{S2})$, $h'(b_{S2})$, $h'(e_{S3})$, $h'(b_{S3})$, $h'(e_{S4})$, ... and the output sequence of $[I,b_I,h_I,SEQ\rangle$ is the same as that of $[S,b_S,h_S,SEQ\rangle$

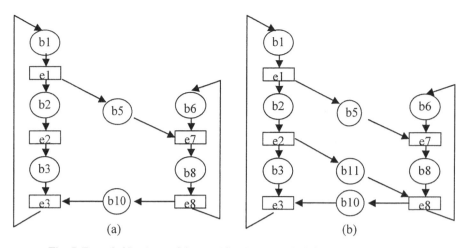

Fig. 5. Extended Petri net of the specification (a) and of the implementation (b)

"←"

From Lemma1: $[I,b_I,h_I,SEQ\rangle$ is represented by $h_I(c(b_I))$, e_{I1}, b_{I1}, e_{I2}, b_{I2}, e_{I3},..., and $[S,b_S,h_S,SEQ\rangle$ is represented by $h_S(c(b_S))$, e_{S1}, b_{S1}, e_{S2}, b_{S2}, e_{S3},... . $[I,b_I,h_I,SEQ\rangle = (B_{Ki},E_{Ki},R_{Ki},I,O,PCO,PO,\iota_{Ki},o_{Ki})$ and $[S,b_S,h_S,SEQ\rangle = (B_{Ks},E_{Ks},R_{Ks},c_{Ks}, I,O,PCO,PO,\iota_{Ks},o_{Ks})$. The output sequences of $[I,b_I,h_I,SEQ\rangle$ and $[S,b_S,h_S,SEQ\rangle$ are the same: $o(e_{Ii}) = o(e_{Si})$. Because I and S are sequential Petri nets, $|PCO|=1$ and SEQ contains exactly one element seq. That implies that $\iota(e_{Ii}) = \iota(e_{Si})$. A homomorphism h': $[S,b_S,h_S,SEQ\rangle \to [I,b_I, h_I,SEQ\rangle$ can be defined as follows: h'$(e_{Si})= e_{Ii}$, h'$(b_{Si})= b_{Ii}$, h'$(h_S(c(b_S)))= h_I(c(b_I))$. This homomorphism defines that $[I,b_I,h_I,SEQ\rangle$ is weak simulating $[S,b_S,h_S,SEQ\rangle$ and b_I is weak simulating b_S.

Theorem 2: The conformance relation on strongly connected FSMs is the same as that on strongly connected sequential Petri nets.

Proof: Follows directly from Lemma 2 and Theorem 1.

Thus, the answer to the first question from above is affirmative. For answering the second question we have to define the global FSM of an extended Petri net. For communicating finite state machines (CFSM) the Cartesian product of the states is used to generate a FSM. On this FSM the conformance relation is defined.

Definition 12: The global FSM $G(P)=(I,O,S,\lambda,\delta)$ of an extended strongly connected Petri Net $P = (B,E,R,I,O,PCO,PO,\iota,o)$ contains the set S of all sets $c \subseteq B$, such that (B,E,R,c) is contact free, there exists an event e with $^\bullet e \subseteq c$, and $\delta(c,i)=c'$ and $\lambda(c,i)=o'$ if it exists an event e with $^\bullet e \subseteq c$, $e^\bullet$ and c' are disjoint, $c'=(c-^\bullet e) \cup e^\bullet$ and $\iota(e)=i$ and $o(e)=o'$.

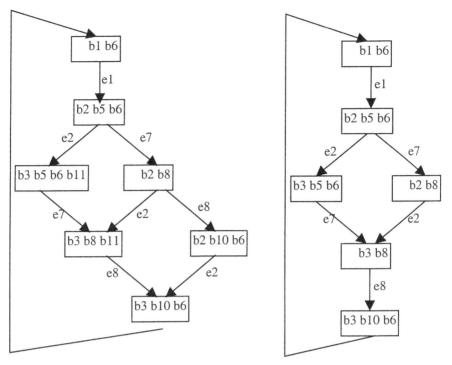

Fig. 6. Global FSMs of the specification (a) and of the implementation (b) as defined in figure 5.

We demonstrate the difference between the conformance relation on extended Petri nets and global FSM of extended Petri nets by an example. Figure 5 shows a specification and an implementation. Both extended Petri-Nets have two PCOs defined by ({i1},{o1}) and ({i2},{o2}). The input and output function is defined by $\iota(e)=i1$ $o(e)=o1$ for e= e1,e2,e3 and $\iota(e)=i2$ $o(e)=o2$ for e= e7,e8.

The initial markings of the specification are for condition b1 {b1, b6}, for condition b2 {b2, b8}, for condition b3 {b3, b10, b6}, for condition b5 {b5, b6, b3}, for condition b6 {b6, b5, b3}, for condition b8 {b8, b3} and for condition b10 {b10, b3, b6}.

The initial markings of the implementation are for condition b1 {b1, b6}, for condition b2 {b2, b8}, for condition b3 {b3, b10, b6}, for condition b5 {b5, b6,b3, b11}, for condition b6 {b6, b5, b3, b11}, for condition b8 {b8, b3, b11}, for condition b10 {b10, b3, b6} and for condition b11 {b11, b8, b3}.

Thus for each condition b in the specification a condition b' in the implementation exists with b=b' and b' is weakly simulating b. Also for each condition b'=b1, b2, b3, b5, b6, b8, b10 a condition b in the specification exists with b=b' and b' is weakly simulating b. Condition b11 is weakly simulation condition b8. The implementation Petri net conforms to the specification Petri net.

The corresponding global FSMs of the extended Petri nets of Figure 5 are shown in figure 6.

In state b2b8 in the specification the sequence of input symbols in_seq=i2, i1, i1, i1, i2, i1 produces the output sequence out_seq=o2, o1, o1, o1, o2, o1. There is no state in the implementation which will produce out_seq when in_seq is applied. The FSMs do not conform even though the Petri nets conform. That means conformance of global FSM of Petri nets is not equivalent to conformance on Petri nets. One reason for this is the asymmetry of the weak simulation relation on the executions of the extended Petri nets. A discussion which conformance relation is more "intuitive" as well as a comparison of the expressiveness of the two notions will be given in a subsequent paper.

5 Model Checking and Distributed Test Generation

So far we have assumed that models both for the specification and for the implementation are given. In this case, conformance checking is a special form of model checking. In particular, since both specification and implementation are deterministic, we can use a classical partition refinement algorithm to establish conformance.

Let H_0 be the relation consisting of all pairs $(b_I, b_S) \in B_I \times B_S$. H_{i+1} is constructed from H_i as follows:

$(b_I, b_S) \in H_{i+1}$ iff
1. $(b_I, b_S) \in H_i$, and
2. $\forall\, e_I \in b_I^\bullet, e_S \in b_S^\bullet\; (\iota_{KI}(e_I) = \iota_{KS}(e_S) \rightarrow o_{KI}(e_I) = o_{KS}(e_S))$, and
3. $\forall\, e_S \in b_S^\bullet\; \forall\, b_S' \in e_S^\bullet\; \exists\, e_I \in b_I^\bullet\; \exists\, b_I' \in e_I^\bullet : (b_I', b_S') \in H_i$

Since we are dealing with finite Petri nets, the iteration must reach a fixed point. Let H be the relation reached upon stabilization. Then I conforms to S if

1. $\forall\, b_S \in B_S\; \exists\, b_I \in B_I : (b_I, b_S) \in H$, and
2. $\forall\, b_I \in B_I\; \exists\, b_S \in B_S : (b_I, b_S) \in H$.

Often, the internals of the implementation are not accessible in a systems validation process. In this case, we have to resort to black box testing techniques. In our approach, each execution of the specification can serve as the basis for the generation of test cases. We use the following algorithm to generate executions from a given extended Petri net P:

- Start with an arbitrary condition b and let c(b) be $\cup\{^\bullet e \mid e \in b^\bullet\}$
- The initial part of the execution is a copy of all conditions in c(b)
- Put a mark on all conditions in c(b)
- Repeat indefinitely
  - Choose a maximal set of events which are either enabled in P, or can be enabled by putting a token on a condition which is not marked, such that the inputs of these events contain at most one input from each PCO and PO, respectively.

- Put a mark on all conditions which have received a token, as well as on all conditions in the pre- and postset of an enabled transition.
- Fire the chosen events in P, and extend the execution by appending a copy of all chosen events and their postsets to it.

During the execution of the Petri Net the initial marking c(b) is generated on the fly. Starting with the marking containing all conditions in the preset of the succeeding events of b, conditions are added when a state is reached in which no event can be fired. Only conditions of the preset of an event are added such that the event get fireable. Each condition can only be added once to the initial state. In comparision to FSMs, an execution is described by a combination of the input sequence and the output sequence together with the reached states of the FSM.

Since each execution represents a class of different interleavings, it is sufficient to generate only a few representatives from it which cover the intended behavior. The particular interleaving is chosen according to some specific heuristics, e.g., firing all enabled events always according to some specific ordering. Another alternative is to map the execution to a TTCN test case description and let the TTCN scheduler do the linearization of concurrent events.

In general, this technique can reduce the number of test cases by an exponential factor. If each potential error is preserved by this reduction technique, then the testing coverage in the specification is the same as that without partial order reduction; i.e., the same test termination and measurement criteria can be used.

6 Conclusion and Further Work

We have proposed a conformance relation between specification and implementation of distributed systems. This relation is based on a local view of the distributed system which allows to avoid or reduce the state explosion problem. As distributed model we use an extension of one-safe Petri nets. The conformance relation on extended Petri nets is based not only on the equivalence between transitions but also on the equivalence between conditions which are local states. The equivalence between conditions depends only on the structural properties of the Petri net and is independent of any specific initial marking.

We compared the conformance relation on Petri nets with the conformance relation on FSMs. The result is that for sequential systems both conformance relations are equivalent. For distributed systems the conformance relations are not equivalent.

Further work includes a more detailed comparison between both conformance relations for distributed systems. In particular, we plan to investigate a variant of the conformance relation on Petri nets using the strong simulation relation and compare it with the relation on FSMs. Furthermore, we plan to adopt methods for generation of test to the conformance relation on Petri nets and to define a coverage measure which determines the extent to which a black box implementation conforms to a specification. Finally, we will implement concrete testcases in TTCN-3 for telecommunication networks. For this, a concrete algorithm for the generation of a module from causal nets has to be developed. For practical applications in telecommunications, we also plan to develop methods to handle the different parameters and information elements sent within the messages across the network and to incorporate these into the generation of TTCN-3 testcases.

References

[GSM03.02] ETSI: ETSI TS 100 522 Digital cellular telecommunications system (Phase 2+): Network architecture (GSM 03.02 version 7.1.0 Release 1998): 2000

[SCCP] ITU-T: Recommendations Q.711-Q.714 Specifications of Signalling System No. 7 - Signalling connection control part: 1996

[GSM04.08] ETSI: ETSI TS 100 940 Digital cellular telecommunications system (Phase 2+); Mobile radio interface layer 3 specification (3GPP TS 04.08 version 7.17.0 Release 1998): 2002

[MAP] ETSI: ETSI TS 129 002 Digital cellular telecommunications system (Phase 2+) (GSM); Universal Mobile Telecommunications System (UMTS); Mobile Application Part (MAP) specification (2GPP TS 29.002 version 4.50 Release 4): 2001

[ISUP] ITU-T: Recommendations Q.761-Q.764 Signalling System No. 7 – ISDN user part signalling procedures: 1999

[TTCN-2] ISO. ISO/IEC 9646-3, Tree and Tabular Combined Notation (TTCN), Second Edition (1997)

[TTCN-3] ETSI: ETSI ES 201 873 Methods for Testing and Specification (MTS); The Tree and Tabular Combined Notation version 3: 2001

[1999_GSBWL] Vassiliou-Gioles, T., Schieferdecker, I., Born, M., Winkler, M., Li, M.: Configuration and Execution Support for Distributed Tests. In: Csopaki, G., Dibuz, S., Tarnay, K. (eds): Testing of Communicating Systems. Kluwer Academic publishers. Boston, Dordrecht, London (1999) 61-76

[DSS1] ITU-T: Recommendation Q.931 Digital Subscriber Signallingsystem No. 1 (DSS 1) – ISDN User-Network Interface Layer 3 Specification For Basic Call Control: 1998

[ISUP 2] ITU-T: Recommendation Q.764-Annex H Specification of Signalling System No.7 – ISDN User Part Signalling Procedures Annex H: State Transition Diagrams: 1995

[SDL] ITU-T: Recommendation Z.100 Languages for Telecommunications Applications – Specification and Description Language, 1999

[Yao Petrenko Bochmann 1993] Yao, M., Petrenko, A., v. Bochmann, G.: Conformance Testing of Protocol Machines without Reset. In: Danthine, A., Leduc, G., Wolper, P. (eds): Protocol Specification, Testing and Verification, XIII.Elsevier Science Publishers B. V. (North-Holland) (1993) 241-253

[Anido Cavalli 95] Anido, R., Cavalli, A.R.: Guaranteeing full fault coverage for UIO-based testing methods. In: Proceedings of the 8th Int. Workshop on Protocol Test Systems, Evry, France (1995) 221-236

[Lee Yannakakis 1996] Lee, D., Yannakakis, M.: Princliples and Methods of testing Finite State Machines – A Survey. Proceedings of the IEEE. Vol. 4 (8), 1996, 1090-1123

[Petrenko Ulrich Chapenko 1998] Petrenko, A., Ulrich, A., Chapenko, V. :Using partial-orders for detecting faults in concurrent systems. In: Proceedings of Workshop on Testing of Communicating Systems (IWTCS'98), Russia, 1998

[Ulrich König 1999] Ulrich, A., König, H.: Architectures for Testing Distributed Systems. In: Csopaki, G., Dibuz, S., Tarnay, K. (eds): Testing of Communicating Systems. Kluwer Academic publishers. Boston, Dordrecht, London (1999) 93-108

[Jard 2001] Jard, C., Principles of Distributed Test Synthesis based on True-Concurrency Models. In: Schieferdecker, I., König, H., Wolisz, A.: Testing of Communicating Systems XIV. Kluwer Academic publishers. Boston, Dordrecht, London (2002) 301-316

[Kim Shin Janson Kang 1999] Kim, M., Shin, S.T., Chanson, S. T., Kang, S.: An Enhanced Model for Testing Asynchronous Communicating Systems. In: Formal Description Techniques and Protocol Specification, Testing and Verification, 19. IFIP (1999) 337-355

[Ramalingam Das Thulasiraman 95] Ramalingam, T., Das, A., Thulasiraman, K.: Fault detection and diagnosis capabilities of test sequence selection methods based on the FSM model. Computer communications, vol.18(2), 1995,113-122

[Reisig 88] Reisig, W.: Petri Nets. Springer 1988

[Tretmans 96] Test Generation with Inputs, Outputs and Repetitive Quiescence. Software–Concepts and Tools, 17(3):103-120, 1996

An Automata-Based Approach to Property Testing in Event Traces

Hesham Hallal[1], Sergiy Boroday[1], Andreas Ulrich[2], and Alexandre Petrenko[1]

[1] CRIM, 550 Sherbrooke West, Suite 100, Montreal, H3A 1B9, Canada
{Hallal,Boroday,Petrenko}@crim.ca
[2] Siemens AG, Corporate Technology SE1, Otto-Hahn-Ring 6, 81739 München, Germany
andreas.ulrich@siemens.com

Abstract. We present a framework for property testing where a partially ordered execution trace of a distributed system is modeled by a collection of communicating automata. We prove that the model exactly characterizes the causality relation between the events in the observed trace. We present the implementation of this approach in SDL, where ObjectGEODE is used to verify properties, and illustrate the approach with an industrial case study.

Keywords: monitoring, passive testing, distributed traces, property checking, SDL

1 Introduction

In the age of the Internet, distributed systems have emerged as a viable option to pave the information highway towards brighter horizons. However, the shine of distributed systems has always been dimmed by the cost associated with their design and validation. Although asynchrony and geographic distribution add to the value of distributed systems; the same characteristics render especially their validation difficult. Meanwhile, the possibility of using formal methods as a reliable means for the development and validation of distributed systems restores the hope that both the time-to-market and the validation cost could be slashed. Once a formal specification of the behavior of a system becomes available, many development activities, such as test derivation could easily be automated. However, the development of distributed systems rarely yields formal specifications of their behavior that would make formal methods fully applicable. Recognizing this fact, research in both academia and industry aims at developing tools for debugging and testing that do not require formal specifications in the first place. Such tools usually rely on monitoring functions of distributed systems and produce logfiles of execution traces that can be analyzed further. This type of analysis is also known as passive testing of distributed systems.

A general approach to trace analysis could be outlined as follows. The distributed system is instrumented in a way that events executed by the distributed processes are collected. Such events typically denote the send and receive of messages, local actions and others. A trace is produced that includes all the events collected during a system run, and an appropriate analysis tool can be applied to check the trace against some user-defined properties. There exists a large body of work on developing various tools to visualize traces, see, e.g. [3, 21, 31]. Their goal is to facilitate efforts of the designer or tester for locating and correcting bugs by filtering out unrelated information and by offering a proper visualization of traces. The analysis is performed manually either online (simultaneously with the system execution) or post mortem. Another group of methods targets the analysis phase by offering means to verify certain properties in the distributed system under test (SUT) [9, 12, 17].

Developing a tool for checking properties in execution traces, one faces the choice either to elaborate algorithms for a specific class of properties and to implement them in specialized tools or to reuse a general-purpose model checker. In the first scenario, the daunting task is to implement the algorithms from scratch. On the other hand, reuse of an off-the-shelf model checker allows the developer of an analysis tool to rely on reliable, highly sophisticated, and versatile products, in which many years of research and development have already been invested.

In this paper, we present an automata-based approach to property checking in distributed systems based on traces collected during a system execution. Our approach relies on a partially ordered set (poset) of events, where the partial order is the traditional *happened-before* relation [5], and its corresponding lattice, known as the lattice of ideals, consistent cuts or, simply, global states. Intuitively, such a lattice represents the joint behavior of the distributed processes observed in a SUT. We propose to model the SUT using finite automata. Then, we prove that the composition of these automata is isomorphic to the lattice of the poset, which allows us to translate the problem of property testing in event traces to a typical model checker problem. Our framework for trace analysis treats both synchronous and asynchronous communications simultaneously.

From the theoretical perspective, our framework can be considered as a generalization of the framework proposed in [6], where event causality is studied in traces with synchronous or asynchronous communications. While the authors mention the possibility of dealing with traces in both communication modes, they do not provide a formal treatment of such traces. The Communication of threads running on a single machine is naturally modeled by rendezvous, while the interaction of threads running on machines that are geographically distributed is essentially asynchronous message passing. Moreover, rendezvous could be used as an abstraction of some message exchange patterns. This explains the mixed nature of the framework developed in this paper. Instead of modeling synchronous communication by a pair of events, as is done in several previous works, we use just a single abstract rendezvous event. This allows us to treat an event trace as a partially ordered set of events, so existing techniques based on partial orders remain applicable. Notice also that [6] as well as [10] mostly concentrate on studying the hierarchy of various classes of traces, while we elaborate on modeling partial orders using automata for the purpose of trace analysis. Compared to [10], as well as other papers on the semantics and model checking of message sequence charts (MSCs) [2, 20], our framework includes local events and rendezvous that are ignored there. In [19] a global state graph (Büchi automaton) is built from an MSC, similar to [12] and [17] which consider the ideal lattice. The paper [17] describes an approach for trace analysis based on a method for building the ideal lattice from a trace and demonstrates that property verification could be performed while the lattice is built. The class of properties is restricted to those allowing automata representation (e.g., LTL), while the use of a general model checker that is advocated in our paper allows us to consider more complex properties. Earlier findings are published in [14] and [30].

This paper is organized as follows. In Section 2, we formally define event traces that include send, receive and local events, as well as rendezvous events, thus generalizing the framework of [6]. In Section 3, we detail our approach of modeling distributed processes by means of automata and show that the composition of automata and the automaton of the lattice of ideals are isomorphic. In Section 4, we present an implementation of our approach based on the commercial tool ObjectGEODE. We discuss the modeling of an event trace in SDL and the use of ObjectGEODE to verify properties of the SUT. Then, in Section 5, we apply our approach to an industrial case study. Finally, Section 6 concludes the paper and mentions directions for further work.

2 Event Traces

A distributed system consists of sequential processes $P_1, \ldots, P_n$ that perform local actions and communicate by exchanging messages and by performing rendezvous (synchronization points). In our framework, a process reports its own actions by generating a non-empty set E_i of events that are classified into *local* (l), *send* (s), *receive* (r), and *rendezvous* (z) events, and are the only events observable outside the system. The local events indicate some change in the local state of a process. Events of types s and r are generated when the corresponding processes exchange a message m, and a rendezvous event indicates that certain processes are involved in a rendezvous v. Let M and R denote the disjoint sets of messages and rendezvous, respectively. Rendezvous events are the only common events between processes; in other words, any non-empty set $E_i \cap E_j$, where $i \neq j$, contains only rendezvous events.

Assuming that each event is generated by a process P_i only once, the events constitute a sequence T_i whose length is $|E_i|$, called the *local trace* of P_i. P_i is sequential, so the events in E_i are totally ordered by a causal relation, $\prec_i$, defined by their occurrence in T_i. Each $\prec_i$ is an irreflexive total order. We define the mappings $\Gamma_1, \Gamma_2, \Gamma_3$ as follows:

- Γ_1 is defined from M into the set of send events, and $\Gamma_1(m) = s$, where s is the send of message m.
- Γ_2 is defined from M into the set of receive events, and $\Gamma_2(m) = r$, where r is the receive of message m.
- Γ_3 is defined from R into the set of rendezvous events, and $\Gamma_3(v) = z$, where z is the event with which a process P_i, such that $z \in E_i$, performs the rendezvous v.

Let $E = \cup E_i$ be the set of all the events generated in the system. In E, matching of receive with send events and identifying rendezvous events is defined by the relation Π that is the smallest subset of $E \times (M \cup R) \times E$ such that

- $(\Gamma_1(m), m, \Gamma_2(m)) \in \Pi$ for all $m \in M$ and
- $(\Gamma_3(v), v, \Gamma_3(v)) \in \Pi$ for all $v \in R$.

We consider in our framework only systems with point-to-point communications, which are formalized as follows.

Definition 1. Given a system of communicating processes $P_1, \ldots, P_n$ that execute the set of events E, exchange messages in the set M, and perform rendezvous in the set R, a relation $\Pi \subseteq E \times (M \cup R) \times E$ is a *point-to-point communication* relation if
1. the mappings Γ_1, Γ_2, and Γ_3 are bijections;
2. if $s, r \in E_i$ then $(s, m, r) \notin \Pi$ for all $m \in M$;
3. if $(z, v, z) \in \Pi$ then $|\{ i \mid z \in E_i \}| \geq 2$.

Conditions 1 and 2 state that all messages are unique and each message is transmitted between only two distinct processes. Similarly, conditions 1 and 3 ensure the uniqueness of rendezvous and restrict them to multi-party rendezvous (a process would not perform a rendezvous with itself). In the system of processes $P_1, \ldots, P_n$, causal dependencies between events from different processes are induced by the point-to-point communication relation.

Definition 2. The *causality* relation $\prec$ on the set of events E of the system $P_1, \ldots, P_n$ with the point-to-point communication relation Π is the smallest subset of $E \times E$ that satisfies the following rules.

1. If $a \prec_i b$ then $a \prec b$.
2. If $(s, m, r) \in \Pi$ then $s \prec r$.
3. If $a \prec b$ and $b \prec c$ then $a \prec c$.

The third condition ensures transitivity. Thus the causality relation is a transitive closure of a relation defined by the local orders and the precedence of a send event over the matching receive event. An event a is said to be a *predecessor* of another event b if $a \prec b$. Thus, a is a predecessor of b if and only if there exists a finite sequence $c_0, c_1, \ldots, c_k$ such that $a = c_0$ and $b = c_k$, and for every j, $1 \leq j \leq k$, either $c_{j-1} \prec_i c_j$ for some i or $(s, m, r) \in \Pi$ for some message m.

For an arbitrary collection, $(T_1, \ldots, T_n)$, of local traces of events generated by the processes of a distributed system to be causally consistent, we require that the causality relation on the set of events in a tuple of local traces be cycle-free.

Definition 3. A tuple $(T_1, \ldots, T_n)$ of local traces produced by a system $P_1, \ldots, P_n$ is said to be an *event trace* of the system if the causality relation $\prec$ on the set of events E is irreflexive.

Since the causality relation is transitive by definition, an event trace is defined when the relation on E is a strict partial order. On the other hand, the reflexive closure of $\prec$, denoted $\preccurlyeq$, ($a \preccurlyeq b$ if and only if $a \prec b$ or $a = b$) is a partial order, and $(E, \preccurlyeq)$ is a poset.

An *ideal* P of poset $(E, \preccurlyeq)$ is a subset of E such that if $e \in P$ and $e' \preccurlyeq e$ then $e' \in P$. Let $I(E, \preccurlyeq)$ denote the set of all the ideals of poset $(E, \preccurlyeq)$. It is known that $I(E, \preccurlyeq)$ and a set inclusion relation form a lattice, which we denote by $(I(E, \preccurlyeq), \subseteq)$ and call it the *ideal* lattice (also called the lattice of consistent cuts or global states). Since in our case, the set of events is finite, the set E is the supremum and the empty set $\varnothing$ is the infimum of the lattice $(I(E, \preccurlyeq), \subseteq)$. The ideal lattice represents all the possible interleavings of events and provides a simple way to check properties of concurrent systems [17].

Here, the question arises how to build the ideal lattice of an event trace. Several methods have already been discussed, see, e.g., [17]. However, the practicality of such methods can be viewed only in light of satisfying the goal of trace analysis: verifying properties based on an event trace. The work in [17] proposes an efficient method to build the ideal lattice and indicates the possibility of using a standard model checker to verify properties on the lattice directly, a task that any well known tool, e.g., SPIN [15], SMV [27], or ObjectGEODE [29], could solve. The use of the obtained lattice as an input to a model checker faces a scalability problem since the model checker could simply reject a specification whose size exceeds its input capacity. We believe that a more efficient solution to the problem lies in applying a model checker to a modular specification instead of a monolith one. Indeed, modular specifications are more compact and allow a model checker to fully exploit sophisticated search techniques avoiding a full state space search inevitable in case of monolith specifications.

3 Modeling Event Traces Using Automata

In this section, we describe how a given event trace can be modeled by a system of communicating automata and demonstrate that the system exactly characterizes the causality relation of the event trace, i.e., the composition of the automata and the lattice of ideals are isomorphic. We first recall a few definitions from automata theory.

An *automaton* A is a tuple $<\Sigma, Q, q_0, \rightarrow_A, Q_f>$, where Σ is a finite set of actions, Q represents a finite set of states, q_0 is denoted the initial state; $\rightarrow_A \subseteq Q \times \Sigma \times Q$ is a transition relation, and $Q_f \subseteq Q$ is a set of final states. An automaton is *deterministic* if $q -a\rightarrow q'$ and $q -a\rightarrow q''$ imply that $q' = q''$ for all $a \in \Sigma$ and $q \in Q$. A state q of an automaton is said to be *reachable* from state q' if there exists a sequence of transitions $q' -a_1\rightarrow q_1 -a_2\rightarrow q_2...-a_i\rightarrow q_i$, where $q_i = q$. We use q ***after*** σ to denote the state q_i reached from the state q after a finite sequence of transitions, $q -a_1\rightarrow q_1 -a_2\rightarrow q_2...-a_i\rightarrow q_i$, where $\sigma = a_1a_2...a_i$. We assume that each state is reachable from itself with an empty word ε, i.e., q *after* $\varepsilon = q$. An automaton is *initially connected* if all its states are reachable from the initial state. In the following, we consider only automata that are deterministic and initially connected. We define the set *events*(σ) to be the set of all events in a given word σ. We call $L_{acc}(q)$ the language accepted by an automaton initialized in state q. The language accepted by the automaton starting in the initial state is denoted $L_{acc}(A)$. An initially connected automaton A is *minimal* if all the final states of A are reachable from every state, and $L_{acc}(q) \neq L_{acc}(q')$ for each two states q and q', i.e., states are pairwise distinguishable.

We use the operator $\|$ to compose automata. Given $A_1 = <E_1, Q_1, q_{01}, \rightarrow_{A1}, Q_{f1}>$ and $A_2 = <E_2, Q_2, q_{02}, \rightarrow_{A2}, Q_{f2}>$, the *composition automaton*, denoted $A_1 \| A_2$, is a tuple $< E, Q, q_0, \rightarrow, Q_f>$, where $E = E_1 \cup E_2$, $q_0 = (q_{01}, q_{02})$; $Q \subseteq Q_1 \times Q_2$, $\rightarrow$, and Q_f are the smallest sets obtained by applying the following rules.

- If $q_1 -a\rightarrow_1 q'_1$ and $a \notin E_2$ then $(q_1, q_2) -a\rightarrow (q'_1, q_2)$.
- If $q_2 -a\rightarrow_2 q'_2$ and $a \notin E_1$ then $(q_1, q_2) -a\rightarrow (q_1, q'_2)$.

- If q_1 -a$\rightarrow_1 q'_1$ and q_2 -a$\rightarrow_2 q'_2$ then (q_1, q_2) -a$\rightarrow (q'_1, q'_2)$.
- $Q_f = Q \cap (Q_{f1} \times Q_{f2})$.

The composition is associative; it can be applied to finitely many automata. The composition of n automata $C_1 \ldots C_n$ is an automaton $C = C_1 \parallel \ldots \parallel C_n$ over the alphabet $E = E_1 \cup \ldots \cup E_n$. Let $\omega \in E^*$, we use $\omega_{\downarrow Ei}$ to denote the projection of ω onto the set E_i. The following properties of the composition automaton can directly be established from the definition.

Proposition 4. If $(q_1, \ldots, q_p)$ **after** ω is defined then $(q_1, \ldots, q_p)$ **after** $\omega = (q_1$ **after** $\omega_{\downarrow E1}, \ldots, q$ **after** $\omega_{\downarrow Ep})$. If $\omega \in L_{acc}(C)$ then $\omega_{\downarrow Ei} \in L_{acc}(C_i)$.

3.1 Ideal Lattice Automaton

An ideal lattice is usually visualized as a graph, called the covering digraph [17], in which nodes represent the elements of the poset and edges represent the relations between the elements (with the omission of transitive edges). We represent the ideal lattice by an automaton, called the *ideal lattice automaton*, which corresponds to the covering digraph and is defined as follows.

Definition 5. Given an event trace $(T_1, \ldots, T_n)$ with the causality relation $\prec$, the *ideal lattice automaton*, denoted T, is a tuple $<E, I(E, \prec), \varnothing, \rightarrow_T, \{E\}>$, where E is the set of events, $I(E, \prec)$ is the set of states and $\varnothing$ is the initial state, $\rightarrow_T = \{(P, a, R) \mid P, R \in I(E, \prec), R = P \cup \{a\}, \text{ and } R \neq P\}$, $\{E\}$ is the set of final states.

Each word accepted by the ideal lattice automaton contains exactly one instance of an event in E and thus defines a total order that is a linearization of the causality relation $\prec$, i.e., the word is one possible interleaving of the events in the trace. The accepted language $L_{acc}(T)$ of the ideal automaton consists exactly of all linearizations of the strict partial order $\prec$ [17, 4]. In the following, we will not distinguish between a total order on E and the word it defines, thus $\sigma \in L_{acc}(T)$ also denotes a total order such that $\sigma \supseteq \prec$. The states of the ideal lattice automaton represent all possible states of the system that generated the event trace; thus we can use the ideal lattice automaton to check properties of the system exhibited during its execution.

Proposition 6. The ideal lattice automaton is minimal.

Proof. By construction, T is initially connected. Then, by definition of a lattice, an infinum and a supremum exist for every subset of the lattice. This implies that in the ideal lattice automaton there exists a sequence of transitions from every state to the final state that is the supremum E of the lattice. This means that the final state of the ideal lattice automaton is reachable from every state. We prove now that the states of T are pairwise distinguishable. Let I, J be two different states (ideals) of T. $L_{acc}(I)$ and $L_{acc}(J)$ are the sets of the linearizations of the posets $(E \setminus I, \prec)$ and $(E \setminus J, \prec)$, respectively. From $I \neq J$ it follows that $E \setminus I \neq E \setminus J$. Hence $L_{acc}(I) \neq L_{acc}(J)$. QED.

3.2 Composition of Automata

Given an event trace $(T_1, ..., T_n)$ with the causality relation $\prec$, we define $p = n + |M|$ communicating automata, where n automata model the processes $P_1, ..., P_n$ and accept the corresponding local traces $T_1, ..., T_n$, respectively, and $|M|$ automata that model message delays inherent to asynchronous communications. A delay automaton of message $m \in M$ is a three state automaton that accepts the word $s.r$ such that $\Gamma_1(m) = s$ and $\Gamma_2(m) = r$. A similar approach is taken in [10] where communications via individual message channels are used instead of message delay automata. In our work, we consider a more general case of traces of systems with mixed synchronous and asynchronous communications instead.

Definition 7. Given a process P_i with its local trace T_i, the *local trace* T_i automaton is the tuple $<E_i, I(E_i, \preccurlyeq), \emptyset, \rightarrow_{Pi}, \{E_i\}>$, where E_i is the set of events, $I(E_i, \preccurlyeq)$ is the set of states, $\emptyset$ is the initial state, $\rightarrow_{Pi} = \{(P, a, R) \mid P, R \in I(E_i, \preccurlyeq), R = P \cup \{a\}, \text{ and } R \neq P\}$, and $\{E_i\}$ is the set of final states.

According to Definition 5, the local trace automaton corresponding to local trace T_i is the ideal lattice automaton of the local (total) order $\prec_i$. Thus, according to Proposition 6 a local trace automaton is minimal. The accepted language of such an automaton contains exactly one word that is the corresponding local trace.

Definition 8. Given a message $m_i \in M$, let $\Gamma_1(m_i) = s_i$, and $\Gamma_2(m_i) = r_i$, the *message* m *delay automaton* is the tuple $<E_{i+n}=\{s_i, r_i\}, \{\emptyset, \{s_i\}, \{s_i, r_i\}\}, \emptyset, \rightarrow_{mi}, \{\{s_i, r_i\}\} >$, where $\{s_i, r_i\}$ is the set of events, $\{\emptyset, \{s_i\}, \{s_i, r_i\}\}$ is the set of states, $\emptyset$ is the initial state, $\rightarrow_{mi} = \{(\{\emptyset\}, s_i, \{s_i\}), (\{s_i\}, r_i, \{s_i, r_i\})\}$, and $\{\{s_i, r_i\}\}$ is the set of final states.

Each message delay automaton is an ideal lattice automaton of a total order of two events, reflecting the precedence of a send event over the matching receive event and thus is minimal.

Let $C_1, ..., C_p$ be the set of (component) automata that model n communicating processes $P_1, ..., P_n$ and $|M|$ message delay automata, where $C_i = <\Sigma_i, Q_i, q_{0i}, \rightarrow_i, \{q_{fi}\}>$, such q_{0i} is the initial state of C_i that corresponds to an empty ideal, and q_{fi} is the final state of C_i. Consider the composition $C_1 \parallel ... \parallel C_p = C$. The set of events of C is the union of all Σ_i, where Σ_i is either E_i or $\{s_i, r_i\}$ for an appropriate $m_i \in M$. Hence the set E of all events of the given distributed system constitutes the set of events of C. The initial state of C is $q_0 = (q_{01}, ..., q_{0p})$. The set of states Q_C and the transition relation $\rightarrow$ of C are defined by the semantics of the composition operator $\parallel$. The set of final states of the composition automaton is $Q_f = Q_C \cap \{(q_{f1}, ..., q_{fp})\}$, which means that Q_f contains only one state or is empty when $(q_{f1}, ..., q_{fp})$ is not reachable in C. Thus, $C = <E, Q_C, q_0, \rightarrow, Q_f>$.

3.3 Isomorphism

Next we show that the composition of the local trace automata and message delay automata is isomorphic to the automaton of the ideal lattice based on the fact that two minimal automata that accept the same language are isomorphic. We already proved that the ideal automaton is minimal. We shall first demonstrate that the composition automaton is minimal and then that the two automata accept the same language. By definition of the composition operator, the composition automaton is initially connected. So, we need to show that the state $(q_{f1}, ..., q_{fp})$ is reachable from all states of the composition automaton and that all states are pair wise distinguishable.

Proposition 9. Given an event trace $(T_1, ..., T_n)$ with the causality relation $\prec$, the state $(q_{f1}, ..., q_{fp})$ is reachable from every state of the composition automaton C.

Proof. Assume that state $(q_{f1}, ..., q_{fp})$ is not reachable from some state in Q_C. Two cases are possible here: the state belongs to a cycle or there is a deadlock state with no outgoing transitions. There is no cycle in the composition automaton, since each component automaton is acyclic, see Proposition 4. Then let $q \in Q_c$ be a state with no outgoing transitions. The automaton C is initially connected, thus there exists a word $\sigma \in E^*$ such that $q = q_0$ *after* $\sigma = (q_{01}$ *after* $\sigma_{\downarrow E1}, ..., q_{0p}$ *after* $\sigma_{\downarrow Ep})$ and there exists a component automaton C_i such that q_{0i} *after* $\sigma_{\downarrow Ei} \neq q_{fi}$. Then *events*$(\sigma_{\downarrow Ei}) \subset E_i$ and there exists an event $e \in E_i \setminus$ *events*$(\sigma_{\downarrow Ei})$ such that all the predecessors of e (in E_i) are in q_{0i} *after* $\sigma_{\downarrow Ei}$. Hence e cannot be a local event, so there exists at least one component automaton C_j ($j \neq i$) such that $e \in C_i \cap C_j$. Moreover, state q_{0j} *after* $\sigma_{\downarrow Ej}$ has no transition on e, otherwise the composition automaton would have a transition from q. There are two cases possible: either $e \in q_{0j}$ *after* $\sigma_{\downarrow Ej}$ or $e \notin q_{0j}$ *after* $\sigma_{\downarrow Ej}$. In the first case, $e \in \sigma$ and, according to Proposition 4, $e \in \sigma_{\downarrow Ei}$ and thus $e \in q_{0i}$ *after* $\sigma_{\downarrow Ei}$. But, $e \in E_i \setminus$ *events*$(\sigma_{\downarrow Ei})$, a contradiction. Consider now the second case, $e \notin q_{0j}$ *after* $\sigma_{\downarrow Ej}$, this implies that q_{0j} *after* $\sigma_{\downarrow Ej} \neq q_{fj}$. Then there should exist a transition from state q_{0j} *after* $\sigma_{\downarrow Ej}$ on another event, say e', that is a predecessor of e. Similar to the above considered cases, e' cannot be a local event and therefore there should exist another component automaton that shares this event and so on. Due to a finite number of events in the composition, an event must reoccur in this sequence. This means that a reoccurring event is a predecessor of itself. This is impossible due to irreflexivity of the causality relation $\prec$ on the event trace. So, $(q_{f1}, ..., q_{fp})$ is the final state reachable from all the states in Q_c the composition automaton for an event trace. QED.

Proposition 10. States of C are pairwise distinguishable.

Proof. Assume that there exist two states $q = (q_1, ..., q_i, ..., q_p)$ and $q' = (q'_1, ..., q'_i, ..., q'_p)$ that are not distinguishable, in other words, $L_{acc}(q) = L_{acc}(q')$. According to Proposition 4, for each i it holds that $L_{acc}(q_i) = L_{acc}(q'_i)$ since $L_{acc}(q_i)$ and $L_{acc}(q'_i)$ are both singletons. Each component automaton is minimal, thus, $q_i = q'_i$. This implies $q = q'$. QED.

Thus, the following statement holds.

Lemma 11. The composition automaton C is minimal.

Next, we show that the two automata, C and T, accept the same language, i.e., $L_{acc}(T) = L_{acc}(C)$.

Proposition 12. $L_{acc}(C) \subseteq L_{acc}(T)$.

Proof. Proposition 4 states that each projection of an arbitrary word ω, accepted by C, takes the corresponding component automata to the final state. Thus, all events of E are present in ω. In addition, any event occurs exactly once in ω, otherwise this would mean that events reoccur in component automata. Moreover, by construction, words accepted by local trace and message delay automata define the corresponding local orders and the send-receive precedence, respectively. Hence the word ω satisfies these total orders. Since the causality relation is just a transitive closure of local orders along with send-receive precedence, ω represents a linearization of the causality relation. Therefore, the ideal lattice automaton accepts ω. QED.

To prove that $L_{acc}(T) \subseteq L_{acc}(C)$, we need an auxiliary proposition, which establishes a relation between words of $L_{acc}(T)$ and states of the composition automaton C. Let $E\!\downarrow_\sigma$ denote $E \cap \mathit{events}(\sigma)$ and let σ_k be a prefix of length k of a word $\sigma \in L_{acc}(T)$.

Proposition 13. For any k, $0 \le k \le |E|$, prefix σ_k of $\sigma \in L_{acc}(T)$, q_0 *after* $\sigma_k = (E_1\!\downarrow_{\sigma_k}, \ldots, E_p\!\downarrow_{\sigma_k})$.

Proof. Base of induction. For $k = 0$, σ_0 is an empty word, and q_0 *after* $\sigma_0 = q_0 = (q_{01}, \ldots, q_{0p}) = (q_{01}$ *after* $\varepsilon, \ldots, q_{0p}$ *after* $\varepsilon)$.

Inductive step. Let $\sigma_k = e_1 e_2 \ldots e_k$ and $\sigma_{k+1} = e_1 e_2 \ldots e_k e_{k+1}$, where $e_1, e_2, \ldots, e_k, e_{k+1} \in E$. Assume q_0 *after* $\sigma_k = (E_1\!\downarrow_{\sigma_k}, \ldots, E_p\!\downarrow_{\sigma_k})$, we need to show that for word σ_{k+1}, q_0 *after* $\sigma_{k+1} = (E_1\!\downarrow_{\sigma_{k+1}}, \ldots, E_p\!\downarrow_{\sigma_{k+1}})$. So, it is sufficient to prove that q_k *after* $e_{k+1} = (E_1\!\downarrow_{\sigma_{k+1}}, \ldots, E_p\!\downarrow_{\sigma_{k+1}})$. According to the properties of a partial order, $\sigma_{k+1} = \sigma_k e_{k+1}$ is a linearization of an ideal, so all the predecessors of e_{k+1} are in σ_k. Then $E_i\!\downarrow_{\sigma_k}$ and $E_i\!\downarrow_{\sigma_{k+1}}$ are different ideals of the total order $(E_i, \preccurlyeq)$ for any E_i that contains e_{k+1}. This means that the automaton C_i has a transition $(E_i\!\downarrow_{\sigma_k}, e_{k+1}, E_i\!\downarrow_{\sigma_{k+1}})$. Thus $E_i\!\downarrow_{\sigma_{k+1}} = E_i\!\downarrow_{\sigma_k} \cup \{e_{k+1}\}$; if $E_i\!\downarrow_{\sigma_{k+1}} = E_i\!\downarrow_{\sigma_k}$ then $e_{k+1} \notin E_i$. According to the definition of the composition operator, there exists a transition $(E_1\!\downarrow_{\sigma_k}, \ldots, E_p\!\downarrow_{\sigma_k}), e_{k+1}, (E_1\!\downarrow_{\sigma_{k+1}}, \ldots, E_p\!\downarrow_{\sigma_{k+1}})$. QED.

Proposition 14. $L_{acc}(T) \subseteq L_{acc}(C)$.

Proof. According to Proposition 13, each word σ accepted by T takes C to state $(E_1\!\downarrow_\sigma, \ldots, E_p\!\downarrow_\sigma) = (E_1, \ldots, E_p)$, which, according to the construction of C, is its final state. QED.

Lemma 15. $L_{acc}(T) = L_{acc}(C)$.
Lemma 11 and 15 imply the following.

Theorem 16. The composition automaton C and the ideal lattice automaton T are isomorphic.

The latter theorem shows that the composition of the automata representing local traces and the ideal lattice automaton of the distributed system under test can be used interchangeably to reason about the system. In particular, model checking technology can be applied to check properties on the collected traces of distributed systems.

4 Using SDL Tools in Trace Analysis

By following the approach to trace analysis outlined in the previous sections, we translate the problem of trace analysis into a typical model checking problem. So, given a logfile of events, implementation of the approach is reduced to the following tasks:

1. Extracting an event trace from a logfile produced by a monitoring tool.
2. Choosing an appropriate model checking tool and describing a system of local trace and message delay automata in the specification language of this model checker.

4.1 Event Trace Extraction

Monitoring a distributed system yields a *logfile* that comprises a not necessarily ordered collection of events indicating the actions executed. To verify whether the logfile describes an event trace, we have to analyze each event individually and its relation to others. We assume that every event is recorded in a certain structure, the *event record*, that contains information sufficient to identify the type of the event, the process that generated it, and to determine properties of the causality relation. Moreover, it must be possible to verify whether all events satisfy the point-to-point communication relation. In our framework, we assume that every event record in the logfile has the following fields:

- Type of event: Communication (Send, Receive, or Rendezvous), Local
- Name of the issuing process
- Local ordinal number
- Partner for communication events
- Message / rendezvous parameters
- Local parameters of the issuing process (variables, ...).

In a logfile, each event is uniquely identified by its complete event record structure. Local ordinal numbers determine the local order of each process in the system. As explained in Sect. 2, matching of receive with send events and identifying rendezvous events is defined by the point-to-point communication relation. So communication events should be verified for the existence of their matching events. The final check should make sure that the causal relation between the events in the logfile is a strict partial order.

4.2 Choosing a Model Checker

A system of communicating automata can be extracted from an event trace and described in the language of a chosen model checker fully automatically. Therefore, the choice is dictated by the richness of the language used for property specification and not by the system to be analyzed. User friendliness is another important selection criterion. Indeed, to be accepted by practitioners, any integrated environment for trace analysis should hardly require test engineers to master, for example, a temporal logic.

From this viewpoint, ObjectGEODE (OG for short) of Telelogic [29] meets the above requirements. OG supports modeling systems in SDL extended with synchronous communication and performing specification simulation and property checking. Moreover, OG provides the SDL-like language GOAL (GEODE Object Automata Language) to specify properties [1, 29]. Once an SDL specification is built from a logfile, the OG simulator performs the verification of specified properties. To deploy OG, we need to translate the automata model into an SDL system, addressing the aspects of structure, behavior, communication, and data.

The structure of an SDL system is modeled using a hierarchy of *system / block / process / procedure* statements. In our case, it is sufficient to define a system with a single block that is composed of several processes. The obvious approach is to map each local trace and message delay automaton into a designated SDL process. The communication between processes, which is point-to-point and represents multi-party rendezvous, is then achieved via signals that represent data exchanged in SDL channels. To do so, the representation of an automaton can be obtained by projecting the collected trace onto the set of events of this automaton: send, receive, rendezvous, and local events. Each send event of the automaton translates to an OUTPUT statement in SDL with the signal name showing the name of the message. Similarly, each receive event is mapped to an INPUT statement, and a local event is translated to TASK statements. As for rendezvous events, SDL does not support synchronous communications, but OG extends standard SDL by allowing the definition of rendezvous channels, over which signals are exchanged without any delay. However, this extension allows only two-party rendezvous to be directly modeled, thus we restrict our implementation of the approach to cover only this type of rendezvous. Subsequently, states are inserted between the events. The overall behavior of the system is modeled by the joint behavior, represented as a global state graph, of all communicating processes in SDL.

Such a straightforward solution requires the definition of as many SDL processes as there are processes and messages in the given event trace. This leads to a large system specification and could represent a significant obstacle to the model checker in handling the model during syntax and semantic checking and simulation. A better solution consists in replacing message delay automata by individual channels, where a separate channel with associated individual input queue is used to carry every message. These input queues play the role of the message delay automata. An advantage of this approach is the drastic reduction of the number of processes defined in the model. However, individual channels are not supported in standard SDL.

The need to further reduce the size of the resulting SDL specification motivates the search for another solution to the problem of translating message delay automata to SDL. We decided to use the standard SDL input queue. However, to breach FIFO order of consumption and prevent discarding messages during simulation, we use an SDL feature, the asterisk SAVE construct, which prohibits a process in its current state from discarding those signals which could not be consumed by an explicitly defined transition, as suggested in [25] and [22, 18]. This prevents signal loss when signals arrive at the input queue of a process out of the expected order. The saved signals are kept for future consumption, which means that messages are delivered finally in the order as stored in the collected event trace.

4.3 Workflow of the Approach Using SDL and OG

The approach based on OG is summarized in the workflow shown in Figure 1. The implementation efforts are reduced to building just a front-end to OG featuring:

1. A system specification tool TRAYSIS that builds an SDL specification from the collected trace.
2. A property specification tool called Property Manager that eases the process of writing properties for trace verification.

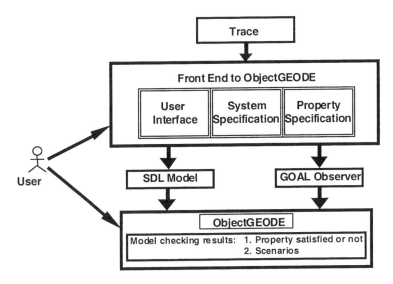

Fig. 1. The workflow of the approach.

4.3.1 The System Specification Tool

We present here the Trace Analysis (TRAYSIS) tool, Figure 2, which currently accepts logfiles with either synchronous or asynchronous communications. Its main task is to automate the process of producing an SDL specification from a logfile of an event trace specified in XML syntax. The tool offers the following main features:

- XML logfile treatment and model construction. The tool checks the logfile as discussed in Section 4.1. In case of missing communication events, the tool informs the user and converts the unmatched communication events into local events of the processes. The tool also checks for flaws in the logfile that would render the generated SDL model syntactically incorrect, e.g. the presence of SDL keywords in the logfile and the use of some illegal characters. Once a logfile is successfully checked, the tool generates the SDL model of the system. Causality cycles are then detected by OG during simulation of the generated SDL model as they cause deadlock prior to the termination of the processes.
- Customized model generation. This feature enables the user to cope with large logfiles. The tool offers three types of filtering: process filter, signal filter, and a filter that extracts a segment from the initial logfile.
- Statistics about the SDL model. The tool offers listings of processes, variables, and signals occurring in the model to help the user better understand the generated model.

4.3.2 Pattern Specification

There is a wide range of properties that can be sought in distributed systems. These properties are classified into two types: state-based and event-based properties. Event-based properties are expressed in terms of atomic or composite events of the distributed system, while state-based properties formulate assertions on state variables of processes of the system. The work of Dwyer et al. to build a repository of typical and frequently used specification patterns (similar to design patterns) [8], [24] is an effort to consider both types of properties. For our project, we rely on the repository from [24] to build a library [11] of specification property templates in GOAL in which we also consider mixed properties that are expressed in terms of both events and state predicates. However, we needed to modify the original patterns, with the addition of a termination predicate, to count for the fact that we deal only with finite traces in our post mortem analysis.

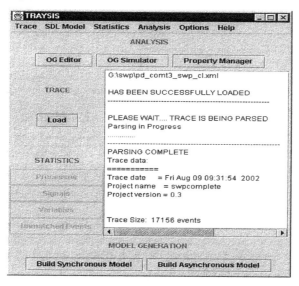

Fig. 2. The TRAYSIS tool.

GOAL is an automata-based language to specify properties of SDL models. The structure of GOAL, which is similar to SDL [1], allows one to specify any event-based or state-based properties as well as mixed properties as long as they are expressible by automata [13]. A property presentation in GOAL is called an *observer* that implements an extended finite automaton with accepting states. Observers are usually described in terms of entities (objects, signals, variables etc.) of the system. In addition, GOAL allows the declaration of two types of designated states: *success* and *error* states. Entering a success state (error state) indicates that the system respects (violates) the property expressed in the observer. However, the use of the success/error convention is completely up to the user and has no formal meaning. In our observers, reachability of error states indicates that the property is not satisfied, and absence of error states indicates that the property is satisfied (independently of interleavings). Presence of success indicates that there are possible interleavings on which the property is satisfied. During simulation, OG builds the synchronous product of the observer and the specification of the system. The output includes a report on the number of error and success states encountered along with scenarios that lead to these observer states. Scenarios can be replayed as counter or supporting examples later.

The Property Manager is a tool that supports the process of property specification for trace analysis in OG by offering the user a library of predefined patterns [11]. Every pattern of the library is a parameterized template of a GOAL observer. The tool helps the user customize the observer to specific settings of the observed event trace. A screenshot of the Property Manager tool is shown in Figure 3.

5 Case Study

We applied the trace analysis approach in an industrial case study. The task of a development team at Siemens' Power Transmission and Distribution Division was to implement a *sliding window protocol* (SWP) mechanism on top of the factory automation protocol suite PROFIBUS. This extension of PROFIBUS is required to support communication among distributed power control devices within regional power grids. A protocol analyzer observes the exchange of PROFIBUS messages within the communication network and produces an execution trace of the behavior of the system.

The SWP is a well-anticipated means of flow control for a reliable data transfer taking place between two communication partners and is well studied [28]. The protocol in our case study offers a symmetrical service for both communication partners, i.e., partners, server and client, can send data, which is flow-controlled. The protocol uses the two messages *TVoid* to send data to a partner and *TWindowAck* to send an acknowledgement back. Both messages carry the two parameters *Win_Send* and *Win_Quit* indicating the sequence numbers of the latest message sent and the latest acknowledgement received, respectively. The sliding window protocol requires that no more than a certain, pre-defined number of messages (the maximum window size) be sent in a row before an acknowledgement is received. A single acknowledgement can confirm the receipt of up to *mws* unacknowledged messages simultaneously, where *mws* is the maximum window size.

In our case study, we verified various properties of the behavior of the network of communicating power control devices. These properties include the check whether the maximum window size is always respected and the following distributed properties. The first property requires that the total number of unacknowledged messages in the whole network do not exceed a certain threshold at anytime. The requirement is specifically of interest in systems where concurrency control is required. The second property states that the total number of messages in transit does not exceed a certain limit. These properties help the developers of the system evaluating the efficiency and reliability of the communication media and checking whether their occurrence contributes to undesired behavior. It is clear that these properties cannot be checked locally at the server side since the server is unaware of any messages in transit.

We formalized the last property in GOAL using the "global response" property template taken from our template library. Due to lack of space we present a simplified version of the observer in Figure 3. The observer uses three probes $pb1$, $pb2$, and $pb3$ to get access to variables that are otherwise internal to the SDL system. The SDL processes $a2$, $a40$, and $a36$ are the communication partners that were identified from the recorded trace, where $a2$ is the server and the remaining processes act as clients. Their behaviors are modeled based on the trace information and the model construction rules for asynchronous communication as outlined before. The observer waits for the occurrence of an output message at any of the client processes (left branch) and stores the message number of the sent message (*Win_Send*) in an array to keep track of unacknowledged messages. Every time a message is sent by a client, the observer checks whether the total number of messages in transit exceeds a certain limit (2 messages) that is specified by the variable *limit*. In this case the observer reaches the error state. The number of messages in transit in the system is the difference between the number of messages sent by the clients and the number of acknowledgement messages issued by the server (detected in the right branch). However, the observer does not signal a violation before the first acknowledgment is detected which indicates that the server is alive. Once the observer detects that an acknowledgement message is received by any of the clients (the middle branch), it compares the acknowledgement number (*Win_Quit*) with the message numbers stored in the array of this client and resets the array and the counters appropriately by calling the *updtarray* procedure.

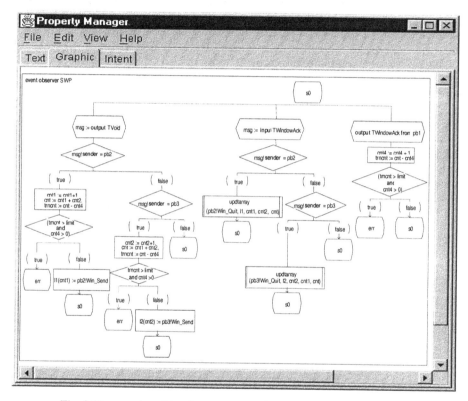

Fig. 3 Observer describing the property on the number of messages in transit.

During verification in OG, the observer is triggered by the occurrences of *TVoid* and *TWindowAck* messages. If the error state *err* is reached, the verification process terminates. Then, an error scenario is created that depicts the history of the system behavior. This error scenario can be visualized as an MSC. The visualization helps test engineers and developers understand better the cause of the property violation. In our case study, the verification of a trace containing about 1000 messages exchanged between the involved processes was done relatively fast since the resulting state space remained rather small (20800 states). The maximum number of messages in transit detected in the observed network helped assessing the bandwidth of the network.

6 Conclusion

We presented a general formal framework for trace analysis of distributed systems that applies to a logfile of events collected during execution of the system. We formally defined an event trace to consist of message send and receive events, rendezvous events, and local events; partially ordered by an irreflexive causality relation. We modeled an event trace by a system of communicating automata that correspond to the processes observed in the SUT and demonstrated that the system exactly characterizes the causality relation of the event trace. We proved that the composition of the automata is isomorphic to the ideal lattice automaton defined by the event trace.

We used SDL and the general-purpose model checker ObjectGEODE to implement our approach. We implemented the front-end tool TRAYSIS to carry out the translation from a logfile to an SDL specification of the SUT. We also built a library of property patterns in GOAL and interfaced it with the Property Manager tool to customize patterns to specific settings. Last but not least, we illustrated the trace analysis approach on an industrial case study on the verification of sliding window protocol implementations.

Our tools allow us to deal with relatively large traces (we processed traces of about twenty thousand events). However, more work is required to cope with even larger traces from industrial applications. Our case study indicates that further development of user-friendly property specification tools and pattern libraries based on the automata approach is also needed.

References

1. B. Algayres, Y. Lejeune, E. Hugonnet, "GOAL: Observing SDL behaviors with GEODE", in *SDL'95 with MSC in CASE (ed. R. Braek, A. Sarma), Proc. of the 7th SDL Forum*, Oslo, Norway, Sept. 1995, Elsevier Science Publishers B. V. (North Holland), pp. 359-372.
2. R. Alur and M. Yannakakis. "Model Checking of Message Sequence Charts". *In CONCUR'99: Concurrency Theory, Tenth International Conference*, LNCS 1664, pages 114--129, 1999.
3. J. P. Black, M. H. Coffin, D. J. Taylor, T. Kunz, A. A. Basten, "Linking Specifications, Abstraction, and Debugging", *CCNG Technical Report E-232, Computer Communications and Network Group*, University of Waterloo, Nov. 1993.
4. R. Bonnet and M. Pouzet. "Linear Extensions of Ordered Sets". *In I. Rival, editor, Ordered Sets*, pages 125-170, D. Reidel Publishing Company, 1982.
5. K. Chandy, L. Lamport, "Distributed Snapshots: Determining Global States of Distributed Systems", *ACM Transactions on Computing Systems* 3(1), pp. 63-75, Feb. 1985.
6. B. Charron-Bost, F. Mattern, and G. Tel. "Synchronous, Asynchronous, and Causally Ordered Communications". *Distributed Computing*, 1995.
7. F. Dietrich, X. Logean, S. Koppenhoefer, J.-P. Hubaux, "Testing Temporal Logic Properties in Distributed Systems", *In Proc. of the 11th International Workshop on Testing of Communicating Systems*, Tomsk, Russia, Aug. 1998.
8. M. Dwyer, G. Avrunin, and J. Corbett, "Patterns in Property Specifications for Finite-state Verification", *In Proc. 21st International Conference on Software Engineering*, May 1999.
9. M. Dwyer, L. Clarke, "Data Flow Analysis for Verifying Properties of Concurrent Programs", *In Proc. of ACM SIGSOFT'94*, New Orleans, LA, USA, 1994.

10. A. Engels, S. Mauw, and M.A. Reniers. "A Hierarchy of Communication Models for Message Sequence Charts". *In T. Mizuno, N. Shiratori, T. Higashino, and A. Togashi, editors, FORTE/PSTV'97*, pages 75-90, Osaka, Japan, Nov. 1997. Chapman & Hall. To appear in Science of Computer Programming, 2002.
11. Q. Fan. "Formalizing Properties for Distributed System Testing", *Master Thesis in Preparation*.
12. E. Fromentin, M. Raynal, V. Garg, and A. Tomlinson, "On the Fly Testing of Regular Patterns in Distributed Computations", *Internal Publication # 817*, IRISA, Rennes, France, 1994.
13. R. Groz, "Unrestricted Verification of Protocol Properties on a Simulation Using an Observer Approach", *Protocol Specification, Testing and Verification, VI*, Montréal, Canada, North-Holland, 1986, pp. 255-266.
14. H. Hallal, A. Petrenko, A. Ulrich, S. Boroday, "Using SDL Tools to Test Properties of Distributed Systems", *In proc. of Workshop on Formal Approaches to Testing of Software (FATES) in affiliation with CONCUR'01*; Aalborg, Denmark; BRICS Technical Re-port NS-01-4, Aug. 2001.
15. G.J. Holzmann. "The Model Checker SPIN". *IEEE Transactions on Software Engineering*, 23(5):279-295, 1997.
16. B. E. Jackl, "Event-Predicate Detection in the Debugging of Distributed Applications", *Master's Thesis*. Department of Computer Science, University of Waterloo, 1996.
17. C. Jard, T. Jeron, G. V. Jourdan, and J. X. Rampon, "A General Approach to Trace-checking in Distributed Computing Systems", *In Proc. IEEE Int. Conf. on Distributed Computing Systems*, Poznan, Poland, Jun. 1994.
18. KLOCwork. Corporate website, http://www.KLOCwork.com/.
19. P. B. Ladkin and Stefan Leue. "Interpreting Message Flow Graphs". *Formal Aspects of Computing 7(5)*, p. 473 - 509, Sept./Oct. 1995.
20. Leue and P.B. Ladkin. "Implementing and Verifying MSC Specifications Using Promela/Xspin". *In J.-C. Grégoire, G. Holzmann and D. Peled (eds.), Procs of the DIMACS Workshop SPIN96, the 2nd Intl Workshop on the SPIN Verification System. DIMACS Series Volume 32, American Mathematical Society*, Providence, R.I., 1997
21. D. Luckham and B. Frasca, "Complex Event Processing in Distributed Systems", *Stanford University Technical Report* CSL-TR-98-754, Mar. 1998, 28 pages.
22. N. Mansurov, D. Zhukov, "Automatic Synthesis of SDL models in Use Case Methodology". *In R. Dssouli, G. v. Bochmann, and Y. Lahav (eds.), SDL'99: The Next Millenium, Proc. of the ninth SDL Forum*, Montreal, Québec, Canada, Jun. 21 - 25, 1999.
23. B. Miller, J. Choi, "Breakpoints and Halting in Distributed Programs", *In Proc. of the 8th IEEE Int. Conf. on Distributed Computing Systems*, San Jose, Jul. 1988.
24. Pattern Specification System, http://www.cis.ksu.edu/santos/spec-patterns.
25. G. Robert, F. Khendek and P. Grogono, "Deriving an SDL Specification with a Given Architecture from a Set of MSCs", *In A. Cavalli and A. Sarma (eds.), SDL'97: Time for Testing - SDL, MSC and Trends, Proc. of the eight SDL Forum*, Evry, France, Sept. 22 - 26, 1997.
26. R. L. Smith, G. S. Avrunin, L. Clarke, and L.J. Osterweil. "An Approach Supporting Property Elucidation". *In Proc. of the 24th International Conference on Software Engineering*, Orlando, FL, May 2002, pages 11-21.
27. The SMV System, www-2.cs.cmu.edu/~modelcheck/smv.html.
28. A. Tanenbaum. "Computer Networks". *Prentice Hall*, 1996.
29. Telelogic, "ObjectGEODE SDL Simulator Reference Manual".
30. A. Ulrich, H. Hallal, A. Petrenko, S. Boroday, "Verifying Trustworthiness Requirements in Distributed Systems with Formal Log-file Analysis". *In Proc. of the thirty-sixth Hawaii International Conference on System Sciences (HICSS-36)*, 2003.
31. P. A. S. Ward, "A Framework Algorithm for Dynamic Centralized Dimension-Bounded Timestamps", *In Proc. of CASCON 2000*, Mississauga, Ontario, Canada.

Fault Diagnosis in Extended Finite State Machines

Khaled El-Fakih[1], Svetlana Prokopenko[2],
Nina Yevtushenko[2], and Gregor v. Bochmann[3]

[1] Department of Computer Science, American University of Sharjah, UAE
kelfakih@aus.ac.ae
[2] Tomsk State University, Russia
{prokopenko,yevtushenko}@elefot.tsu.ru
[3] School of Information Technology and Engineering, University of Ottawa, Canada
bochmann@site.uottawa.ca

Abstract. In this paper, we propose a method for the derivation of an adaptive diagnostic test suite when the system specification and implementation are given in the form of an extended finite state machine. The method enables us to decide if it is possible to identify the faulty transitions in the system when faults have been detected in a system implementation. If this is possible, the method also returns test cases for locating the faulty transitions. An example is used to demonstrate the steps of the method.

1 Introduction

The purpose of diagnostic testing is to locate the differences between a specification and its implementation, when the implementation is found to be faulty. In the software domain where a system is represented as an FSM, some work has already been done on diagnostic testing [2][4]. However, no work has been reported for systems represented as Extended FSMs.

In [1] we considered the problem of the derivation of an adaptive diagnostic test suite for a system of two communicating FSMs. It is known that we can not always locate the difference between the system specification and its implementation, due to the fact that different faults can result in the same behavior of a system/implementation under test (SUT). In [1], we presented a method that enables us to decide if it is possible to locate a faulty component machine, and if this is possible then tests for locating the fault(s) are derived. In this paper, we use a similar approach to the diagnostic testing of a single Extended FSM (EFSM). We assume that either predicate, transfer or assignment faults may occur in an EFSM implementation. Moreover, none of these faults increases the number of states in the implementation of the system. Similar to [1], we present a method for the derivation of an adaptive diagnostic test suite that enables us to decide whether it is possible to identify the faulty transitions in the given system, and if this is possible then tests for locating the faulty transitions are derived. We assume that the specification domain of each variable is finite and therefore, an EFSM can be represented as a classical FSM. We use a non-deterministic FSM, called *Fault Function* (FF) [6], for the compact representation of transfer, predicate and assignment faulty transitions. The fault domain is the union of sub-machines of three FFs, *FF-predicate*, *FF-transfer* and *FF-assignment*. In this paper, we present a new method how a FF can be reduced by deleting sub-machines that do not agree with the observed Input/Output behavior of an SUT. We also describe two strategies for the derivation of diagnostic tests that differentiate between different implementations without the explicit enumeration of sub-machines of the FFs. In order to reduce the number of superfluous or infeasible sub-machines of FFs that do not correspond to possible implementations we study how a faulty transition of an EFSM affects transitions of the corresponding FSM.

This paper is organized as follows. Section 2 includes necessary preliminaries needed for understanding the diagnostic problem introduced in Section 3 and solved in subsequent sections. Section 6 concludes the paper and includes some insights for future work.

2 Preliminaries

A non-deterministic finite state machine (FSM) is an initialized non-deterministic complete machine that can be formally defined as follows [7]. A *non-deterministic finite state machine* A is a 5-tuple $\langle S, I, O, h, s_0 \rangle$, where S is a finite nonempty set of states with s_0 as the initial state; I and O are finite nonempty sets of inputs and outputs, and $h: S \times I \rightarrow 2^{S \times O} \setminus \varnothing$ is a behavior function where $2^{S \times O}$ is the set of all subsets of the set $S \times O$. The behavior function defines the possible transitions of the machine. Given present state s_i and input symbol i, each pair $(s_j, o) \in h(s_i, i)$ represents a possible transition to the next state s_j with the output o. This is also written as a transition of the form $s_i \xrightarrow{i/o} s_j$. If for each pair $si \in S \times I$ it holds that $|h(s,i)| = 1$ then FSM A is said to be *deterministic*. In the deterministic FSM A instead of behavior function h we use two functions, transition function $\delta: S \times I \rightarrow S$ and output function $\lambda: S \times I \rightarrow O$. FSM $B = (S', I, O, g, s_0)$, $S' \subseteq S$, is a *sub-machine* of A if for each pair $si \in S' \times I$, it holds that $g(s,i) \subseteq h(s,i)$.

Sometimes a behavior of an FSM is not defined for some pairs $si \in S \times I$. In this case, the transition is said to be *undefined* at state s under input i. An FSM where some transitions are undefined is called *partial*. If a partial FSM has a complete submachine, i.e. a submachine where each transition is defined, then the FSM has the largest complete submachine [3]. The latter can be obtained by iteratively deleting states where at least one transition is undefined. If the initial state has an undefined transition then the FSM has no complete submachine. Otherwise, the procedure is terminated when no state can be deleted.

As usual, function h can be extended to the set I^* of finite input sequences. Given state $s \in S$ and input sequence $\alpha = i_1 i_2 \ldots i_k \in I^*$, output sequence $o_1 o_2 \ldots o_k \in h(s, \alpha)$ if there exist states $s_1 = s, s_2, \ldots, s_k, s_{k+1}$ such that $(s_{j+1}, o_j) \in h(s_j, i_j)$, $j = 1, \ldots, k$. Input/Output sequence $i_1 o_1 i_2 o_2 \ldots i_k o_k$ is called a *trace* of A if $o_1 o_2 \ldots o_k \in h(s, i_1 i_2 \ldots i_k)$. We let the set $h^o(s, \alpha) = \{\gamma | \exists s' \in S \ [(s', \gamma) \in h(s, \alpha)]\}$ denote the *output projection* of h, while denoting $h^s(s, \alpha) = \{s' | \exists \gamma \in Y^* \ [(s', \gamma) \in h(s, \alpha)]\}$ the *state projection* of h. If for each prefix β of α the set $h^s(s, \beta)$ is a unique state then we say that the state $h^s(s, \alpha)$ is *deterministically reachable* from the initial state via the sequence α.

Given state s of FSM $A = (S,I,O,h,s_0)$ and state t of FSM $B = (T,I,O,g,t_0)$, states s and t are *equivalent*, written $s \cong t$, if for each input sequence $i_1 i_2 ... i_k \in I^*$, it holds that $h^o(s,i_1 i_2...i_k) = g^o(t,i_1 i_2...i_k)$. If states s and t are not equivalent then they are *distinguishable*, written $s \neq t$. A sequence $\alpha \in I^*$ such that $h(s, \alpha) \neq g(t, \alpha)$ is said to *distinguish* states s and t. States s_1 and s_2 of FSM A are said to be *separable* [7] if there exists an input sequence α such that the sets $h^o(s_1,\alpha)$ and $h^o(s_2,\alpha)$ are disjoint. Sequence α is called a *separating* sequence for the states s_1 and s_2.

FSMs $A = (S,I,O,h,s_0)$ and $B = (T,I,O,g,t_0)$ are *equivalent*, written $A \cong B$, if their sets of traces coincide. A sequence α that distinguishes the initial states of non-equivalent FSMs A and B is said to *distinguish* FSMs A and B.

Given a deterministic FSM $A = (S, I, O, \delta, \lambda, s_0)$, a non-deterministic FSM FF defined over the same input and output alphabets is called a *Fault Function* (FF) for A if it includes A as a sub-machine. Fault functions were introduced in [6] to represent in a compact way all implementations of a given FSM with a given type of error. In this case, the set of all deterministic sub-machines of FF is considered as a *fault domain* of A.

A sequence $s_0 \xrightarrow{i_1/o_1} s_1, ..., s_{k-1} \xrightarrow{i_k/o_k} s_k$ of consecutive transitions is called a *path* of FSM A. The path is said to be *deterministic* if it has no transitions with different next states and/or outputs for the same state-input combination. In other words, for any $j, p < k$ it holds: if $s_j = s_p$ and $i_{j+1} = i_{p+1}$ then $o_{j+1} = o_{p+1}$ and $s_{j+1} = s_{p+1}$. In this paper, we are interested in deterministic sub-machines of a given FSM. Since a path of a deterministic sub-machine is always deterministic we consider only deterministic paths of a given FSM. Moreover, a deterministic sub-machine has a family of deterministic paths if and only if these paths are *deterministically compatible*, i.e. for any two paths of the family there are no transitions with different next states and/or outputs for the same state-input combination.

Given two FSMs $A = \langle S, I, O, h, s_0 \rangle$ and $B = \langle T, I, O, g, t_0 \rangle$, the intersection $A \cap B$ is the largest initially connected sub-machine of FSM $\langle S \times T, I, O, H, s_0 t_0 \rangle$ where for each pair (st, i), $st \in S \times T$, $i \in I$,

$H(st, i) = \{(s't', o) | (s', o) \in h(s, i), (t', o) \in g(s, i)\}$.

The function $H(st, i)$ is undefined if there is no $o \in O$ such that $(s', o) \in h(s, i)$ and $(t', o) \in g(s, i)$ for appropriate $s' \in S$ and $t' \in T$. The intersection represents the set of common output responses of FSMs A and B to each input sequence and generally, the intersection is a partial FSM. It is known [8] that if the intersection has no complete submachine then any two deterministic sub-machines of A and B are non-equivalent. Moreover, in this case, there exists a set of input sequences, a so-called *distinguishing set* [3], such that any two sub-machines of A and B have different sets of output responses to sequences of this set. In [3] an algorithm is given for determining the distinguishing set.

2.1 The EFSM Model

The EFSM model extends the FSM model with input and output parameters, context variables, operations and predicates defined over context variables and input parameters.

Formally [5], an extended finite state machine M is a pair (S,T) of a set of states S and a set of transitions T between states from S, such that each transition $t \in T$ is a tuple (s,x,P,op,y,up,s'), where:

$s,s' \in S$ are the initial and final states of a transition;

$x \in X$ is an input, where X is the set of inputs, and D_{inp-x} is the set of possible input vectors, associated with the input x, i.e. each component of an input vector corresponds to an input parameter associated with x;

$y \in Y$ is output, where Y is the set of outputs, and D_{out-y} is the set of possible output vectors, associated with the output y, i.e. each component of an output vector corresponds to an output parameter associated with y;

P, op, and up are functions, defined over input parameters, and context variables, namely:

$P: D_{inp-x} \times D_V \to \{\text{True, False}\}$ is a predicate, where D_V is a set of context vectors $\mathbf{v}$;

$op: D_{inp-x} \times D_V \to D_{out-y}$ is an output parameter function;

$up: D_{inp-x} \times D_V \to D_V$ is a context update function.

As in [5], we use the following definitions:

Given an input x and an input vector from D_{inp-x}, the pair of input x and vector from D_{inp-x}, is called a *parameterized input*. A sequence of parameterized inputs is called a *parameterized input sequence*. A context vector $\mathbf{v} \in D_V$ is called a *context* of M. A *configuration* of M is a pair of a state s and a context vector $\mathbf{v}$. Given a parameterized input sequence of an EFSM we can calculate the corresponding parameterized output sequence by simulating the behavior of the EFSM under the input sequence starting from the initial state and initial values of the context variables.

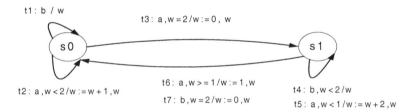

Fig. 1. The EFSM EM_1

Consider the EFSM in Fig 1. It has two states, two non-parameterized inputs a and b and a context variable w. The value of the variable w is an output of the machine. When an input a is applied to the machine at the state s_0 and w is equal to 2, the predicate of the transition t_3 is valid. The machine moves to state s_1, updates w according to action $w:=0$ and produces output 0.

Definition [5]: An EFSM is said to be:
- *Consistent* if for each transition t with input x, every element in $D_{inp-x} \times D_V$ evaluates exactly one predicate to true among all predicates guarding the different transitions with the same start state and input x; in other words, the predicates are mutually exclusive and their disjunction evaluates to true.
- *Completely specified* if for each pair $(s, x) \in S \times X$, there exists one transition leaving state s with input x.

In this paper, we consider a class of specification and implementation EFSMs that are consistent and completely specified. A behavior of such EFSM is defined under each parameterized input sequence. Moreover, for each parameterized input sequence there exists a single parameterized output sequence that is produced by the EFSM for the given input sequence. Two EFSMs are said to be equivalent if their parameterized output responses to each parameterized input sequence coincide.

2.2 Unfolding a Given EFSM to an Equivalent FSM and a Fault Model for EFSMs

When the specification domain of each context variable and input parameter is finite an EFSM can be unfolded to an equivalent FSM by simulating its behavior with respect to all possible values of context variables and input vectors. The equivalence relation means the set of traces of the FSM coincides with the set of parameterized traces of the EFSM. Given a state s of EFSM A, a context vector $\mathbf{v}$, an input x and vector $\mathbf{\rho}$ of input parameters, we derive the transition from configuration $s\mathbf{v}$ under input $x\mathbf{\rho}$ in the corresponding FSM. We first determine the outgoing transition (s,x,P,op,y,up,s') from state s where the predicate P is true for input vector $\mathbf{\rho}$ and context vector $\mathbf{v}$, update the context vector to the vector $\mathbf{v}'$ according to the assignment up of this transition, determine the parameterized output $y\omega$ and add the transition $(s\mathbf{v},x\mathbf{\rho},y\omega,s'\mathbf{v}')$ to the set of transitions of the FSM. The obtained FSM has the same number of states as the number of different configurations $(s,\mathbf{v})$ of the EFSM that are reachable from the initial state. It is known that the simulation can lead to a state explosion problem.

As an example, consider the EFSM EM_1 presented in Fig. 1. At state s_0 two inputs can be applied to the machine; these inputs a and b are not parameterized. The variable w is a context variable and its domain is equal to $\{0,1,2\}$. When a is applied to the machine and the value of context variable w is equal to 0, the machine updates w according to an assignment $w:=w+1$ and moves from the initial configuration $(s_0,0)$ to the configuration $(s_0,1)$ and produces the output $w=1$, because in our example, an output coincides with the value of the context variable w. So, the corresponding FSM has a transition labeled with $a/1$ from the state $(s_0,0)$ to the state $(s_0,1)$. The FSM M_1 that corresponds to the EM_1 of Fig. 1 is shown in a tabular form in Fig. 2.

	$(s_0;0)$	$(s_0;1)$	$(s_0;2)$	$(s_1;0)$	$(s_1;2)$	$(s_1;1)$
a	$(s_0;1)/1$	$(s_0;2)/2$	$(s_1;0)/0$	$(s_1;2)/2$	$(s_0;1)/1$	$(s_0;1)/1$
b	$(s_0;0)/0$	$(s_0;1)/1$	$(s_0;2)/2$	$(s_1;0)/0$	$(s_0;0)/0$	$(s_0;0)/0$

Fig. 2. The FSM M_1 that corresponds to the EFSM EM_1 of Fig. 1

In our example, a reference behavior at the state $(s_1,1)$ is presented in Fig. 2. However, by direct inspection, one can see that this configuration is not reachable from the initial state. We include the configuration in the description, since it becomes reachable if a fault occurs in the EFSM.

It is known that if an EFSM is consistent and completely specified, the corresponding FSM is complete and deterministic. Two EFSMs are equivalent iff their corresponding FSMs are equivalent.

In this paper, we consider a fault model based on transfer, predicate, and assignment faults of a consistent and completely specified EFSM. Consider a transition $t=(s,x,P,op,y,up,s')$ of an implementation EFSM EM. Transition t has a transfer fault if its final state is different from that specified by the reference EFSM, i.e. an implementation machine has a transition (s,x,P,op,y,up,s'') instead of t. Moreover, t has a predicate fault, if the predicate of the transition in the implementation EFSM is different from P, i.e. an implementation EFSM has the transition (s,x,Q,op,y,up,s') instead of t, where $Q \neq P$, i.e. Q and P return different results for some value(s) of input and context vectors. Transition t has an assignment fault if it has an action other than that specified by the reference EFSM. That is after the execution of the wrong transition, the context and/or output vector will have a value different than expected by the reference assignment statement. We note that an implementation with a predicate fault is in general not consistent, unless certain assumptions can be made about the implementation, as explained in subsequent sections.

3 Fault Diagnosis for EFSMs

Let $EM\text{-}RS$ be an EFSM representing the specification of the given system while EM' is its implementation. We denote $M\text{-}RS$ the corresponding FSM. Both specification and implementation EFSMs are complete and consistent. We also assume that the implementation belongs to the finite set of machines $E\Re = \{EM_0=EM\text{-}RS, EM_1,..., EM_l\}$. If the implementation at hand (SUT) say EM' does not pass a given test suite TS, then our objective is to recognize EM' (or identify the set of faulty implementations that are equivalent to EM').

In order to solve the given problem, we work at the FSM level, rather than at the EFSM level. That is, we unfold the given EFSM specification $EM\text{-}RS$ and obtain an FSM $M\text{-}RS$. Due to our assumptions, the implementation at hand is a deterministic complete FSM of the finite set $\Re = \{M_0=M\text{-}RS, M_1,..., M_l\}$, where each FSM M of $\Re$ corresponds to an EFSM EM of $E\Re$. The set $\Re$ is further called the *fault domain* of RS.

A set of input sequences is called a *diagnostic test suite* $Diag_{TS}$ w.r.t. the given fault domain $\Re$ if for each two non-equivalent machines of $\Re$, there exists a sequence that distinguishes them. The set $Diag_{TS}$ can be considered as a distinguishing set of the fault domain $\Re$. When the test cases of $Diag_{TS}$ are applied to the SUT, we can always identify the SUT up to the subset of equivalent machines. The problem is that the number of machines in the fault domain and correspondingly the size of the $Diag_{TS}$ are usually huge. However, to identify a machine of the set $\Re$, it is enough to use its identifier that is a subset of a distinguishing set, i.e. usually is much shorter. Moreover, since we do not know the implementation at hand we have to derive the identifier only on-the-fly, i.e. by the use of an adaptive experiment with the SUT. In this case, the set of diagnostic test cases is not given a priori but is incrementally derived throughout the experiment.

In this paper we use non-deterministic FSMs, called *Fault Functions* (FFs), for the compact representation of the fault domain. A detailed description of these FFs is given in the following subsections.

Let $M\text{-}RS = (S, I, O, \delta, \lambda, s_0)$ be a complete deterministic specification FSM of a given EFSM and FF is a fault function of $M\text{-}RS$. As a consequence of the compact representation, a FF is known to contain *infeasible sub-machines* that do not correspond to possible faulty implementations. In order to reduce the number of infeasible sub-machines we define three FFs for different types of the considered faults instead of a unique FF. Moreover, since a single fault of a given EFSM usually implies multiple faults in the corresponding FSM, for each fault of the EFSM, we determine the set of corresponding transitions of the FSM affected by the fault and we obtain a corresponding partition of input-state combinations of FF. The partition is also used to reduce the number of infeasible sub-machines. We denote *FF-Transfer*, *FF-Predicate*, and *FF-Assignment* these three FFs that we consider for the compact representation of transfer, predicate, and assignment faults. The fault domain is the union of all the deterministic feasible sub-machines of these FFs.

3.1 An Overview of the Diagnostic Approach

Let *RS* be the specification FSM obtained by unfolding the given EFSM, while *TS* is a conformance test suite. If the SUT of the given system produces unexpected output responses to the test suite *TS*, then the SUT is not equivalent to *RS*, i.e. the SUT is a faulty implementation. Our objective is to determine what machine of the fault domain is equivalent to the SUT.

First we derive the three FFs of the specification system, *FF-Predicate*, *FF-Transfer*, and *FF-Assignment*. Here we notice that we use the unfolding procedure only once. The FFs are derived explicitly from the obtained specification FSM based on the types of faults. The set of all deterministic sub-machines of a *FF* includes each implementation FSM that corresponds to the specification EFSM with predicate, transfer, or assignment faults.

Since our implementation system is deterministic we do not take into account non-deterministic paths of FFs. Similar to [1], we first remove from a FF a behavior that does not agree with the observed outputs to the applied test suite *TS*. In Section 4, we describe a novel algorithm that removes from a non-deterministic FSM those sub-machines whose output responses to the test suite do not agree with those obtained by applying the test suite *TS* to the SUT.

Our diagnostic algorithm works under the assumption that the SUT has either predicate, transfer or assignment faults. If the SUT is equivalent to a deterministic sub-machine of a single FF, then there exist corresponding predicate, transfer, or assignment faults, and we try to locate them. However, if the SUT is equivalent to the deterministic sub-machine of two different FFs, or two sub-machines of the same FF, then the faulty machine cannot be uniquely identified. This is due to the fact that there are different possible faults that cause the same observable behavior of the SUT. However, in this case, if needed, we can determine the subset of these possible faults. If none of the above cases occurs, we conclude that the implementation has faults that are not captured by the considered fault model.

In order to draw one of the above conclusions, we should have test cases such that by observing the output responses of the SUT to these test cases, we can distinguish the SUT from other sub-machines of the FFs. If the FFs after deleting sub-machines with the output responses which do not agree with those observed with the conformance test suite, have no equivalent deterministic sub-machines then there exists a distinguishing set of input sequences such that, given the set of output responses to these input sequences, we always can determine a single FF such that the machine under test is a sub-machine of the FF. Otherwise, we generate for the two FFs a set of sequences that distinguish some deterministic sub-machines of these FFs, then we apply these sequences to the SUT and we reduce the FFs based on the observed behavior of the SUT. We repeat the latter process as much as possible. Afterwards, we try to further reduce the remaining FFs, by repeating a process similar to the above, but for each FF alone. Here we note that a single fault in an EFSM can imply several faulty transitions in the corresponding FSM. To derive the FFs we consider each transition of the specification EFSM, insert a corresponding fault and determine the transitions of the corresponding FSM affected by the fault.

A submachine of the *FF-Transfer* (*FF-Predicate* or *FF-Assignment*) is said to be *feasible* if it corresponds to an EFSM that can be obtained from the specification EFSM through transfer (predicate or assignment) faults. In this paper, we define appropriate properties that restrict the fault domain of each FF, and thus can be used to reduce the number of the infeasible sub-machines. For this purpose, for each possible single transfer (assignment or predicate) fault in the EFSM, we determine the cluster of its corresponding faults in the FSM. The clusters obtained, for a particular FF, form a partition of its transitions.

3.2 *FF-Transfer*, *FF-Predicate*, and *FF-Assignment* and Their Properties

A transition of an implementation EFSM EM has a *transfer fault* if its final state is different from that specified by the reference EFSM. Consider a transition $t=(s,x,P,op,y,up,s')$ of EM under input x from state s to the tail state s'. When a transfer fault occurs we have a wrong ending state s''. Therefore, for each configuration $s\mathbf{v}$ of EM and parameterized input $x\mathbf{\rho}$ such that the predicate of the transition t is true for the pair $\mathbf{\rho v}$, i.e. $P(\mathbf{\rho},\mathbf{v})=\text{true}$, and there is a transition $(s\mathbf{v},x\mathbf{\rho},y\mathbf{\omega},s'\mathbf{v'})$ in the FSM corresponding to EM, the *FF-transfer* contains the transition $(s\mathbf{v},x\mathbf{\rho},y\mathbf{\omega},s''\mathbf{v'})$ for each state s'' of the EM.

The set of pairs $(s\mathbf{v},x\mathbf{\rho})$ such that the predicate P_t of the transition t from the state s under input x is true for the pair $\mathbf{\rho v}$ is denoted $Dom(P_t)$ and called "domain of P_t". Since an EFSM is consistent, the set of domains $Dom(P_t)$ over all transitions of the EM is a partition of the set of all pairs "state-input" of the corresponding FSM.

As an example, consider transition t_2 of EM_1 in Fig. 1. Under the input a, the predicate of t_2 is true for the configurations $(s_0,0)$ and $(s_0,1)$, thus the pairs $)s_0 0,a($ and $(s_0 1,a)$ form a $Dom(P_2)$ of the partition. For configuration $(s_0,0)$, there is the transition $(s_0 0,a,1,s_0 1)$ to the configuration $(s_0,1)$. Therefore, we add transition $(s_0 0,a,1,s_1 1)$ to *FF-Transfer*. Moreover, from configuration $(s_0,1)$, there is transition $(s_0 1,a,2,s_0 2)$ to the configuration $(s_0,2)$. Therefore, we add transition $(s_0 1,a,2,s_1 2)$ to *FF-transfer*. Due to our restrictions, from configurations $(s_0,0)$ and $(s_0,1)$ and under the input a, in any feasible submachine there can only be transitions to either $(s_0,1)$ and $(s_0,2)$ or to $(s_1,1)$ and $(s_1,2)$. We consider each transition of the EM_1 in Fig. 1 and obtain the *FF-Transfer* shown in Fig. 3.

Proposition 1. A submachine of *FF-transfer* is *feasible* only if for each two transitions of the same predicate domain the state parts of the next configurations coincide.

	$(s_0;0)$	$(s_0;1)$	$(s_0;2)$	$(s_1;0)$	$(s_1;2)$	$(s_1;1)$	Domains
A	$(s_0;1)/1$ $(s_1;1)/1$	$(s_0;2)/2$ $(s_1;2)/2$	$(s_1;0)/0$ $(s_0;0)/0$	$(s_1;2)/2$ $(s_0;2)/2$ $(s_1;1)/1$	$(s_0;1)/1$ $(s_1;1)/1$	$(s_0;1)/1$ $(s_1;1)/1$	$\{(s_00a),$ $(s_01a)\}$ $\{(s_00a)\}$ $\{(s_02a)\}$ $\{(s_12a),$ $(s_11a)\}$
B	$(s_0;0)/0$ $(s_1;0)/0$	$(s_0;1)/1$ $(s_1;1)/1$	$(s_0;2)/2$ $(s_1;2)/2$ $(s_0;0)/0$	$(s_1;0)/0$ $(s_1;0)/0$	$(s_0;0)/0$	$(s_1;1)/1$ $(s_1;1)/1$	$\{(s_00b),$ $(s_01b),$ $(s_02b)\}$ $\{(s_00b),$ $(s_11b)\}$ $\{(s_12b)\}$

Fig. 3. The *FF-Transfer* that corresponds to transfer faults of EM_1 of Fig. 1

Transition $t = (s,x,P,op,y,up,s')$ of a specification EFSM *EM* has a *predicate fault*, if the predicate of the transition of an implementation EFSM is different from P, i.e. an implementation EFSM has the transition (s,x,Q,op,y,up,s') instead of t. Since an implementation EFSM *EM* is consistent, the subset of $D_{inp-x} \times D_V$ that evaluates the faulty predicate Q to true has to be a subset of that of the reference predicate, i.e. $Dom(Q) \subseteq Dom(P)$. As usual, in order to keep an implementation EFSM complete, we use the completeness assumption. For any item of the set $D_{inp-x} \times D_V$ where P is true while Q is false, the implementation EFSM has a loop at the state s with the identity update function up_{id} and the *Null* output (written as '-') that has no parameters. In other words, the implementation EFSM remains at the current state s and does not execute the specified action up; therefore, the context vector is not changed. Moreover, the implementation EFSM produces the *Null* output, i.e it does not produce any output.

Thus, if the predicate fault occurs for the transition $t = (s,x,P,op,y,up,s')$ of a specification EFSM *EM* where the predicate P is changed to predicate Q, then a corresponding implementation EFSM has transitions (s,x,Q,op,y,up,s') and $(s,x,R,-,up_{id},s)$ where the predicate R is true for each item of the set $D_{inp-x} \times D_V$ where P is true while Q is false,. Therefore, the FSM corresponding to the faulty implementation has transitions $(sv,x\rho,-,sv)$ for each pair $(sv,x\rho) \in Dom(R)=Dom(P)\backslash Dom(Q)$.

Thus, when deriving *FF-Predicate*, for each transition $(sv,x\rho,y\omega,s'v')$ of the FSM obtained from the transition $t = (s,x,P,op,y,up,s')$ of the specification EFSM we add the transition $(sv,x\rho,-,sv)$ for each $(sv,x\rho) \in Dom(P)$.

As an example, consider transition t_2 of EM_1 in Fig. 1. Under the input a, the predicate of t_2 is true for the configurations $(s_0,0)$ and $(s_0,1)$. For configuration $(s_0,0)$, there is transition $(s_00,a,1,s_01)$. Therefore, we add transition $(s_00,a,-,s_00)$ to *FF-Predicate*. For configuration $(s_0,1)$, there is transition $(s_01,a,2,s_02)$. Therefore, we add transition $(s_01,a,-,s_01)$. We consider each transition of EM_1 in Fig. 1 and obtain *FF-Predicate* shown in Fig. 4.

A transition of an implementation EFSM *EM* has an *assignment fault* if it has an action other than that specified by the transition. Consider a transition $t = (s,x,P,op,y,up,s')$ of the specification EFSM *EM*. If t has an assignment fault, then after the execution of t, the context vector can be any vector of the set D_V except that specified by t. Thus, the corresponding FSM has a transition $(s\mathbf{v},x\rho,y\omega´,s´\mathbf{v}´)$ for each pair $(s\mathbf{v}, x\rho) \in Dom(P)$ and each $\mathbf{v}´\in D_V$ and $\omega´ \in D_{out\text{-}y}$ [1].

	$(s_0;0)$	$(s_0;1)$	$(s_0;2)$	$(s_1;0)$	$(s_1;2)$	$(s_1;1)$
a	$(s_0;1)/1$ $(s_0;0)/\text{-}$	$(s_0;2)/2$ $(s_0;1)/\text{-}$	$(s_1;0)/0$ $(s_0;2)/\text{-}$	$(s_1;2)/2$ $(s_1;0)/\text{-}$	$(s_0;1)/1$ $(s_1;2)/\text{-}$	$(s_0;1)/1$ $(s_1;1)/\text{-}$
b	$(s_0;0)/0$	$(s_0;1)/1$	$(s_0;2)/2$	$(s_1;0)/0$ $(s_1;0)/\text{-}$	$(s_0;0)/0$ $(s_1;2)/\text{-}$	$(s_1;1)/1$ $(s_1;1)/\text{-}$

Fig. 4. The *FF-Predicate* that corresponds to the predicate faults of EM_1 of Fig. 1

Thus, when deriving the *FF-Assignment*, for each transition $(s\mathbf{v},x\rho,y\omega´,s´\mathbf{v}´)$ of the FSM corresponding to the specification EFSM we include into *FF-Assignment* a transition $(s\mathbf{v},x\rho,y\omega´,s\mathbf{v}´)$ for each possible context vector $\mathbf{v}´$ and each possible output vector $\omega´$. Similar to the *FF-transfer*, the set of domains $Dom(P_t)$ over all transitions of the *EM* is a partition of the set of all pairs "state-input" of the corresponding FSM.

As an example, consider transition t_2 of EM_1 in Fig 1. Under the input a, the predicate of t_2 is true for the configurations $(s_0,0)$ and $(s_0,1)$. For configuration $(s_0,0)$, $w=up(0)=1$ and other possible values of w are 0 and 2. Thus, we add transitions $(s_0 0,a,0,s_0 0)$ and $(s_0 0,a,2,s_0 2)$ to *FF-assignment*. Moreover, for $(s_0,1)$, $w = up(1)=2$ and other possible values of w are 0 and 1. Thus, we add transitions $(s_0 1,a,0,s_0 0)$ and $(s_0 1,a,1,s_0 1)$. Domain $Dom(P_2)$ comprises the pairs $(s_0 0,a)$ and $(s_0 1,a)$. We consider each transition of EM_1 of Fig. 1 and obtain *FF-Assignment* shown as in Fig. 5 below.

	$(s_0;0)$	$(s_0;1)$	$(s_0;2)$	$(s_1;0)$	$(s_1;2)$	$(s_1;1)$
a	$(s_0;1)/1$ $(s_0;0)/0$ $(s_0;2)/2$	$(s_0;2)/2$ $(s_0;1)/1$ $(s_0;0)/0$	$(s_1;0)/0$ $(s_1;1)/1$ $(s_1;2)/2$	$(s_1;2)/2$ $(s_1;0)/0$ $(s_1;1)/1$	$(s_0;1)/1$ $(s_0;0)/0$ $(s_0;2)/2$	$(s_1;1)/1$ $(s_0;0)/0$ $(s_0;2)/2$
b	$(s_0;0)/0$	$(s_0;1)/1$	$(s_0;2)/2$	$(s_1;0)/0$	$(s_0;0)/0$ $(s_0;1)/1$ $(s_0;2)/2$	$(s_1;1)/1$

Fig. 5. The *FF-Assignment* that corresponds to assignment faults of EM_1 of Fig.1

Proposition 2. A submachine of *FF-assignment* is *feasible* only if for each two transitions of the same predicate domain the context parts of the next configurations and the output vectors are updated in the same way.

4 Removing Sub-machines from a Given Fault Function

In this section, we present a new method for reducing the FFs. This is done by removing from each FFs those sub-machines (i.e. possible faulty implementations) whose output responses to the test cases of the given *TS* disagree with those observed by applying these test cases to the SUT.

---

[1] Here we assume the specification domain of each output parameter is finite.

Algorithm 1. Removing from FSM A those sub-machines that do not match the set V of deterministic I/O sequences

Input: A non-deterministic FSM $A = (S, I, O, h, s_0)$ representing a given *FF*, and a set V of deterministic sequences over alphabet $(IO)^*$.

Output: The smallest sub-FSM $A' = (S, I, O, h, s_0)$ of A, that contains all sub-machines of A whose traces include V, if such an A' exists.

Step-1. Given FSM A and the set V of deterministic sequences over alphabet $(IO)^*$, we derive, using A and the sequences of V, the set of all families of deterministically compatible paths labelled with the sequences of V that satisfy the feasibility requirements of the FF A. Each family is represented as a path in a tree *Tree*. We build *Tree* as follows: Starting from the initial state s_0 of A (the root node of *Tree*), we select a sequence from V and derive all deterministic paths of FSM A that agree with the selected sequence and satisfy the feasibility requirements of A. We then, select a new sequence from V, and instead of deriving all deterministic paths for the new sequence starting again from the root node of *Tree*, as done in [1], we start from the leaf node of each path that is associated with the initial state s_0. Moreover, each new path has to be deterministically compatible with the path it appends. If a family of deterministically compatible paths does not exist for a given set V, then there is no sub-machine in A that has V as a subset of its set of traces. In this case, A' does not exist; end of Algorithm 1.

Step-2. For each deterministic path $Path_j$ of the *Tree*, we derive the corresponding FSM B_j by copying the transitions of this path. Moreover, for each pair (s,i) such that the path has no transition from state s under input i, we copy into B_j all corresponding transitions from the original FF A. Then, we obtain the FSM A' as the union of all FSMs B_j over all deterministic paths of *Tree*.

Step-3: Remove from the obtained machine A' all states that are not reachable from the initial state of A'.

As an application example, consider the *FF-Transfer* shown in Fig. 3, and the set $V=\{a/1a/2a/0a/2a/2, b/0a/1\}$ of I/O sequences of a given SUT. At Step-1, using the observed behavior of the SUT and the set V, we derive a Tree of all deterministic compatible paths that satisfy the *FF-Transfer* feasibility constraints.

For example, according to *FF-Transfer*, if t_2 has a transfer fault, then from the root node $(s_0,0)$ and under input a, the machine moves to state $(s_1,1)$ and produces the output 1. However, we do not consider the paths from the ending node $(s_1,1)$ since the observed output of the SUT after applying the second input symbol a is 2, while according to *FF-Transfer*, any submachine at state $(s_1,1)$ produces 1 to the input a. For the same reason, after obtaining the output 1202 to the input sequence $aaaa$, we do not consider the paths from the ending node of the transition $(s_0,2)$-$a/0$->$(s_0,0)$. Finally, we observe that according to *FF-Transfer*, a sub-machine of the FF that produces 1202 to the input sequence $aaaa$ can be either at state $(s_1,2)$ or $(s_0,2)$. At each of these states only the output 1 can be produced for the next input a, while due to the observed behavior the SUT produces the output 2. Thus, we eliminate *FF-Transfer* since there is no sub-machine in *FF-Transfer* that has V as a subset of its traces, i.e. the SUT has no transfer faults. We apply Algorithm 1 to *FF-Predicate* and assure the SUT has no predicate faults. It is therefore expected to have assignment fault(s).

Then, we apply Algorithm 1 to *FF-Assignment* of Figure 5 and we obtain the machine shown in Fig. 6 below. Now we locate the faulty transition $(s_1,2,a,2,s_0,2)$ that

	$(s_0;0)$	$(s_0;1)$	$(s_0;2)$	$(s_1;0)$	$(s_1;2)$
a	$(s_0;1)/1$	$(s_0;2)/2$	$(s_1;0)/0$	$(s_1;2)/2$	$(s_1;2)/2$
b	$(s_0;0)/0$	$(s_0;1)/1$	$(s_0;2)/2$	$(s_1;0)/0$	$(s_0;0)/0$ $(s_0;1)/1$ $(s_0;2)/2$

Fig. 6. Machine obtained after applying Algorithm 1 to *FF-Assignment* of Fig. 5

corresponds to the assignment fault of transition t_6 of the reference EFSM. The context variable is updated to $w:=2$ instead of $w:=1$.

5 Other Steps of the Diagnostic Method

In this section we derive an identifier of a given SUT based on an adaptive experiment with the SUT. We refer to an implementation identifier as an *adaptive diagnostic test suite* and we propose a method for the derivation of an identifier of a given SUT. We recall that the identifier can be derived up to the set of equivalent implementations that are equivalent to the SUT. In this case, we know that the fault corresponds to one of remaining equivalent implementations.

According to [3], if there exists a distinguishing set, say *DIS*, for two given FFs, say *FF-predicate* and *FF-assignment*, then after applying the sequences of *DIS* to the SUT and observing corresponding output responses, we can always determine if the SUT is a submachine of *FF-predicate* or *FF-assignment*, or if it is not a submachine of any of them. Therefore, if the machines *FF-predicate* and *FF-transfer* and *FF-assignment* are pair-wise distinguishable we can always identify the type of the fault.

However, we do not need to derive a complete distinguishing set based on the intersection of two FFs. Instead, we state simpler sufficient conditions for the derivation of test cases that allow us to reduce two given FFs. These conditions are also applicable when the two FFs have no distinguishing set.

Theorem 1. Given two FFs FF_1 and FF_2, let state *st* be a deterministically reachable state of the intersection $FF_1 \cap FF_2$ through an input sequence β, such that the sets of outputs of FF_1 and FF_2 at states *s* and *t* under some input *a* do not coincide. Then after observing the output response of the SUT to the last input symbol *a* of $r.\beta.a$ we can delete from FF_1 and FF_2 the outgoing transitions of states *s* and *t* with inputs *a* and outputs that do not match the observed output of the SUT. If for *s* or *t* all outgoing transitions with an appropriate input label *a* are removed we can delete FF_1 or FF_2.

Hereafter, we assume that the type of a faulty transition is already identified, i.e. only one fault function FF remains. Now, to locate a faulty sub-machine (implementation) of the remaining FF, we distinguish between the different deterministic sub-machines of FF. This can be done by deriving input sequences as described in the following theorem. These sequences then are applied to the SUT in order to reduce the FF according to the observed behavior. Afterwards, if only a single implementation remains, then the fault is uniquely located, else if a number of equivalent implementations remain, then we can locate the fault up to the set of equivalent implementations (equivalent faults).

Theorem 2. Given an *FF*, let state *s* of *FF* be deterministically reachable through an input sequence β. If there exists an input *a* such that *FF* has at least two different output responses to *a* then after observing the response of the SUT to the last input symbol a of $r.\beta.a$ we can delete the outgoing transitions of s with input labels a and outputs different than the observed one. Otherwise, if there exist some outgoing transitions of s labeled with the same input/output a/o such that in FF the destination states of these transitions are separable by the sequence γ, then after observing the output responses of the SUT to the sequence γ of $r.\beta.a.\gamma$ we can delete the outgoing transitions of s that are labeled with a/o and have destination states with outputs to γ different than the observed ones.

As an application example, we consider the *FF-assignment* of Figure 6 obtained after applying Algorithm 1. The state $(s_1,2)$ is deterministically reachable through the I/O sequence $a/1a/2a/0a/2$. Moreover, there are different output responses to the input b at state $(s_1,2)$. Therefore, the second fault can be identified after applying the input sequence $aaaa.b$ to the SUT. If the output response of the SUT to the last input symbol of this sequence is 2, then the second faulty transition is $(s_1 2,b,2,s_0 2)$. This transition corresponds to the assignment fault of t_7, where the context variable is updated as $w:=2$ instead of $w:=0$.

We note that if the FFs cannot be further reduced based on the above two theorems, then similar to [1], we suggest to divide these FFs into several FFs by fixing some of its transitions as deterministic transitions. In the worst case, we may need to explicitly enumerate all deterministic sub-machines of a given FF. However, our preliminary experiments show that the FFs obtained after applying Algorithm 1 are almost deterministic.

6 Conclusions

In this paper we have proposed an original method for the fault diagnosis in Extended Finite State Machines (EFSMs). The method assumes that an implementation EFSM is complete, consistent and it can have either predicate, transfer, or assignment faults. However, our ability to locate the fault is known to essentially depend on the observability (i.e. outputs produced) of the implementation EFSM. In software implementations, it is easy to increase the observability by reading some internal variables (eg. some context and state variables). Currently we are investigating the problem of determining the minimum number of variables that we need to observe in order to locate the faulty transitions. Moreover, we are adapting the method for inconsistent implementation EFSMs and we are experimenting with the method to assess its applicability to realistic application examples.

References

1. El-Fakih, K., Yevtushenko, N., Bochmann, G.: Diagnosing Multiple Faults in Communicating Finite State Machines. In Proc. of the IFIP 21st FORTE, Korea (2001) 85-100
2. Ghedamsi, A. Bochmann. G.: Test Result Analysis and Diagnostics for Finite State Machines. Proc. of the 12-th ICDS, Yokohama Japan (1992)
3. Koufareva, I.: Using Non-Deterministic FSMs for Test Suite Derivation. Ph.D. Thesis, Tomsk State University, Russia (2000)

4. Lee, D., Sabnani, K.: Reverse Engineering of Communication Protocols. Proc. of ICNP, October (1993) 208-216
5. Petrenko, A., Boroday, S., Groz, R.: Confirming Configurations in EFSM'. Proc. of the IFIP joint conference on FORTE XII and PSTV XIX, China (1999)
6. Petrenko, A., Yevtushenko, N.: Test Suite Generation for a FSM with a Given Type of Implementation Errors. Proc. of the 12th Int. Workshop Protocol Specification, Testing and Verification, (1992)
7. Starke, P.: Abstract Automata. North-Holland/American Elsevier (1972)
8. Yevtushenko, N., Koufareva, I.: Relations Between Non-Deterministic FSMs. Tomsk, Spectrum (2001)

A Guided Method for Testing Timed Input Output Automata

Abdeslam En-Nouaary and Rachida Dssouli

Department of Electrical and Computer Engineering
Concordia University, 1455 de Maisonneuve W., Montréal
Québec H3G 1M8, Canada
{ennouaar,dssouli}@ece.concordia.ca

Abstract. Real-time systems are those systems whose behaviors are time dependent. Reliability is one of the characteristics of such systems and testing is one of the techniques that can be used to ensure reliable real-time systems. This paper presents a method for testing real-time systems specified by Timed Input Output Automata (TIOA). Our method is based on the concept of test purposes. The use of test purposes helps reduce the number of test cases generated since an exhaustive testing of a TIOA causes the well-known state explosion problem. The approach we present in this paper consists of three main steps. First, a synchronous product of the specification and test purpose is computed. Then, a sub-automaton (called Grid Automata) representing a subset of the state space of this product is derived. Finally, test cases are generated from the resulting grid automata. The test cases generated by our method are executable and can easily be represented in TTCN (Tabular Tree Combined Notation).

Keywords: Real-Time Systems, Timed Input Output Automata, Testing, Test Purposes.

1 Introduction

Testing plays a key role in software life cycles. It consists of executing a physical implementation of a computer system with the intention of finding and discovering errors. This is done by submitting a set of test cases (also called test suite) to the implementation and observing its reactions. If the outputs of the implementation for a test case do not match those derived from the specification, the implementation is said faulty (i.e., a fault is detected). A test case is a sequence of input actions allowed by the environment. The test cases we apply to the implementation of a system are systematically generated from the formal specification of that system. A test cases generation algorithm should be practical in the sense that it must derive few test cases while ensuring good fault coverage. The term fault coverage refers to the ability of a test cases generation method to detect the potential faults in the implementation under test.

Over the last three decades, many algorithms have been developed for testing untimed specification models such as Finite State Machines (FSMs) and

Extended FSMs (EFSMs). However, testing real-time systems is still a new research field since researchers have started investigating the issue only at the mid-nineties. Even some algorithms have been devised for testing real-time systems (see for example [ENDKE98,SVD01,MMM95,CL97,COG98,KAD+00, KENDA00], [FAUD00,SPF01,KLC98,HNTC99,NS98,ENDK02,EN02,Hog01]), most of these methods suffer from the state space explosion problem and generate a great number of test cases. So, the necessity for the development of new techniques that are practical and with good fault coverage still exists.

In this paper, we present a framework for testing real-time systems using test purposes. A test purpose is a precise representation of the functionality to be tested. Thus, test purposes allow us to reduce the number of test cases generated and incrementally carry out the testing process. The formal model we use to describe both the specification and test purposes is *Timed Input Output Automaton (TIOA)* [AD94,NSY92,LA92]. To generate test cases from TIOA, our approach proceeds in three steps. First, a synchronous product of the specification and test purpose is computed. Then, an automaton representing a subset of the state space of this product is constructed. Finally, test cases are derived from the resulting grid automata.

The remainder of this paper is structured as follows. Section 2 introduces the TIOA model and the test purpose concept as well as the theoretical results needed for the rest of the paper. Section 3 presents our approach for timed test cases generation. Section 4 discusses the results and concludes the paper.

2 Backgrounds

This section presents the TIOA model and the test purpose concept as well as the theoretical ingredients we need for the subsequent sections. All these concepts and results are illustrated with simple examples so that the reader later later understands each step of our approach.

Definition 1. *Timed Input Output Automaton*
A TIOA A is a tuple $(I_A, O_A, L_A, l_A^0, C_A, T_A)$, where:

- I_A is a finite set of input actions. Each input action begins with "?".
- O_A is a finite set of output actions. Each output action begins with "!".
- L_A is a finite set of locations.
- $l_A^0 \in L_A$ is the initial location.
- C_A is a finite set of clocks all initialized to zero in l_A^0.
- $T_A \subseteq L_A \times (I_A \cup O_A) \times \Phi(C_A) \times 2^{C_A} \times L_A$ is the set of transitions.

A transition in a TIOA, denoted by $l \xrightarrow{\{?,!\}a,G,\lambda}_A l'$, consists of a source location l, an input or output action $\{?,!\}a$, a clock guard G, which should hold to execute the transition, a set of clocks λ to be reset when the transition is fired, and a target location l'. We assume that the transitions are instantaneous and the clock guards over C_A are conjunctions of formulas of the form $x \; op \; m$, where: $x \in C_A$, $op \in \{<, \leq, =, >, \geq\}$ and m is a natural. Moreover, we suppose that

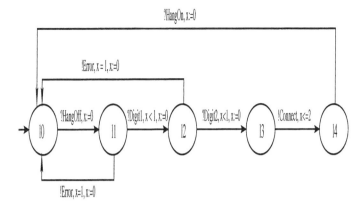

Fig. 1. An Example of TIOA.

each clock $x \in C_A$ has a bounded domain $[0, B_x] \cup \{\infty\}$ [SV96], where B_x is the largest integer constraint appearing in the constraints over x in the automaton. This means that each clock x is relevant only under the integer constant B_x, and all the values of x greater than B_x are represented by ∞.

Example 1. Figure 1 shows an TIOA with one clock. It is a specification of an hypothetical telephone system somewhat similar to that presented in [CL97]. The task of the telephone system is to issue an output *Connect* whenever a user hangs off (input $Hangoff$ on the Figure) and composes two digits ($Digit1$ and $Digit2$ on the Figure) forming the number to be called. After the connection is established, the user hangs on and the system goes back to its idle state and starts waiting for another connection request. The behavior of the system should respect the following time constraints. First, the user should type the first digit within 1 time-unit after having hanged off. Moreover, the amounts of time separating $Digit1$ and $Digit2$ must be no more than 1 time-unit. Finally, the system must respond with *Connect* within 2 time-units after the last digit has been typed. Whenever an input's time constraint is not respected by the user, the system times out, issues *Error* and goes back to its idle state.

The TIOA introduced so far is an abstract model because it doesn't explicit all the possible executions. Such executions, called the operational semantics, can informally stated as follows. The TIOA starts at its initial location with all clocks initialized to zero. Then, the values of clocks increase synchronously and measure the amount of time elapsed since the last initialization or reset. At any time, the TIOA can make a transition $l \xrightarrow{\{?,!\}a,G,\lambda} l'$ provided that the current location is l and the values of clocks satisfy the clock guard G. In this case, all the clocks in λ are reset and the TIOA changes its location to l'. To formalize the operational semantics of TIOA, we need the following definitions.

Definition 2. *Clock valuation*
Let $A = (I_A, O_A, L_A, l_A^0, C_A, T_A)$ be a n−clocks TIOA (i.e., a TIOA with n clocks).

- A clock valuation of A (or over C_A) is an application $v : C_A \to [\mathbf{R} \cup \{\infty\}]^n$, which assigns a non-negative real number or ∞ to each clock $x \in C_A$. We represent a clock valuation by a vector $(v_{x_1}, v_{x_2}, .., v_{x_n})$ and denote the set of all clock valuations by $V(C_A)$.
- For any clock valuation $v \in V(C_A)$ and any non-negative real number d, $v + d$ is also a clock valuation that assigns the value $v(x) + d$ to each clock $x \in C_A$. $v + d$ is the clock valuation reached from v by letting time elapses by d time units.
- For any clock valuation $v \in V(C_A)$ and any subset of clocks $X \in C_A$, $[X := 0]v$ is also a clock valuation that assigns the value 0 to each clock $x \in X$ and agrees with v on the rest of clocks. $[X := 0]v$ is the clock valuation obtained from v by resetting clocks X.
- A clock valuation $v \in V(C_A)$ satisfies a clock guard G, denoted by $v \models G$, if and only if G holds under v.

Definition 3. *States of TIOA*
Let $A = (I_A, O_A, L_A, l_A^0, C_A, T_A)$ be a TIOA.

- A state of A is a pair (l, v) consisting of a location $l \in L_A$ and a clock valuation $v \in V(C_A)$.
- The initial state of A is the pair (l_A^0, v_0), where $v_0(x) = 0$ for each clock $x \in C_A$. We denote the set of states of A by $S(A)$.

The operational semantics of a TIOA A is formally given by a timed labeled transition system, called the regions graph. The latter is constructed using the equivalence relation $\sim$ [AD94] on the set of clock valuations $V(C_A)$.

Definition 4. *Equivalence between Clock Valuations*
Let $A = (I_A, O_A, L_A, l_A^0, C_A, T_A)$ be a TIOA, and v and $v' \in V(C_A)$. We say v and v' are equivalent, written $v \sim v'$, iff:

- $\forall x \in C_A, \lfloor v(x) \rfloor = \lfloor v'(x) \rfloor$
- $\forall x, y \in C_A$ such that $((v(x) \neq \infty) \wedge (v(y) \neq \infty)), (fract(v(x)) \leq fract(v(y)) \Leftrightarrow fract(v'(x)) \leq fract(v'(y)))$
- $\forall x \in C_A$ such that $v(x) \neq \infty, (fract(v(x)) = 0 \Leftrightarrow fract(v'(x)) = 0)$
  Here, $\lfloor t \rfloor$ and $fract(t)$ denote the integer and fractional parts of t respectively.

Definition 5. *Clock region*
Let $A = (I_A, O_A, L_A, l_A^0, C_A, T_A)$ be a TIOA. A clock region of A (or over C_A) is an equivalence class generated by the relation $\sim$ given in definition 4. We denote the clock region of a clock valuation v by $[v]$ and the set of all clock regions of A by $Reg(A)$.

Example 2. The set of regions for the TIOA of Figure 1 is given in Figure 2. For instance, the clock valuations $v_1 = \frac{1}{2}$ and $v_2 = \frac{1}{10}$ have the same behaviors when time progresses and so are equivalent (i.e., they belong to the same region). This means that if a state (l, v_2) accepts a trace then the state (l, v_1) also does.

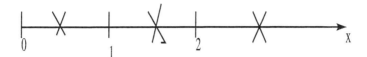

Fig. 2. An Example of Clock Regions.

Table 1. Number of Clock Regions in TIOA.

Formula	1-clock	2-clocks	3-clocks	4-clocks
[AD94]	8	128	3072	98304
[EEN98]	4	13	88	474

An upper bound of the number of regions in a $n-$clocks TIOA A has been given in [AD94]. However, the exact number of these regions is given in [EEN98]. Table 1 gives the number of regions resulting from applying both formulas to three TIOAs with different numbers of clocks having all the same domain $[0,1] \cup \{\infty\}$.

An important property of clock valuations is that each clock valuation can be represented by an equivalent one with all coordinates having the form $\frac{m}{n+1}$, where m is a non-negative integer and n is the number of clocks in TIOA. The following proposition formalizes this property.

Proposition 1. *Relationships between Clock Valuations*
Let $A = (I_A, O_A, L_A, l_A^0, C_A, T_A)$ be a $n-$clock TIOA.
For all $v \in V(C_A)$, there exists non-negative integers, $m_1, m_2, .., m_n$, and $v' \in V(C_A)$ such that $v' = (\frac{m_1}{n+1}, \frac{m_2}{n+1}, .., \frac{m_n}{n+1})$ and $v' \sim v$.

Proof. For the sake of space, the proof is given in the technical report of this paper.

The following proposition generalizes the result of proposition 1 to clock regions in a TIOA.

Proposition 2. *Characterization of Clock Regions*
Let $A = (I_A, O_A, L_A, l_A^0, C_A, T_A)$ be a $n-$clock TIOA.
For every $R \in Reg(A)$, there exists at least one clock valuation $v = (\frac{m_1}{n+1}, \frac{m_2}{n+1}, .., \frac{m_n}{n+1})$ ($m_1, m_2, .., m_n$ are non-negative integers) that characterizes R (i.e., $[v] = R$).

Proof. For the sake of space, the proof is given in the technical report of this paper.

Definition 6. Regions Graph
Let $A = (I_A, O_A, L_A, l_A^0, C_A, T_A)$ be a TIOA. The regions graph of A is an automaton $RG = (\Sigma_{RG}, S_{RG}, s_{RG}^0, T_{RG})$ where:

- $\Sigma_{RG} = I_A \cup O_A \cup \mathbf{R}^{>0}$,
- $S_{RG} = \{\langle l, [v] \rangle \mid l \in L_A \wedge v \in V(C_A)\}$,

- $s_{RG}^0 = \langle l_A^0, [v_0] \rangle$, where $v_0(x) = 0$ for all $x \in C_A$,
- RG has a transition $s \xrightarrow{\{?,!\}a} s'$, from $s = \langle l, [v] \rangle$ to $s' = \langle l', [v'] \rangle$ on action $\{?,!\}a$ iff there is a transition $\xrightarrow{\{?,!\}a,G,\lambda}_A l'$ such that $v \models G$ and $v' = [\lambda := 0]v$.
- RG has a delay transition $s \xrightarrow{d} s'$, from $s = \langle l, [v] \rangle$ to $s' = \langle l, [v'] \rangle$ on time increment $d > 0$, iff $[v'] = [v + d]$.

Since the regions graph can be seen as a timed reachability analysis graph of TIOA, it is heavily used for the verification and testing of timed dependant systems. Moreover, it is clear from the definition 6 above that each state of the regions graph has a delay transition labeled with the symbol d. Here, the value of d is in the interval $]0, 1[$. The delay transitions on d greater than or equal to 1 are obtained using the following rule: if $s_1 \xrightarrow{d_1} s_2$ and $s_2 \xrightarrow{d_2} s_3$ then $s_1 \xrightarrow{d_1+d_2} s_3$.

As the symbol d in the definition 6 takes an infinite number of values between 0 and 1, we can use propositions 1 and 2 to instantiate the delay transitions with the same value and obtain a finite sub-automaton of the regions graph. This operation is termed *sampling the regions graph* and the resulting sub-automaton is called *Grid Automaton (GA)*[LY93,ENDKE98,SVD01,ENDK02]. The following proposition formalizes the result.

Proposition 3. *Sampling the Region Graph*
Let $A = (I_A, O_A, L_A, l_A^0, C_A, T_A)$ be a $n-$clock TIOA.
There exists a sub-automaton of the regions graph of A with all delay transitions labeled with the same delay $\frac{1}{n+1}$.

Proof. For the sake of space, the proof is given in the technical report of this paper.

After introducing the TIOA model and all theoretical results we need for subsequent sections, we now define the concept of test purpose [ISO91,Tre92,KLC98], [KC00,GHN93,BGRS01,SPF01] that plays a key role in our contribution.

Definition 7. Timed Test Purpose
A timed test purpose TP is an acyclic TIOA $(I_{TP}, O_{TP}, L_{TP}, l_{TP}^0, C_{TP}, T_{TP})$ with a special set of accepting locations.

Informally, the test purpose represents the property we want to verify. For a real-time system implementation, this property is a set of interactions with the environment as well as the time constraints of these interactions. For example, a test purpose may consist of checking whether a sequence of interactions is permitted by an implementation of a real-time system within a certain time interval. The accepting locations of a test purpose represent the locations where the test verdict should be "Pass".

Example 3. To illustrate the concept of test purpose, consider again the telephone system of Figure 1. Figure 3 shows an example of test purposes, which consists of checking whether the implementation accepts $Digit1$ within 1-time

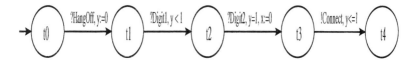

Fig. 3. An Example of Test Purpose.

unit and $Digit2$ 1-time unit after the user has hanged off, and responds with $Connect$ at the latest 1-time unit after $Digit2$ is dialed. Here, instead of generating test cases for the whole system, the user needs just some of them to verify whether or not the implementation permits his/her test purpose. The advantage of using test purposes is therefore the saving of time and money needed for testing implementations.

To test an implementation against a test purpose, the implementation must, at the same time, respect the specification and satisfy the test purpose. Therefore, we should compute a special composition [Kan95](called synchronous product in [KLC98]) of the specification and test purpose before generating test cases.

Definition 8. Synchronous Product
Let Let $A = (I_A, O_A, L_A, l_A^0, C_A, T_A)$ be a specification and $TP = (I_{TP}, O_{TP}, L_{TP}, l_{TP}^0, C_{TP}, T_{TP})$ be a timed test purpose. The synchronous product of A and TP is the TIOA $SP = (I_{SP}, O_{SP}, L_{SP}, l_{SP}^0, C_{SP}, T_{SP})$ such that:

- $I_{SP} = I_A \cup I_{TP}$ and $O_{SP} = O_A \cup O_{TP}$.
- $L_{SP} \subseteq L_A \times L_{TP}$.
- $l_{SP}^0 = (l_A^0, l_{TP}^0)$.
- $C_{SP} = C_A \cup C_{TP}$.
- L_{SP} and T_{SP} are the smallest relations defined by the following two rules:
  - $(l_1, l_2) \in L_{SP} \wedge l_1 \xrightarrow{\{?,!\}a, \lambda_1, G_1}_A l_1' \in T_A \wedge l_2 \xrightarrow{\{?,!\}a, \lambda_2, G_2}_A l_2' \notin T_{TP} \implies (l_1', l_2) \in L_{SP} \wedge (l_1, l_2) \xrightarrow{\{?,!\}a, \lambda_1, G_1}_A (l_1', l_2) \in T_{SP}$.
  - $(l_1, l_2) \in L_{SP} \wedge l_1 \xrightarrow{\{?,!\}a, \lambda_1, G_1}_A l_1' \in T_A \wedge l_2 \xrightarrow{\{?,!\}a, \lambda_2, G_2}_A l_2' \in T_{TP} \implies (l_1', l_2') \in L_{SP} \wedge (l_1, l_2) \xrightarrow{\{?,!\}a, \lambda_1 \cup \lambda_2, G_1 \& G_2}_A (l_1', l_2') \in T_{SP}$.

Example 4. Figure 4 shows the synchronous product of the specification in Figure 1 and the test purpose in Figure 3. As we can see from this figure, some executions of the specification are not allowed by the synchronous product because the time constraints of a transition in the specification and its corresponding one in the test purpose might not be simultaneously satisfied. For example, the execution $?HangOff.\frac{2}{3}.?Digit1.\frac{2}{3}.?Digit2.!Connect$ is allowed by the specification but not by the synchronous product.

3 Test Cases Generation

This section is devoted to test cases generation from a TIOA using the definitions and results of the previous section. Our approach is based on test purposes to

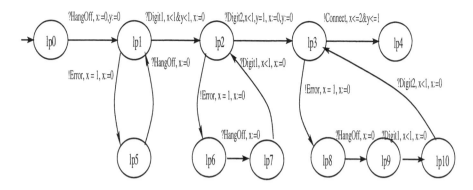

Fig. 4. Synchronous Product of Figures 1 and 3.

test only the critical parts of the system and therefore minimize the number of test cases generated. The proposed method consists of three main steps:

- The construction of a synchronous product.
- The sampling of the regions graph of the synchronous product.
- The traversal of the resulting sub-automaton.

In what follows, we will explain each of these steps and illustrate it using Figures 1 and 3.

3.1 Construction of a Synchronous Product

Since our approach is based on test purposes, the goal of this step is to construct a synchronous product of the specification of the system to be tested and the test purpose to be verified. This construction is based on the definition 8, which is transformed into the algorithm presented in Figure 5.

The algorithm takes a specification and a test purpose as inputs and returns their synchronous product as output. The synchronous product is a TIOA representing somehow an intersection between the TIOA of the specification and that of test purpose. This is needed because when we want to test an implementation using test purposes, we would like to check whether or not the implementation satisfies both the specification and test purpose. The test verdict is given based on the following three rules. If the implementation satisfies both the specification and test purpose, the verdict is "Pass". However, if the implementation respects the specification but does not satisfy the test purpose, the verdict is "Inconclusive". Finally, if the implementation does not respect the specification, the verdict is "Fail".

To construct the synchronous product, the algorithm first creates the initial location of the TIOA by concatenating the initial location of the specification and that of the test purpose. Then, the algorithm incrementally constructs the transitions of the synchronous product and adds the resulting states to the set of states. The transitions and states of the synchronous product are created according to the three rules stated in definition 8.

INPUT: - A specification's TIOA $S = (I_S, O_S, L_S, l_S^0, C_S, T_S)$.
  - A test purpose TIOA $TP = (I_{TP}, O_{TP}, L_{TP}, l_{TP}^0, C_{TP}, T_{TP})$.
OUTPUT: - A synchronous product $SP = (I_{SP}, O_{SP}, L_{SP}, l_{SP}^0, C_{SP}, T_{SP})$.
  $l_{SP}^0 \leftarrow (l_S^0, l_{TP}^0)$.
  Add l_{SP}^0 to L_{SP}.
  $C_{SP} \leftarrow (C_S \cup C_{TP})$.
  $RL \leftarrow l_{SP}^0$ // the set of reachable locations.
  $HL \leftarrow \emptyset$ // the set of handled locations.
  While $(RL \backslash HL \neq \emptyset)$ do
    Get a location $l = (l_1, l_2)$ from $RL \backslash HL$.
    Add l to HL.
    If $l_1 \xrightarrow{\{?,!\}a, G_1, \lambda_1}_S l_1' \in T_S$ and $l_2 \xrightarrow{\{?,!\}a, G_2, \lambda_2}_{TP} l_2' \notin T_{TP}$ then
      Add (l_1', l_2) to RL.
      Add $(l_1, l_2) \xrightarrow{\{?,!\}a, G_1, \lambda_1}_{SP} (l_1', l_2)$ to T_{SP}.
    EndIf
    If $l_1 \xrightarrow{\{?,!\}a, G_1, \lambda_1}_S l_1' \in T_S$ and $l_2 \xrightarrow{\{?,!\}a, G_2, \lambda_2}_{TP} l_2' \in T_{TP}$ then
      Add (l_1', l_2') to RL.
      Add $(l_1, l_2) \xrightarrow{\{?,!\}a, G_1 \& G_2, \lambda_1 \cup \lambda_2}_{SP} (l_1', l_2')$ to T_{SP}.
    EndIf
  EndWhile

Fig. 5. Synchronous Product's Construction Algorithm.

The complexity of the algorithm is $\theta((|L_S| \times |L_{TP}|) \times |T_S|)$. Indeed, the loop *while* in step3 of the algorithm executes for each reachable location. At the worst case, the number of locations in the synchronous product is $(|L_S| \times |L_{TP}|)$ (see definition 8). For each iteration of the loop *while*, we have to traverse all the transitions of both the specification and test purpose to see if there is any transition leaving from l_1 and l_2 respectively. The complexity of this traversal is $\theta(|T_S| + |T_{TP}|)$, which is equivalent to $\theta(|T_S|)$ since $|T_{TP}|$ is less than $|T_S|$.

3.2 Sampling the Regions Graph of the Synchronous Product

Since the synchronous product obtained in the previous step is a TIOA, its operational semantics is given by its regions graph. However, as one can see from the definition 6, each state in the regions graph consists of an infinite number of clock valuations and has a delay transition labeled with the generic symbol d. To generate test cases from the regions graph of the synchronous product, we have to choose a set of representatives for each state and accordingly instantiate the delay transition d. The objective of this step of our approach is to construct a sub-automaton of the regions graph of the synchronous product according to proposition 3. This operation is called the sampling of the regions graph and the resulting sub-automaton is called the *Grid Automata (GA)* [LY93,ENDKE98,SVD01,ENDK02]. Figure 6 shows the algorithm used to construct such sub-automaton.

INPUT : - A synchronous product $SP = (I_{SP}, O_{SP}, L_{SP}, l_{SP}^0, C_{SP}, T_{SP})$.
OUTPUT: - A sub-automaton GA of the regions graph of SP.
   $s^0 \leftarrow (l_{SP}^0, 0)$.
   granularity $\leftarrow \frac{1}{k+1}$.
STEP3: Initialize the sets to be used
   $RS \leftarrow l_{SP}^0$ // the set of reachable states.
   $HS \leftarrow \emptyset$ // the set of handled states.
   While $RS \backslash HS \neq \emptyset$ do
      Get a state $s = (l, v)$ from $RS \backslash HS$
      Add s to HS
      For each transition $l \xrightarrow{\{?,!\}a, G, \lambda} l'$ in TIOA do
         If $v \models G$ then Add $(l, v) \xrightarrow{\{?,!\}a} (l', [\lambda := 0]v)$ to GA if it does not exist
         Add $(l', [\lambda := 0]v)$ to RS if it does not exist
         EndIf
      EndFor
      Add $(l, v) \xrightarrow{granularity} (l, v + granularity)$ to GA if it does not exist.
      Add $(l, v + granularity)$ to RS if does not exist
   EndWhile

Fig. 6. Sampling Algorithm.

The algorithm takes a synchronous product as input and constructs a grid automaton of its regions graph. The algorithm proceeds in many steps. Given the TIOA of the synchronous product, we first calculate the granularity of sampling. This granularity is equal to $\frac{1}{k+1}$, where k is the number of clocks in the TIOA. In a second step, we create the initial state formed with the initial location of the TIOA and a clock valuation that sets all clocks to zero. In a third step, we create all reachable states from the initial state with repetitive $\frac{1}{k+1}$ delay transitions. Then, for each reachable state (l, v), we create a transition $(l, v) \xrightarrow{\{?,!\}a} (l', [\lambda := 0]v)$ for each transition $l \xrightarrow{\{?,!\}a, G, \lambda} l'$ in TIOA such that v satisfies G. Afterwards, we repeat the same process starting with state $(l', [\lambda := 0]v)$.

Example 5. The granularity used to sample the regions graph of the synchronous product of Figure 4 is $\frac{1}{3}$. The resulting grid automata containing only the executions, which lead to a verdict "Pass" is shown in Figure 7. Notice that each state of this automata has an outgoing delay transition labeled with the same delay $\frac{1}{3}$ (the transition between $(lp3, (0,0))$ and $(lp3, (1,1))$ on delay 1 is the sum of three consecutive transitions on delay $\frac{1}{3}$).

The complexity of the algorithm is exponential on the number of clocks of TIOA. Indeed, the outer loop *while* of the algorithm executes for each reachable state. At the worst case, the number of states in a n-clocks TIOA is exponential on the number of clocks n(see [EEN98]). For each iteration of the loop *while*, we have to traverse all the transitions of the TIOA leaving from l (l is the location of the state we choose in the current iteration). The complexity of this traversal is $\theta(|T_S|)$.

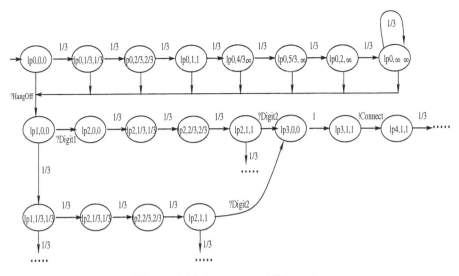

Fig. 7. Grid Automata of Figure 4.

3.3 Traversal of the Sub-automaton

This step of our approach consists of traversing the automaton derived during the previous step to obtain test cases for the system being tested. The algorithm used here depends on the data structure adopted to represent the sub-automaton. An important representation of the sub-automaton is to use a graph. Therefore, timed test cases can be derived from the graph using a depth traversal. So, each traversal represents a test case that starts at the initial state of the grid automaton and finishes when a leaf is reached. Figure 8 shows the algorithm used to generate test cases.

The algorithm takes the sub-automaton extracted during the second step of our approach as input and produces test cases as outputs. The algorithm proceeds as follows. After initializing all the variables to be used (VS and TC), we traverse all the states of GA, one by one, starting from the initial state and we add the chosen state to the set VS to indicate that it has been visited. Then, we add all the neighbors of the chosen state to the set NS and we recursively handle them all before going back to choose another state from the set of previous states.

Example 6. The test cases resulting from applying step3 on Figure 7 are given in Table 2. To lessen the length of test cases, we have summed up all consecutive delays in each of them. Here, the test case $?HangOff.?Digit1.1.?Digit2.1$ means that when testing the implementation, the tester should apply the input $?HangOff$ followed by the input $?Digit1$, waits 1 time unit, applies the input $?Digit2$ and observes the output $!Connect$ within 1 time unit.

The complexity of the algorithm is $\theta(max(a,n))$, where a and n are respectively the number of transitions and the number of states in GA. Indeed, the

INPUT : - A sub-automaton of the regions graph of SP.
OUTPUT: - A set of timed test cases.
  $VS \leftarrow \emptyset$. (the set of visited states)
  $TC \leftarrow \emptyset$. (a test case)
  While ($S_{SP}\backslash VS \neq \emptyset$) do
    Choose a state s from $S_{SP}\backslash VS$ (the first time, the initial state is chosen).
    Add s to VS.
    $NS \leftarrow$ all the neighbors of s. (the set of neighbor states)
    While ($NS \neq \emptyset$) do
      Choose and remove a state s_1 from NS such that $s \xrightarrow{\{?,!\}a} s_1 \in T_{GA}$.
      Concatenate $\{?,!\}a$ with TC.
      Add all the neighbors of s_1 to NS.
      If s_1 has no outgoing transition then
        Print TC.
        $TC \leftarrow TC\backslash\{?,!\}a$.
      EndIf
    EndWhile
  EndWhile

Fig. 8. Timed Test Cases Generation Algorithm.

Table 2. Timed Test Cases Generated.

$?HangOff.?Digit1.1.?Digit2.1.!Connect$
$\frac{1}{3}.?HangOff.?Digit1.1.?Digit2.1.!Connect$
$\frac{2}{3}.?HangOff.?Digit1.1.?Digit2.1.!Connect$
$1.?HangOff.?Digit1.1.?Digit2.1.!Connect$
$\frac{4}{3}.?HangOff.?Digit1.1.?Digit2.1.!Connect$
$\frac{5}{3}.?HangOff.?Digit1.1.?Digit2.1.!Connect$
$2.?HangOff.?Digit1.1.?Digit2.1.!Connect$
$\frac{7}{3}.?HangOff.?Digit1.1.?Digit2.1.!Connect$
$\frac{8}{3}.?HangOff.?Digit1.1.?Digit2.1.!Connect$
$?HangOff.\frac{1}{3}.?Digit1.\frac{2}{3}.?Digit2.1.!Connect$
$\frac{1}{3}.?HangOff.\frac{1}{3}.?Digit1.\frac{2}{3}.?Digit2.1.!Connect$
$\frac{2}{3}.?HangOff.\frac{1}{3}.?Digit1.\frac{2}{3}.?Digit2.1.!Connect$
$1.?HangOff.\frac{1}{3}.?Digit1.\frac{2}{3}.?Digit2.1.!Connect$
$\frac{4}{3}.?HangOff.\frac{1}{3}.?Digit1.\frac{2}{3}.?Digit2.1.!Connect$
$\frac{5}{3}.?HangOff.\frac{1}{3}.?Digit1.\frac{2}{3}.?Digit2.1.!Connect$
$2.?HangOff.\frac{1}{3}.?Digit1.\frac{2}{3}.?Digit2.1.!Connect$
$\frac{7}{3}.?HangOff.\frac{1}{3}.?Digit1.\frac{2}{3}.?Digit2.1.!Connect$
$\frac{8}{3}.?HangOff.\frac{1}{3}.?Digit1.\frac{2}{3}.?Digit2.1.!Connect$

traversal of all the states of GA (i.e., the outer *while* loop) takes a time of $\theta(n)$ and for each state the inner *while* loop executes a number of times equal to the number of transitions outgoing from that state. By summing up all the transitions outgoing from all states, one can easily see that the complexity of the inner *while* loop is $\theta(a)$.

4 Conclusion and Discussion

We presented a timed test cases generation method based on test purposes and using TIOA model. Our approach consists of three steps. First, we construct a synchronous product of the specification and test purpose. Then, we sample the regions graph of the synchronous product in order to construct an automaton (Grid Automata) easily testable in the sense that each of its state has an outgoing delay transition labeled with the same delay. Finally, we traverse the grid automaton to extract test cases for the system.

Our method generates few test cases even for huge specification. Moreover, the test cases derived by our approach are executable in that the predicate of each transition traversed by each test case is satisfied. To study in much more details the scalability and the test coverage of our method, we are currently implementing it in order to apply it on different examples with different sizes.

Comparing our approach to timed test purpose based methods we are aware of [KLC98,SPF01], we point out the following similarities and differences. Castanet et al. [KLC98] construct a synchronous product of the specification and test purpose, as defined in this paper, for tests generation. However, the authors use Timed Input Output Machine ($TIOM$) model to describe both the specification and test purpose. $TIOM$ is different from TIOA in that the time constraints of the transitions in $TIOM$ are given as intervals and so the number of clocks used is just one. Moreover, the testing of $TIOM$ consists of calculating two types of intervals for each transition: the final potential interval and the success interval. The former defines the lower and upper bounds for the execution of the transition with respect to the beginning of the test (i.e., the initial state). The latter narrows the final potential interval by taking into account the minimum and maximum delays between the transition and the preceding one. The test verdict is *fail* whenever a transition is executed outside its final potential interval; it is *Pass* when all transitions are executed within their success intervals; and it is *inconclusive* if a transition is fired within its final potential interval but outside its success interval.

On the other hand, Fauchal et al. [SPF01] use timed automata (TA), as in our approach, to describe the specification and test purposes. However, timed test cases are generated based on a synchronous product of the regions graphs of the specification and test purpose rather than their TAs. This is done in three steps. First, the set of specification transitions sequences containing the same actions and in the same order as the test purpose are extracted. Then, the regions graph of each of these sequences and that of the test purpose are constructed. Finally, the regions graph of each sequence is synchronized with that of test purpose to generate test cases. For each transition in the resulting synchronous product, the authors generate two test cases to cover the borders of each clock region.

References

[AD94]   R. Alur and D. Dill. A Theory of Timed Automata. *Theoretical Computer Science*, 126:183–235, 1994.

[BGRS01] P. Baker, J. Grabowski, E. Rudolph, and I. Schieferdecker. A Message Sequence Chart-profile for Graphical Test Specification, Development and Tracing . In *18th International Conference and Exposition on Testing Computer Software, Washington, DC. USA* , June 2001.

[CL97] D. Clarke and I. Lee. Automatic Generation of Tests for Timing Constraints from Requirements. In *Proceedings of the Third International Workshop on Object-Oriented Real-Time Dependable Systems*, Newport Beach, California, February 1997.

[COG98] Rachel Cardell-Oliver and Tim Glover. A Practical and Complete Algorithm for Testing Real-Time Systems. In *FTRTFT1998 - Formal Techniques for Real-Time Fault Tolerant Systems, Lyngby, Danmark*, 1998.

[EEN98] A. Elqortobi and A. En-Nouaary. Dénombrement du Nombre des Régions dans un Automate Temporisé. Technical Report TR-1116, Département IRO, Universite de Montréal, Montréal, Canada, January 1998.

[EN02] A. En-Nouaary. Testing Real-Time Systems using Test Purposes. In *International Workshop on Communication Software Engineering (IWCSE),Marrakech, Morocco*, December 2002.

[ENDK02] A. En-Nouaary, R. Dssouli, and F. Khendek. Timed Wp-Method: Testing Real-Time Systems. *IEEE Transactions on Software Engineering, November*, November 2002.

[ENDKE98] A. En-Nouaary, R. Dssouli, F. Khendek, and A. Elqortobi. Timed Test Cases Generation Based on State Characterisation Technique. In *19th IEEE Real-Time Systems Symposium (RTSS'98), Madrid, Spain*, December, 2-4 1998.

[FAUD00] M. A. Fecko, P. D. Amer, M. U. Uyar, and A. Y. Duale. Test Generation in the presence of Conflicting Timers. In *TESTCOM Ottawa, Canada*, August-September 2000.

[GHN93] J. Grabowski, D. Hogrefe, and R. Nahm. Test Case Generation with Test Purpose Specification by MSCs . In *SDL'93*, October 1993.

[HNTC99] T. Higashino, A. Nakata, K. Taniguchi, and A. Cavalli. Generating Test Cases for a Timed I/O Automaton Model. In *Proceedings of the International Workshop on Testing Communica ting Systems (IWTCS'99), Budapest, Hungary*, 1999.

[Hog01] D. Hogrefe. Some Implications of MSC, SDL and TTCN Time Extensions for Computer-aided Test Generation . In *10th SDL-Forum, Copenhagen, Danemark*, June 2001.

[ISO91] ISO. Conformance Testing Methodology and Framework. International Standard IS-9646 9646, International Organization for Standardization — Information Technology — Open Systems Interconnection, Genève, 1991.

[KAD+00] A. Khoumsi, M. Akalay, R. Dssouli, A. En-Nouaary, and L. Granger. An Approach For Testing Real-Time Protocols. In *TESTCOM Ottawa, Canada*, August-September 2000.

[Kan95] I. Kang. *CTSM A Formalism for Real-Time System Analysis based on State-Space Exploration*. PhD Thesis, University of Pennsylvania, 1995.

[KC00] Osmane Koné and Richard Castanet. Test Generation for Internetworking Systems. *Computer Communications*, 23:642–652, 2000.

[KENDA00] A. Khoumsi, A. En-Nouaary, R. Dssouli, and M. Akalay. A New Method for Testing Real-Time Systems. In *RTCSA , Cheju Island, South Korea*, December 2000.

[KLC98] Osmane Koné, Patrice Laurencot, and Richard Castanet. On the Fly Test Generation for Real-Time Protocols. In *International Conference on Computer Communications and Networks, Louisiane, USA*, 1998.

[LA92] N.A. Lynch and H. Attiya. Using Mappings to Prove Timing Properties. *Distributed Computing*, 6(2):121–139, 1992.

[LY93] K.G. Larsen and W. Yi. Time Abstracted Bisimulation: Implicit Specification and Decidability. In *Proceedings Mathematical Foundations of Programming Semantics (MFPS 9)*, volume 802 of *Lecture Notes in Computer Science*, New Orleans, USA, April 1993. Springer-Verlag.

[MMM95] D. Mandrioli, S. Morasca, and A. Morzenti. Generating Test Cases for Real-Time Systems from Logic Specifications. *ACM Transactions on Computer Systems*, 13(4):365–398, November 1995.

[NS98] Brian Nielsen and Arne Skou. Automated Test Generation from Timed Automata. In *5th International Symposium on Formal Techniques in Real-Time and Fault Tolerant Systems FTRTFT'98*, September 1998.

[NSY92] Xavier Nicollin, Joseph Sifakis, and Sergio Yovine. Compiling Real-Time Specifications into Extended Automata. *IEEE transactions on Software Engineering*, 18(9):794–804, September 1992.

[SPF01] Sébastien Salva, Eric Petitjean, and Hacène Fouchal. A Simple Approach to Testing Timed Systems. In *Proceedings of the Workshop on Formal Approaches to Testing of Software, (FATES'01), Aalborg, Denmark*, Aug 2001.

[SV96] J. Springintveld and F. Vaandrager. Minimizable Timed Automata. In B. Jonsson and J. Parrow, editors, *Proceedings of the 4th International School and Symposium on Formal Techniques in Real Time and Fault Tolerant Systems*, Uppsala, Sweden, volume 1135 of *Lecture Notes in Computer Science*. Springer-Verlag, 1996.

[SVD01] J. Springintveld, F. Vaadranger, and P. Dargenio. Testing Timed Automata. *Theoretical Computer Science*, 254:225–257, 2001.

[Tre92] J. Tretmans. *A Formal Approach to Conformance Testing*. PhD Thesis, University of Twente, August 1992.

Interoperability Testing Based on a Fault Model for a System of Communicating FSMs

Vadim Trenkaev*, Myungchul Kim, and Soonuk Seol

Information and Communications University
58-4 Hwaam-Dong, Yuseong-Gu, Daejon, 305-732, Korea
vad@elefot.tsu.ru, {mckim,suseol}@icu.ac.kr

Abstract. This paper presents a fault model for interoperability testing of communication protocols that are modeled by communicating finite state machines, and proposes a technique that extends an initial interoperability test suite, which is given by another existing method, to be a test suite that can detect "almost all" interaction faults based on the fault model. We start with an interoperability test suite derived by a known method and develop a technique for the fault coverage analysis and a technique for the extension of the test suite in order to achieve high fault coverage. We illustrate the proposed techniques with TCP protocol. The fault coverage analysis concludes that the test suite has 100% fault coverage with respect to the proposed fault domain and does not need to be extended. It is shown that our method is applicable to practical protocols and can be used to make interoperability test suites have high fault-detecting capability.

1 Introduction

In the field of communication protocols testing there exist a number of types of testing, in particular, conformance and interoperability testing. Conformance testing is used when a single protocol entity is tested. The objective of conformance testing is to establish whether an implementation under test conforms to its protocol specification. Interoperability testing is used for checking if two or more protocol entities can operate as a system. The objective of interoperability testing is to check whether two or more implementations that usually pass the conformance testing can interact and if so whether they communicate with each other correctly providing the expected services.

In the case of conformance testing, there are many research works in which test suite derivation methods are based on a formal fault model [1]. A Labeled Transition System (LTS) and a Finite State Machine (FSM) can serve as examples of a formal model of the system that consists of a single entity. To evaluate the quality of a test suite, two well-known criteria are used: a fault coverage and total length of a test suite, i.e., a high-quality test suite must detect all or almost all faults of a given fault

---

* On leave from Tomsk State University

domain and must be short enough. The fault domain including such types of faults as any output and/or transfer fault in the implementation at hand is widely used.

Here we notice in the case of interoperability testing, we need a more complex formal model of a system under test (SUT), for example a system of communicating finite state machines (SCFSM). There are a number of papers that deal with testing a SCFSM [2-5] with respect to a formal fault model [6]. In this paper, we use this model for interoperability testing. Namely, we assume that a joint behavior of the protocol entities is described by a SCFSM and at most one component FSM can be faulty. To be more rigorous, the following is assumed. Communicating FSMs have been thoroughly tested in isolation and found conforming to their specifications, and interact through a media that can be modeled by an FSM with a single state. However, the component FSMs can be incompatible. For example, some options are implemented only for a single component FSM; the component implementations can have different codes for the same variable. Such faults can be modeled as output faults of component FSMs, i.e., when the output of an appropriate transition is wrong. Thus we consider interoperability faults to be output faults in component implementations. It is assumed that the access to internal interfaces of protocol entities is not granted.

In this paper, we develop a technique for a test suite derivation w.r.t. single output faults of component transitions which are responsible for interacting between communicating FSMs. The technique includes the fault coverage analysis of a given interoperability test suite is derived by another existing method and an algorithm for extending the initial test suite to increase its fault coverage.

We illustrate the proposed fault coverage analysis with TCP protocol, using the initial interoperability test suite which is derived by the method from [8]. The test suite for the TCP has 100% fault coverage w.r.t. the above fault domain.

The rest of the paper is structured as follows. Section 2 presents the related work while preliminaries are given in Section 3. In Section 4, we present a fault model for interoperability testing. We explain in Section 5 how a given test suite can be expanded in order to have a higher fault coverage and propose a technique for such extension. In Section 6, we apply the proposed fault coverage analysis to the TCP protocol. Section 7 concludes the work.

2 Related Work and Motivation

There are a number of papers that deal with interoperability testing [7-12]. In most of them when deriving an interoperability test suite, the technique is based on a reachability analysis. Depending on how the protocol implementations to be tested interact, and depending on the points of observation and control, different interoperability testing architectures are used. Some papers present a framework for interoperability testing [9,10]. Viho et al [10] show how the existing concepts of conformance testing can be used for interoperability testing. Griffeth et al [12] present a method for automatic generation of test cases which cover all the required interoperations and contain a number of test cases with total length close to minimum. We also notice that various kinds of interactions can occur between protocol entities; the latter can communicate synchronously or asynchronously, based on the single or multi stimulus principles [11].

However, the authors of most existing papers do not discuss the fault coverage of an interoperability test suite. The goal of interoperability testing often is to execute all possible system interactions not caring which interoperability faults can be detected. Therefore, the fault coverage of a derived interoperability test suite remains unknown, i.e., we usually can say nothing about faults that are detected by the test suite. To discuss the fault coverage we need to formally describe the set of faults to be detected. The fault coverage of a given test suite is then calculated as the ratio of number of faults that are detected by the test suite, to total number of possible faults. The fault coverage shows the quality of a test suite in terms of its fault-detecting capabilities; the higher is the fault coverage the higher is the quality of a test suite [13]. Thus, the evaluation of the fault coverage of a given test suite is an important issue in the protocol testing. In this paper we propose a model of interoperability faults and a method of the calculation of the fault coverage.

3 Preliminaries

3.1 Finite State Machine

We define a *deterministic* FSM A as a 6-tuple $(S, X, Y, \psi, \varphi, s_0)$, where S is a finite set of states with s_0 as the initial state, X is a finite set of inputs, Y is a finite set of outputs, $\psi: S \times X \to S$ is the *transition* function and $\varphi: S \times X \to Y$ is the *output* function. The 4-tuple (s, x, y, s'), $s, s' \in S$, $x \in X$, $y \in Y$, is called a *transition*, where s, s' are, respectively, the initial and final states of the transition, and x, y are, respectively, the input and the output symbols.

We assume that a FSM has a *reset capability*, i.e., there is a special reset input "r" that takes the FSM from any state to the initial state. The output "*null*" is produced when performing this transition. Moreover, we assume that the transition $(s, r, null, s_0)$, $s \in S$, is always correctly implemented.

In this paper we use only complete FSMs. However, usually a protocol specification is not completely specified. For each state and each undefined input at the state, we augment the description of an FSM by use of a completeness assumption [1]. The corresponding complete FSM remains at the state while producing external output "*null*". In a real situation the "*null*" output can be detected by waiting a specified time-out period.

As usual, we extend the output function to input sequences. We assign $\psi(s, \varepsilon) = s$ and $\varphi(s, \varepsilon) = \varepsilon$, for any $s \in S$, where ε is so-called empty symbol. Given states $s, s' \in S$, input sequence $\alpha = x_1 x_2 \ldots x_k \in X^*$, and output sequence $\beta = y_1 y_2 \ldots y_k \in Y^*$, $\varphi(s, \alpha) = \beta$ if there exist states $s_1 = s, s_2, \ldots, s_k, s_{k+1} = s'$ such that $\psi(s_i, x_i) = s_{i+1}$, $\varphi(s_i, x_i) = y_i$, $i = 1, \ldots, k$.

The equivalence relation between two states s and s' of FSM A holds if $\forall \alpha \in X^*$ [$\varphi(s,\alpha) = \varphi(s',\alpha)$], otherwise, the states are non-equivalent. Two FSMs A and B are said to be *equivalent* if their initial states are equivalent; this means that the machines have one and the same behavior. For non-equivalent FSMs there is a sequence such that the responses of the machines do not coincide. A sequence is said *to distinguish* FSM A and FSM B if FSMs have different output responses to this sequence.

Let $X = X_1 \cup X_2 \cup \ldots \cup X_n$. Given a sequence $\alpha \in X^*$ and $k=1,\ldots,n$, let a sequence $\beta \in X_k^*$, is obtained from the sequence α by deleting all the symbols that are not in X_k. Then the sequence β is called X_k-*projection* of sequence α.

3.2 System of Communicating Finite State Machines

Formal methods can be applied for interoperability testing if a joint behavior of the protocol entities is formally described. In this paper we assume a behavior of the specification system and of a system under test is described by a *system of communicating finite state machines* [2] that exchange symbols over channels. Each protocol entity is a FSM (further a component machine) while each channel is a perfect bounded FIFO (first-in and first-out) queue.

We assume a SCFSM works in *a slow environment* [2], i.e., the next input can be applied only when all the processes in the system under the previous input have been completed. Neither the specification system nor its implementation has a *live-lock*, i.e., component machines execute finite number of transitions under any external input. Each component machine has no input queue and produces a pair of outputs, an external output and an internal output, as a response to a current input.

Let a system under test be specified as a system of communicating finite state machines $A1=(Q,X_1 \cup E_2, Y_1 \times E_1, \psi_1, \varphi_1, q_0)$ and $A2=(T, X_2 \cup E_1, Y_2 \times E_2, \psi_2, \varphi_2, t_0)$ with the structure shown in Fig. 1.

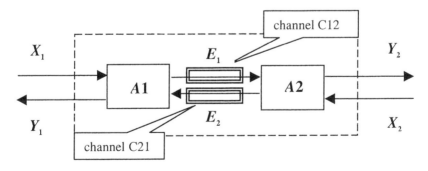

Fig. 1. System under test as system of communicating finite state machines.

In order to deal with the situation when only one output is produced by a component machine we assume that each alphabet Y_1 and E_1 (Y_2 and E_2) has empty symbol ε. Thus output pair (e,ε) or (ε, y), $e \in E_1$, $y \in Y_1$, corresponds the situation when a component machine produces output e or y, respectively. Thus the SCFSM has $X=X_1 \cup X_2$ as the set of external inputs, $Y=Y_1 \cup Y_2$ as the set of external outputs, while E_1 and E_2 are internal alphabets. For the sake of simplicity, the alphabets are assumed to be pairwise disjoint. Note, that formally we ought to consider $Y=Y_1 \times Y_2$ as external output alphabet. However, since there exist only external pairs (ε, y) or (y, ε) then we omit empty symbol ε and consider $Y=Y_1 \cup Y_2$.

The channel C12 (respectively C21) is a perfect FIFO queue linking from the FSM $A1$ to the FSM $A2$ (from the FSM $A2$ to the FSM $A1$, respectively). The content of the channel C12 (respectively C21) is a sequence of symbols of $E1$ (respectively $E2$). Note, for the above assumptions the content of a channel is a symbol, i.e., the channel size is one.

We use 4-tuple (q,t,c_{12},c_{21}) to represent *a global state* of the SCFSM, where $q \in Q$, $t \in T$, c_{12} and c_{21} are contents of channels C12 and C21, respectively. In other words, a global state of the SCFSM consists of a pair of states of the component machines and the content of each channel. A global state is called *stable* if all channel queues are empty. Otherwise, it is called *transient*. Due to the principle of a slow environment, a stimulus from the environment can be submitted only when the system is at a stable state.

A joint behavior of communicating FSMs $A1$ and $A2$ can be represented by means of a reachability graph that is similar to a global state machine [2] and a composition of FSMs [5]. The reachability graph G is pair (V,E), where the set V of vertices represents the set of global states of the system. The set of edges E corresponds to transitions of the system from one global state to another. Each edge is labeled by an input/output pair of an appropriate component machine. The initial vertex is the stable state $(q_0,t_0,\varepsilon,\varepsilon)$.

The reachability graph is constructed by simulation of a behavior of the system under external inputs, i.e., is derived as follows:

- Let $g \in G$ be a current vertex representing a stable state $(q,t,\varepsilon,\varepsilon)$ while $x \in X_1$ being an external input, i.e., the FSM $A1$ has a stimulus from environment. Then there is an outgoing edge from the vertex g labeled by pair x/α, $\alpha = \varphi_1(q,x)$, where α is a pair of symbols (e,y), $e \in E_1$, $y \in Y_1$. The next state is (q',t,e,ε), where $q' = \psi_1(q,x)$. It means that the FSM $A1$ at state q under the input x produces the output α, enters the next state q' and the queue of the channel C12 is filled by the symbol e. Note that e can be equal ε and, in this case, the next state is a stable state. In similar way, the next state is derived when an external input from X_2 is submitted.

- Let $g \in G$ be a vertex representing a transient state (q,t,ε,c_{21}). Therefore, an external input is absent and a stimulus for a component machine is taken from a queue of the channel C21. Let $c_{21}=c$, $c \in E_2$. Then $g=(q,t,\varepsilon,c)$ and there is an outgoing edge from the state g labeled by pair c/α, where $\alpha = \varphi_1(q,c)$ is a pair of symbols (e,y). The next state is (q',t,e,ε), where $q' = \psi_1(q,c)$. The same construction is used when channel C12 is not empty.

The whole reachability graph representing the behavior of SCFSM is constructed until all states reachable from the initial state are derived.

We consider a SCFSM such that any component machine can produce a pair of external and internal outputs as a response to an accepted input. By this reason, given an external input and a stable state, the system may produce a sequence of external outputs in response to the input until the system enters the next stable state. Since we also assume that any internal dialogue is finite, length of this output sequence cannot exceed an appropriate integer l. In this case, an external behavior of the system can be described by an FSM where outputs can be sequences of length up to l. Thus when a SCFSM has the above properties, in particular it has no live-lock and runs in a slow environment, then its behavior can be described by an FSM that is obtained through the reachability analysis. The composed machine is obtained by hiding all internal actions [2], i.e., symbols of the sets E_1 and E_2.

4 Fault Model for Interoperability Testing

An interoperability test suite is a sequence of external inputs, which are applied to a SUT to verify that its parts can interact with each other as expected. However, most interoperability test suite derivation methods do not return which faults can be detected with a derived test suite. The goal of interoperability testing often is to execute appropriate possible internal interactions only.

Obviously, interoperability testing is concerned only with those failures that occur when protocol entities are interacting. Therefore, it is reasonable to concern about failures occurred through the interaction only, i.e., about a failure when a message produced by one protocol entity has been transformed into another message when reaching another peer entity. In other words, the failure changes a transmitted message. Interoperability testing can be used to detect incompatibilities of protocol implementations such as incompatible options, coding, ect. Note, that compatibility testing can be reduced to media testing, for example the case of incompatibilities of protocol implementations when a message is not accepted by protocol entity due to its format can be considered as the case of a corruption or a loss of a message.

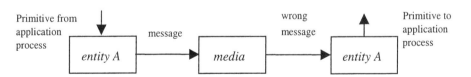

Fig. 2. Communication media with a failure.

Thus we assume that protocol entities are tested separately and found conforming to their specifications, i.e., they are faulty-free. The commutation media including internal interfaces of protocol entities can be faulty; faults of the media correspond to the wrong processing of a given message (Fig. 2).

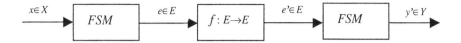

Fig. 3. Media function.

We propose to model a media through which protocol entities interact as a FSM with a single state, i.e., as the mapping $f: E \rightarrow E$, , where E is an internal alphabet (E_1 or E_2) (Fig. 3), and consider that the faults of the media FSM do not increase its number of states. Then the function of correct media is the mapping $f_c(e)=e$ for any $e \in E$ and the function of faulty media is the mapping $f_d(e)=e'$ where $e' \neq e$ for some $e \in E$. When $e' \in E$ then the function of faulty media models the case of a corruption of a message, when $e' = \varepsilon$ then we have the case of a loss of a message.

The replacement of a reference message produced by one component FSM by another wrong message can be modeled as the change of one internal output to another internal output. Remember, that an implementation has a transition with an output fault if the output of the transition is different from that of the specification. Thus the faults of the media can be modeled as output faults of a FSM that represents a behavior of a corresponding protocol entity.

Unfortunately, the number of output faults of an FSM is exponential, i.e., can be very large. On the other hand, the results of performed computer experiments [5] show that a test suite complete w.r.t. single output faults of each component machine usually also detects "almost all" multiple output faults. By this reason, in this paper we only consider single output faults. In addition, we assume that at most one component machine can be faulty. Moreover, since interoperability testing is only concerned about interacting between component machines, we only consider so-called internal output faults. In other words, we consider single output faults at the transitions with a non-empty internal output since only such transitions of component machines are responsible for the interaction between component FSMs. Each transition where the empty word is produced as the internal output is assumed to be faulty-free.

We introduce the formal notion of a single internal output fault of a component machine. Let FSM $A1=(Q, X_1 \cup E_2, Y_1 \times E_1, \psi_1, \varphi_1, q_0)$ be a specification component machine, while FSM $B1=(Q, X_1 \cup E_2, Y_1 \times E_1, \psi_1', \varphi_1', q_0)$ models an implementation of the above specification. We call FSM $A1$ the *reference component* while calling FSM $B1$ an *implemented component*.

We say that $B1$ has *a single internal output fault* if for each pair (q,x), $q \in Q$, $x \in X_1 \cup E_2$, $\psi_1'(q,x)=\psi_1(q,x)$ and there exists exactly one pair (q,x) such that $\varphi_1'(q,x)=(y,e') \neq \varphi_1(q,x)=(y,e)$, $e, e' \in E_1$, $y \in Y_1$. Thus a fault is defined as the replacement of the internal output symbol by a symbol from the same alphabet. A single internal output fault of the component FSM $A2$ is defined in the same way.

5 Interoperability Test Suite Derivation

5.1 Test Architecture

Test architecture is an environment where a system under test is tested. Depending on a testing architecture, a system under test has different points of control and observation. It is known that some test suites require proper points of control and observation for protocol entities in order to recognize all the faulty implementations of an appropriate fault domain. It is shown in [5] that a conformance test suite derived by visiting all transitions of each communicating FSM detects all single output faults when there is an observation point at the internal outputs of component machines. In other words, the test suite detects all faulty implementations if we have a possibility to observe internal output reactions of protocol entities. However, sometimes the access to internal interfaces is not granted. In this paper, when deriving a test suite we consider a more difficult option: the access to internal outputs is not granted.

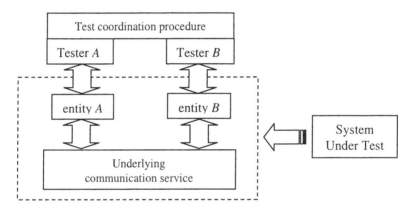

Fig. 4. Test architecture.

Thus we use test architecture in Fig. 4. The SUT is composed of two entities (entity A and entity B). The upper interface of each entity through which the entity communicates with its upper layer is accessible. The low interface is used by entity to interact with the peer entity in the same layer and it is not accessible. Thus only external output sequences can be observed after applying an external input sequence to the implementation. We also assume that a test synchronization procedure exists between Tester A and Tester B.

5.2 Sketch of a Proposed Method

As discussed above, an interoperability test suite visits each component transition concerned on interacting of component machines. However, such test suite does not guarantee detecting all output faults since even traversing all transitions of each component machine it is not enough to detect all output faults when no access is granted to internal outputs [5]. Thus the fault coverage of an interoperability test suite remains unknown. Therefore it is required to make the fault coverage analysis of the interoperability test suite as follows: to check if the test suite can detect "almost all" single internal output faults that can occur in traversed transitions. If the test suite does not detect most such faults then additional test cases should be generated. Here we notice that the fault coverage analysis can also be used to detect redundant test sequences, i.e., can be applied to problem of minimizing of test length. In this paper, we don't consider this problem.

Thus we propose the following test generation process. We first derive an interoperability test suite using an existing method. We then evaluate its fault coverage. Finally, we propose how to extend a non-complete test suite if necessary.

In this paper, to derive an interoperability test suite covering appropriate transitions of the specification component machines we are guided by the approach from [8] that is based on a reachability graph of the specification SCFSM. Thus, we assume an initial interoperability test is given and consider two other steps in details. We first evaluate the fault coverage of the test suite w.r.t. single internal output faults of component machines.

5.3 Fault Coverage Analysis

Let an interoperability test suite be $IT \subset (X_1 \cup X_2)^*$. For the sake of simplicity, we assume that the test suite does not have sequences, which are proper prefixes of other test cases.

As mentioned above, if a SCFSM has no live-lock and runs in a slow environment then its behavior can be described by a composed machine. Let $\mathcal{J}$ be the set of composed machines, which can be obtained from the SCFSM when a reference component is replaced by a machine with a single internal output fault. In other words, the set $\mathcal{J}$ can be derived by explicit simulation of all single internal output faults of each component machine. Let an FSM N denote the composed machine of the specification system. Then the fault coverage analysis can be executed as follows. For each FSM M of set $\mathcal{J}$ that is not equivalent to FSM N check if an interoperability test suite has a sequence that distinguishes the FSMs N and M.

It is known that in general case, the set $\mathcal{J}$ can have an FSM that is equivalent to FSM N. In this case, a behavior of an implemented component with a fault does not influence on a behavior of whole system. Such fault is undetected by any input sequence. We call such fault *undetectable* [5], since it cannot be detected by any test case.

In general there can exist component transitions that are not reachable from the initial state of the component machine. We further assume when an interoperability test does not traverse an appropriate component transition the transition cannot be traversed at all and do not consider output faults for such transitions.

If for each FSM $M \in \mathcal{J}$, that is not equivalent to FSM N, there exists a sequence in the test suite IT distinguishing N and M then the test suite detects all single internal output faults of each component machine and is *complete*. Otherwise, we are required to extend the test suite. A *completeness* (the fault coverage) of a test suite is calculated as the ratio $d/f 100\%$ where f is a number of all detectable faults from the fault domain while d is a number of faults detected with the test suite *IT*.

We below propose how to perform the fault coverage analysis of a given test suite without explicit enumeration of all faults of the fault domain or at least without explicit enumeration of composed FSMs from set $\mathcal{J}$.

When no access is granted to internal outputs a single internal fault can be masked by another component machine. By this reason, the absence of a point of observation at internal outputs implies it is not enough to traverse a transition in order to detect a corresponding output fault. However, for some systems, the absence of the points of observation does not influence the observability of a component behavior. For such systems the fault coverage analysis is reduced to checking appropriate properties of component machines. In particular, this is possible if each component machine is an FSM with so-called *transparent* states.

We say that a state of a FSM is *transparent* if for each internal input the external parts of all outputs produced at the state are not equal to the empty word and are pairwise different.

The definition of the transparent states leads us to the following proposition.

Proposition 1. If each state of FSMs $A1$ and $A2$ is transparent then any interoperability test suite traversing each component transition with a non-empty internal output is complete.

Unfortunately, if the conditions of the proposition are not satisfied then we can say nothing about the completeness of an interoperability test suite.

We further propose the technique for the fault coverage analysis. Given a test case, we derive the reachability graph that besides of reference transitions of appropriate component machine has all transitions with internal output faults. If the corresponding projection of a faulty path is not equal to the expected than the corresponding fault is detected with the test case.

We denote $F(k)=\{(q,x,(e',y),q'): e'\neq e\}$ the set of all transitions that can be obtained from the transition $k=(q,x,(e,y),q')$ via a single internal output fault and call a transition $k'\in F(k)$ as *a faulty (error) transition*, while a transition k is called *a reference (correct) transition*. In similar way, the set of faulty transitions of $A2$ is defined. Each faulty transition $k'\in F(k)$ corresponds to a component machine having a single internal output fault.

Let G' be a reachability graph where a reference transition k of an appropriate component machine is replaced by a faulty transition $k'\in F(k)$. The graph G' is called *a reachability graph for faulty transition k'*.

Fault coverage analysis
Input: interoperability test suite *IT*
Output: verdict about test suite completeness

For each sequence α of the set *IT*, do

1. Construct the path $P(\alpha)$ generated by the sequence α in the reference reachability graph (the reachability graph of specification SCFSM).
2. Obtain the set $RT(\alpha)$ of reference transitions covered by the path $P(\alpha)$.
3. For each reference transition $k \in RT(\alpha)$, derive the set $F(k)$ of faulty transitions.
4. Find a set $W(\alpha)$ as union of sets $F(k)$ over (for all) $k \in RT(\alpha)$.
5. For each faulty transition $k' \in W(\alpha)$ do
   - construct a deterministic path $P'(\alpha)$ generated by the sequence α in the reachability graph for the faulty transition k';
   - compare the external projection of the path $P(\alpha)$ with that of the path $P'(\alpha)$; if they are different then add faulty transition k' to the set $D(IT)$;

Once a faulty transition is detected by a test sequence, it further is not considered when running the algorithm.

Finally, if the set $D(IT)$ has all faulty internal transitions of each component then the interoperability test suite is complete; otherwise, it is incomplete. Each single internal output fault that is not in the set $D(IT)$ is not detected with the test suite.

Here we notice that when a component machine has undetectable faults then the fault coverage analysis algorithm might produce the verdict about incompleteness though the test suite detects any interaction fault. However, it is well known that the problem to eliminate an undetectable fault has the same complexity as the problem of test derivation for this fault. By this reason, in practical situations, a test engineer usually agrees to have a test suite that detects about 95% of the faults of the interest.

Example. We now illustrate the proposed algorithm. Consider the system of communicating finite state machines $A1$ (Fig. 5) and $A2$ (Fig. 6). Such SCFSM models a behavior of a symmetric communication protocol. Each component machine produces a single meaningful output at a time, i.e., we omit the empty item of the output.

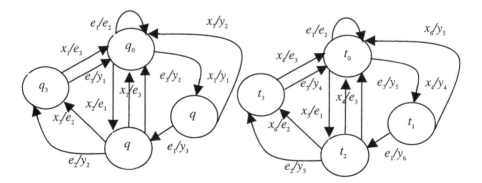

Fig. 5. Component machine $A1$.   **Fig. 6.** Component machine $A2$.

The FSM $A1$ has input alphabet $X_1 \cup E_2$, where $X_1=\{x_1, x_2, x_3\}$, $E_2=\{e_1, e_2, e_3\}$, and output alphabet $Y_1 \times E_1$, where $Y_1=\{y_1, y_2, y_3\}$, $E_1=E_2$, set of states $Q=\{q_0, q_1, q_2, q_3\}$ with q_0 as the initial state. The FSM $A2$ have input alphabet $X_2 \cup E_1$, where $X_2=\{x_4, x_5, x_6\}$, and output alphabet $Y_2 \times E_2$, where $Y_2=\{y_4, y_5, y_6\}$, set of states $T = \{t_0, t_1, t_2, t_3\}$ with t_0 as the initial state. In order to deal with complete FSMs, we augment the machines due to completeness assumption, i.e., for each undefined input the FSM remains at same state while producing external special output "y_7" ("*null*").

The method in [8] returns the interoperability test suite $IT=\{r\ x_1x_5,\ r\ x_2,\ r\ x_4x_2x_1,\ r\ x_4x_2x_3x_1,\ r\ x_4x_2x_3x_4,\ r\ x_4x_2x_4,\ r\ x_4x_2x_6,\ r\ x_5\}$. We use the sequence "$x_2$" to illustrate the algorithm.

The path $P(x_2)$ corresponding to the sequence of transitions "$x_2/e_1,\ e_1/e_2,\ e_2/y_2$" covers the set $RT(x_2)=\{k_1=(q_0, x_2, e_1, q_2),\ k_2=(t_0, e_1, e_2, t_0),\ k_3=(q_2, e_2, y_2, q_3)\}$ of reference transitions. Then the sets of faulty internal transitions covered with the path $P(x_2)$ are as follows: $F(k_1) = \{ (q_0, x_2, e_2, q_2), (q_0, x_2, e_3, q_2)\}$, $F(k_2) = \{ (t_0, e_1, e_1, t_0), (t_0, e_1, e_3, t_0)\}$. We construct the corresponding paths for faulty transitions from $F(k_1) \cup F(k_2)$ (Fig. 7). The only faulty transition $k_2' = (t_0, e_1, e_3, t_0)$ is undetected by the sequence "x_2", since Y-projection of the path $P(x_2)$ coincides with the Y-projection of the path $P'(x_2)$ corresponding to the faulty transition k_2' and is equal to "y_2". By direct inspection, one can assure the faulty transition k_2' is not detected by any sequence from the test suite. Thus, we conclude that the test suite is not complete.

Finally, the set of undetected faulty transitions is $\{(t_0, e_1, e_3, t_0), (q_0, e_1, e_3, q_0), (q_2, x_1, e_2, q_0), (t_2, x_4, e_2, t_0), (t_2, x_6, e_3, t_3)\}$.

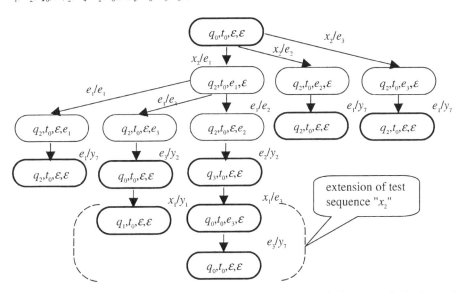

Fig. 7. The reachability graphs corresponding to the test sequence "x_2" that has faulty internal transitions.

Thus the general idea of the algorithm is to enumerate all possible faults and to compare reactions of correct and fault systems under the test suite. Here we notice single internal output faults of each component machine can be explicitly enumerated, since their number is proportional to the size of a component machine. Note, that we treat all the output faults corresponded to a given internal transition in parallel, i.e., we do not derive a faulty composed machine in order to check if the fault is detected with a given test suite.

When a given test suite is not complete we propose a procedure how to augment the test suite in order to detect "almost all" single internal output faults. Note, that if the initial interoperability test suite is complete we do not extend the test suite.

5.4 Extension of the Test Suite

Let $P(\alpha)$ and $P'(\alpha)$ be the paths generated by the sequence α in the reference reachability graph and in the reachability graph for a faulty transition k', respectively.

Random extension algorithm
Input: interoperability test suite *IT*, length *max* of maximum possible extension.
Output: extended interoperability test suite *ET*.

For each sequence α in *IT* do

1. Construct the path $P(\alpha)$.
2. Obtain the set $S(\alpha)$ of undetected faulty transitions covered by the path $P(\alpha)$.
3. For each faulty transition $k' \in S(\alpha)$ assign $\beta = \varepsilon$ and do
   until length of the sequence β is less than *max* do
   - select randomly symbol x from $(X_1 \cup X_2)$;
   - concatenate the sequence β with symbol x;
   - construct paths $P(\alpha\beta)$ and $P'(\alpha\beta)$;
   - if the *Y*-projection of the path $P(\alpha\beta)$ is not equal to that of the path $P'(\alpha\beta)$ then claim the faulty transition k' to be detected and replace the sequence α with the sequence $\alpha\beta$ in the test suite *IT*.
4. If the set of undetected transitions is empty then the extended test suite is complete.

If when performing the algorithm a faulty transition is detected by an extended test sequence then further this faulty transition is not considered.

The idea behind the above algorithm is to extend a test sequence symbol by symbol under a reasonable limit. However, the random extension algorithm does not guarantee the construction of a complete test suite. Moreover, in the worst case any random extension of any test sequence can be ineffective. In fact, an extension algorithm when extending a given test suite must try each external input sequence with length up to *max*. Below we establish an upper bound on length of the sequence *max* after which the extension becomes useless.

As stated in [14], any two FSMs with n and m states, respectively, can be distinguished by a sequence with length up to $n+m-1$. Since an internal output fault does not increase number of states of a component machine then the specification system and its faulty implementation have up to $n_1 n_2$ stable states where n_1 and n_2 are numbers of states of component machines. Therefore, any detectable fault of any component machine is detected by a sequence of length up to $2n_1 n_2 - 1$, i.e., the following proposition can be stated.

Proposition 2. Given a test case $\alpha \in IT$, if a single internal output fault of any component machine is undetected with each sequence $\alpha\beta$, where length of β is $(2n_1n_2 - 1 - |\alpha|)$, then the fault is undetected with any sequence with the prefix α.

However, in practical situations the obtained upper bound is too large. By this reason, usually it is enough to check input sequences of length up to k where k is much lower than $(2n_1n_2 - 1 - |\alpha|)$, for example sequences β of length up to two.

Thus we propose to extend each test case α with each input sequence of length up to two and to estimate the fault coverage of each extended test case $\alpha\beta$ w.r.t. the set of undetected faulty transitions covered by the path $P(\alpha)$. All faults that remain undetected after this procedure usually are undetectable at all.

Example. Consider the above example. The fault coverage analysis of the test suite $IT = \{r\,x_1x_5,\,r\,x_2,\,r\,x_4x_2x_1,\,r\,x_4x_2x_3x_1,\,r\,x_4x_2x_3x_4,\,r\,x_4x_2x_4,\,r\,x_4x_2x_6,\,r\,x_5\}$ returns the following result. The faulty transition (t_0,e_1,e_3,t_0) when traversed is undetected by the sequence "x_2". If we replace the test case "x_2" with the extended sequence "x_2x_1" then the faulty transition becomes detected. Fig. 7 shows how the extension algorithm is executed for the sequence "x_2". By direct inspection, one can assure that the faulty transition (q_0,e_1,e_3,q_0) is detected by the sequence "x_5x_4", the transition (q_2,x_1,e_2,q_0) by "$x_4x_2x_1x_4$", the transition (t_0,x_4,e_2,t_0) by "$x_4x_2x_4x_1$", the transition (t_2,x_6,e_3,t_3) by "$x_4x_2x_6x_4$". Thus the extended interoperability test suite $ET = \{r\,x_1x_5,\,r\,x_2x_1,\,r\,x_4x_2x_1x_4,\,r\,x_4x_2x_3x_1,\,r\,x_4x_2x_3x_4,\,r\,x_4x_2x_4x_1,\,r\,x_4x_2x_6x_4,\,r\,x_5x_4\}$ is complete.

6 Application to TCP

We consider TCP that s shown in Appendix1 where data transfer and a behavior related to a timer are not considered. The interoperability test suite for the TCP derived by the method in [8] is

{listen_1call_2close_1close_2, listen_1listen_2send_data_1,
listen_2call_1close_2close_1, listen_2listen_1send_data_2}.

The index of an external input or output indicates an appropriate component machine.

Note, that TCP specification is not completely specified. We augment the description of the TCP FSM due to the completeness assumption: for each state and each undefined input at the state the TCP FSM remains at the state while producing external output "*null*".

In this paper, we use the model of a slow environment, i.e., multiple stimuli and stimuli during transition are not considered. However, the TCP was initially designed to handle situations when both entities send a message at the same time. By this reason, the TCP FSM has unexecuted transitions due to the limitations of the used formal model. Namely, the transitions (SYN_Sent, SYN, SYN_ACK, SYN_Rcvd), (SYN_Rcvd, SYN_ACK, established, Estab), (SYN_Rcvd, close, FIN, FIN_Wait_1), (FIN_Wait_1, FIN, ACK, Closing), (Closing, ACK, closed, Closed) are not covered by the interoperability test suite. Moreover, since there is no state where FIN_ACK can be sent the transition (FIN_Wait_1, FIN_ACK, {ACK, closed}, Closed) cannot be executed too.

Thus each TCP component has nine transitions with an internal output that are reachable from the initial state (see bold edges in the Appendix). The fault coverage analysis returns the following result. The test suite does not detect all single internal output faults in the transition (FIN_Wait1, ACK, NOTHING, FIN_Wait2) and there is an undetected faulty transition (SYN_Sent, SYN_ACK, {SYN_ACK,established}, Estab). We consider these undetected faults as undetectable since in all cases the correct SCFSM and the SCFSM with the fault have the same final global state. Thus interoperability test suite for the TCP is complete w.r.t. single output faults of the component machines.

7 Conclusion

In this paper, we have proposed a fault model for interoperability testing of communication protocols. Interoperability testing can be used to detect incompatibilities of protocol implementations. In one's turn compatibility testing can be reduced to detecting failures when a message sent by one protocol entity is replaced by another message when transmitting to another peer entity. If a joint behavior of protocol entities is described by a system of communicating finite state machines then such failures can be modeled as output faults of component machines. In order to reduce a number of possible faults in the implementation we limit ourselves with single output faults, since it is well known a test suite detecting all single output faults also captures "almost all" multiple faults. Moreover, since interoperability testing is only concerned about interacting between component machines, we are interested only in internal output faults.

Based on the above fault model, we propose an interoperability test suite derivation method with high fault coverage when the access to internal channels is not granted. We establish a sufficient condition when a test suite visiting all transitions, which are responsible for interacting between communicating FSMs, also detects any internal output fault. When the sufficient condition does not hold for component machines, we propose a technique how to augment a given test suite in order to get higher fault coverage. Undetected faults with the augmented test suite can be listed if necessary.

We illustrate the proposed method with the TCP protocol, using the initial interoperability test suite which is derived by another existing method. The performed fault coverage analysis shows that the test suite is complete and does not need to be extended. We are going to adapt the proposed approach to multi stimulus composition of FSMs as well as to interoperability testing based on other formal models of communication protocols.

References

1. Bochmann, G., Petrenko, A.: Protocol Testing: Review of Methods and Relevance for Software Testing. Sigsoft Software Engineering Notes, spec. issue., USA (1994) 109-124
2. Luo, G., Bochmann G., Petrenko, A.: Test Selection Based on communicating Nondeterministic Finite-State Machines Using a Generalized Wp-Method. IEEE Transactions on S.E., Vol 20. N 2. (1994) 149-162
3. Petrenko, A., Yevtushenko, N., Bochmann, G., Dssouli, R.: Testing in context: framework and test derivation. Computer communications, Vol. 19 (1996) 1236-1249
4. Lee, D., Sabnani, K., Kristol, D., Paul, S.: Conformance testing of protocols specified as communicating finite state machines - a guided random walk based approach. In IEEE Transactions on Communications, vol. 44, N 5 (1996) 631-640
5. Cavalli, A., Prokopenko, S., Yevtushenko, N.: Fault detection power of a widely used test suite for a system of communicating FSMs. Proceedings of the IFIP TC6/WG6.1 13th Inter. Conf. TestCom2000, (2000) 35-59
6. Petrenko, A., Yevtushenko, N., Bochmann, G.: Fault models for testing in context. Proceedings of the IFIP 1st Joint Inter. Conf. FORTE/PSTV96, (1996) 163-178
7. Rafiq, O., Castanet, R.: From Conformance Testing to Interoperability testing. Proceedings of the 3rd International Workshop on Protocol Test Systems, USA (1990) 371-385
8. Seol, S., Kim, M., Kang, S., Park, Y.: Interoperability Test Suite Derivation for the TCP protocol. Proceedings of the IFIP Joint Inter. Conf. FORTE XII/PSTV XIX, (1999) 357-376
9. Kang, S., Shin, J., Kim, M.: Interoperability Test Suite Derivation for Communication Protocols. Computer Networks, 32 (2000) 347-364
10. Viho, C., Barbin, S., Tanguy, L.: Towards a formal framework for interoperability testing. Proceedings of the 21st Inter. Conf. FORTE2001, (2001) 51-68
11. Seol, S., Kim, M., Chanson, S.T.: Interoperability Test Generation for Communication Protocols based on Multiple Stimuli Principle. Proceedings of the IFIP 14th Inter. Conf. TestCom2002, (2002) 151-169
12. Griffeth, N., Hao, R., Lee, D., Sinha R.K.: Integrated system interoperability testing with application to VOIP. Proceedings of IFIP TC6 WG6.1 Joint Inter. Conf. FORTE XIII / PSTV XX, (2000) 69-84
13. Revised Working Draft on "Framework: Formal Methods in Conformance Testing", JTC1/SC21/WG1/Project 54.1 // ISO Interim Meeting / ITU-T on, Paris, (1995)
14. Gill, A.: Introduction to the theory of finite state machine. McGraw-Hill, New York (1962) 207

Appendix: The FSM of Simplified TCP

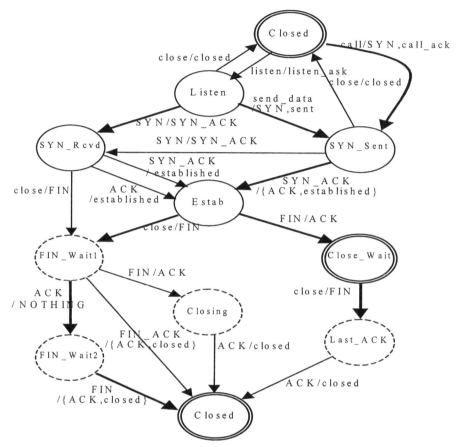

- *Input/Output symbols in capital character: messages passing between peer TCPs (internal messages).*
- *Input/Output symbols in small character: messages accessing tester through PCOs (external messages).*

Framework and Model for Automated Interoperability Test and Its Application to ROHC

Sarolta Dibuz and Péter Krémer

Ericsson Telecommunications Hungary
P.O. Box 107, H-1300, Budapest 3, Hungary
{Sarolta.Dibuz,Peter.Kremer}@eth.ericsson.se

Abstract. In IP world, interoperability testing is heavily used to check the correctness of different implementations. Internet protocols have growing importance in communicating systems. In our paper we show an automatic interoperability test approach and also present its application on an IP-based protocol, ROHC. The primary goal of our work was to define an interoperability testing framework in TTCN-3 that can be used in general. After giving an overview on the ROHC protocol, we also describe the way we have conformance tested it. Then we present MAIT (Model for Automated Interoperability Test) and give detailed explanation on its components and their roles. At the end, we compare the advantages and disadvantages of conventional conformance testing and our interoperability testing model.

1 Introduction

OSI Conformance Testing Methodology [1] can be applied to Internet Protocols, as it has been shown in several papers ([8], [9], [10]). Nevertheless it is not spread in the Internet community. Interoperability testing checks if two different implementations of the same protocol have the capability of inter-working. It is used for testing prototypes built on RFCs and products implementing Internet protocol standards. Interoperability testing is such a well accepted verification method in IETF that a protocol draft can be an IETF standard only if there exists at least two inter-operating implementations for it.

In the telecom world conformance testing is more applied. ETSI, ITU, 3GPP and other standardization bodies develop conformance test suites. Vendors of telecom equipments or operators – the customers – are used to apply these conformance test suites to show conformance of the products or for type approval. Interoperability test is also done after the conformance tests. Its main function is to check if a new network element can inter-operate with the other nodes of the network on the main operation level.

Why is interoperability testing necessary if conformance testing is done or vice-versa? If two IUTs passes the same conformance test suite they can very likely inter-operate, as well. That is how conformance test is defined. But conformance test suites can not cover 100% of the protocol's functionalities. It may happen that interoperability test of two IUTs fails even if the IUTs passed the conformance test. Especially, if the conformance test suite is not a well proven one. So it may happen if the conformance test suite is not a well proven one, that interoperability test of two IUTs fails even if the IUTs passed the conformance test. Protocol definitions contain optional elements. It may happen that the two IUTs implement or don't implement the optional features in a way that in certain cases this leads to problems in interoperability.

If two IUTs can inter-operate using a protocol it also may happen that one of them or both fail on a conformance test. There can be erroneous situations that were not tested during interoperability testing in which the IUTs fail. Erroneous situations can be triggered by conformance test but usually not with interoperability test. Another possibility for non-conforming IUTs, which can inter-operate with each other is that the implementors misunderstood the protocol standard in the same way. So, the two IUTs can inter-operate but conformance test fails for both of them. In this case it is very likely that interoperability test with other IUTs would also fail.

Among others, IETF, ETSI, Sun and the TAHI project used to organize interoperability test events where developers can get together and perform interoperability tests. We have participated on several such events (like *Connectathon*, *IPv6 Bake-off* or *ROHC interop tests*) during the last 3 years. Our experience shows that every developer configures and starts his implementation manually. Moreover, at the end of each interoperability test case the analysis of the logs are also made by hand, which also implies that only the developer of the implementation can perform the interoperability test. We have seen a high need for a method and a tool that can help to automate the testing process.

The aim of our work was to give a framework that automates the testing process by using some parts of CTMF [1] and the flexibility of TTCN-3 [2]. This paper presents a model for interoperability test of Internet Protocols, and also gives an example of its application on ROHC protocol. We also compared the advantages and disadvantages of conventional conformance test and this model of interoperability test in Section 6.

2 Related Work

Conformance testing methodology is well defined and has already proven its stability over the years. In contrary, interoperability testing has no such theoretical base. Several papers have been published related to interoperability testing but most of them deal with test suite generation or derivation [3], [4]. Others [5] perform monitoring and analyzing only, without triggering the IUTs for different actions or testing them actively.

[6] takes into consideration the interoperability test architecture, as well. It proposes three alternatives as possible architectures for interoperability tests. Unfortunately, this article doesn't address practical questions, which become extremely important when somebody wants to perform interoperability tests. For example, even the simplest architecture contains two Testers but the cooperation and synchronization of these Testers are not presented.

Another important issue is the communication between Testers and IUTs. These papers assume that interoperability test can be performed using the same service provider all the way. That is, the Testers send and receive protocol (the same protocol that is under test) messages through their standardized interfaces. Although it can be applied for several protocols but not applicable in general. In case of ROHC, three different interfaces are needed to perform interoperability tests. One is used to provide the necessary input. The second helps to configure the implementation, to trigger the tests and to check if the required action was taken. The third one monitors the exchanged packets. Telecom protocols are very well defined but the same can't be said for Internet Protocols. They usually leave a lot of questions open and the implementations have to make certain decisions (e.g. when to perform a mode change in ROHC). These standards doesn't specify upper and lower interfaces, thus it is not possible to construct an appropriate Upper Tester for these protocols.

Our approach differs in several ways. It gives not only a test architecture but also defines test components, their roles and the communication between them. It doesn't require a standardized upper tester, which is impossible to produce for Internet Protocols in most of the cases. This method also eliminates the problem of testing the states that are not observable from outside. On the other hand, our paper doesn't deal with automatic test suite generation at all.

3 Overview of ROHC

Nowadays, it seems reasonable that IP technology will be the most popular transfer mechanism. Moreover, Voice-over-IP solutions are getting improved and become more and more important. It is possible that mobile networks and cellular phones will also use IP technology in the future. In these systems, the efficient usage of radio bandwidth is crucial in order to serve as much subscribers as possible and to provide acceptable quality at the same time. The main problem with IP when used over wireless links is the large overhead of lower layer protocols. Assuming that RTP is used to transmit speech data, it requires 40 bytes in case of IPv4 and 60 bytes in case of IPv6. (RTP packets are embedded in a UDP packet, which is then carried by an IP packet.) Thus, the size of the payload can be as low as 15-20 octets, depending on the applied speech coder and frame sizes.

Reducing the headers' size is inevitable to improve efficiency but the existing methods ([11], [12]) don't perform well over wireless links ([13]). Basically, radio channels have the following shortcomings:

- high bit error rate ($10^{-3} \ldots 10^{-2}$),
- long round trip time, high latency,
- low bandwidth.

An appropriate header compression scheme must be able to cope with these drawbacks as well as packet loss before the compression point. Since bandwidth is the most expensive resource, the compression ratio and the robustness have far

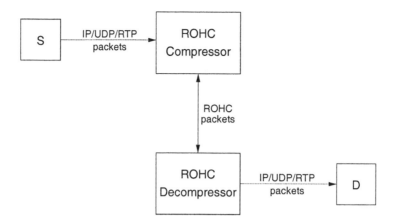

Fig. 1. A typical ROHC configuration

higher importance than the required processing capacity. It is important to note that the redundancy between header field values within a packet and especially between consecutive packets makes header compression possible.

ROHC (RObust Header Compression, [7]) is an IP header compression mechanism designed to perform well over wireless links. It classifies the header fields based on their behavior regarding the correlation to other fields. The conclusion is that most of the header fields never or seldom change. UDP Checksum (if enabled) and RTP Sequence Number (SN) have to be transferred directly, while the other three can be described as a function of SN and some other parameters. That is, in ideal case only the value of two fields (or only one if UDP Checksum is not enabled) have to be communicated explicitly. If a parameter of a function from SN to another field changes (e.g. irregularity in the input RTP stream), additional information is provided to update the necessary parameters of that function.

Figure 1 presents a typical ROHC configuration. The role of the nodes are the following:

- **Source Node (S)**: it produces the input packet stream, which is sent through the Compressor,
- **ROHC Compressor**: it compresses the incoming packets according to its current profile, mode and state, and transmits the resulting ROHC packets,
- **ROHC Decompressor**: restores the original packets (or something similar to the original) and forwards them to the Destination Node,
- **Destination Node (D)**: it is usually a regular application that decodes the original information from the output packet stream.

In ROHC, both the Compressor and the Decompressor has three states that determine the level of compression. States of the Compressor are: Initialization and Refresh (IR), First Order (FO) and Second Order (SO). In the beginning, the Compressor always starts in the lowest compression state (IR) and transits gradually to higher states if it "is sufficiently confident that the decompressor has the information necessary to decompress a header compressed according to that state." [7]. The Compressor can decide the necessary compression state based on the variations in packet headers, on feedback from the Decompressor (positive or negative) or on timeout events. The Decompressor also starts in its lowest compression state (No Context – NC) and transits gradually to higher states: Static Context (SC) and Full Context (FC). Normally, the decompressor never leaves the "Full Context" state once it has reached it. Fallback to lower level states can occur on repeating decompression failures (due to packet loss or bit errors).

Besides states, which determines the level of compression there are also different modes of operation. They control the logic of state transitions and what actions to perform in each state. The current mode of operation can be changed if the Decompressor indicates a mode change request using a Feedback channel. ROHC uses the following modes:

- **Unidirectional (U)**: Packets can be sent in one direction only (from Compressor to Decompressor). Transitions between compressor states are performed only because of timeout events or irregularities in the input stream. Every Compressor and Decompressor start in this mode, mode transitions can be performed if a Feedback channel is available.
- **Bidirectional Optimistic (O)**: The difference to U-mode is that a Feedback channel (from Decompressor to Compressor) is used to send error recovery requests and optional acknowledgments of significant context updates. O-mode aims to maximize compression efficiency with the rare usage of Feedback channel.
- **Bidirectional Reliable (R)**: This mode uses the Feedback channel more intensively and has a stricter logic that ensures more robust context synchronization. R-mode aims to maximize robustness against loss and damage propagation even at high residual bit error rates.

The optimal mode cannot be selected without knowing the characteristics of the environment. The selection depends on the feedback abilities, error probabilities and distributions, etc.

The ROHC RFC defines four different profiles, which can compress different kinds of IP packets. These packet types are (numbers in parentheses denote the number of the profiles): uncompressed packets (0x0000), IP/UDP/RTP packets (0x0001), IP/UDP packets (0x0002), IP/ESP packets (0x0003). In this paper we deal only with profile 0x0001, the same method can be applied to other profiles, as well.

4 Conformance Test of ROHC

In this section we present the test configurations that we have used for the conformance test of ROHC in order to compare it with the interoperability test configuration. A typical configuration (Figure 1) consists of two ROHC implementations: a Compressor and a Decompressor. Since these nodes are working

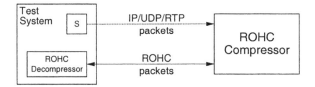

Fig. 2. Conformance test configuration for ROHC Compressor tests

separately and they are connected through a standardized interface, it is possible to test them separately, as well. We have used two different configurations (and test suites) to test Compressors and Decompressors. They are presented in the following two subsections.

4.1 Compressor Tests

To check a particular functionality in the Compressor the writer of the test suite must know how to compress certain RTP packets and how to decompress ROHC packets. This process is defined in the protocol standard, which can be very complicated in some cases. It is a nature of ROHC (and Internet Protocols in general) that an implementation can choose from a high number of possible actions in a certain state. In most of the cases, the RFC leaves it up to the implementation to decide which way to follow. Unfortunately, the same is true for the Decompressor, as well. These properties of the ROHC protocol make its conformance testing more difficult.

The test configuration for Compressor tests can be seen in Figure 2. The configuration consists of a *Test System* and a ROHC Compressor. The *Test System* emulates two nodes:

- **Source Node**, which generates the input IP/UDP/RTP stream for the Compressor
- **ROHC Decompressor**, which can process the compressed ROHC packets.

The two emulated nodes are using separate interfaces. Source Node needs an IP interface in order to send RTP packets. The RTP stream sent by the *Test System* is similar to that of an ordinary application would generate. The emulated Decompressor has a different kind of interface, since it needs to send and receive ROHC packets. Nowadays, two methods are used to transmit ROHC packets: ROHC-over-PPP and ROHC-over-UDP. Our *Test System* supports both kinds of transmission techniques. It is the nature of the protocol that the compressor can't generate a ROHC packet until it receives an RTP packet. Thus, we need to send an RTP packet through the IP interface, first. Then it will be compressed and the *Test System* will receive it through its ROHC interface. All of our tests follow this method and since the protocol assumes this sequential behavior, we didn't have to use parallel test components.

In some cases it is necessary to send a ROHC Feedback packet to the Compressor. The *Test System* uses its ROHC interface to send such packets. For example, to initiate a mode change the Decompressor has to send an appropriate Feedback packet to the Compressor.

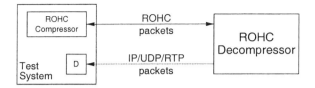

Fig. 3. Test configuration for ROHC Decompressor tests

Our test suite for ROHC Compressor tests consists of 100 TTCN-3 test cases and can check the following functions: mode transitions, context downgrade, changes in the incoming RTP stream, etc.

4.2 Decompressor Tests

Figure 3 shows the conformance test configuration for Decompressor tests. The main difference to the Compressor tests is that in this case the *Test System* emulates a ROHC Compressor and a Destination Node. The *Test System* sends ROHC packets to the Decompressor and processes the reconstructed RTP packets.

Testing a ROHC Decompressor is a bit more difficult than the case of a Compressor. The RFC doesn't define an interface that can be used to remote control a Decompressor. The lack of such an interface prevents us to install an upper tester application. Thus, it is impossible to test mode transitions automatically because the tester can't initiate a mode change remotely.

The test suite for Decompressors contains about 100 TTCN-3 test cases and verifies the correctness of mode transition, context downgrade, irregular changes in the original RTP stream. Unfortunately, these tests can't be executed automatically because of the missing upper tester.

5 Model for Automated Interoperability Test

The Model for Automated Interoperability Test (MAIT) that we have constructed is based on the experience we have gained in interoperability testing of Internet protocols (IPv6, Mobile IPv6, OSPF and ROHC). MAIT uses TTCN-3 [2] as a standardized test description language. This model gives a framework for automated interoperability test suites.

Figure 4 shows the architecture of MAIT. The configuration in this example consists of *IUT 1*, *IUT 2* and the *Test System*. *IUT 1* and *IUT 2* denote the two implementations, which are under test. The *Test System* handles the following tasks:

- remote controlling of *IUT 1* ($PTC\ R1$),
- remote controlling of *IUT 2* ($PTC\ R2$),
- monitoring the network ($PTC\ M$),
- sending protocol messages to IUTs ($PTC\ P$).

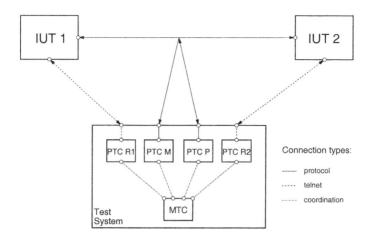

Fig. 4. Sample configuration of interop test

Since these tasks are independent from each other, they are implemented in Parallel Test Components (PTC). The Main Test Component (MTC) is used to coordinate the behavior of PTCs.

$PTC\ R1$ establishes a telnet connection between $IUT\ 1$ and the Test System, this connection is then used to remote control $IUT\ 1$. The test component emulates an ordinary user: configures the IUT, starts an application or triggers a special test by giving the appropriate input. These kind of inputs can't be given through the IUT's standardized interfaces because most of the protocol standards don't specify the upper interface. Basically, it configures, starts and makes everything that only an experienced developer of that particular implementation can do.

$PTC\ R2$ creates the same type of connection but its tasks are slightly different. It also has to start and configure the implementation but then it checks if the test ran correctly and the necessary changes were made. It emulates the user of the other implementation, who follows the tests at the console and sees if everything is working correctly or not. It can be the establishment of a connection, a new record in a database or something else which shows that the test was successful. The messages that are sent on this type of connection are highly depend on the implementations. In order to avoid the re-compilation of the test suite, if a new implementation is tested, these messages are given as test suite parameters.

$PTC\ P$ can send protocol messages to the implementations. There are cases when only a protocol message can trigger a certain action and it also gives the possibility to test inopportune behavior. For example, we can send a mode change request to $IUT\ 1$, which looks like as if it were sent by $IUT\ 2$. In this case, $IUT\ 1$ follows the rules of mode transition and sends a packet according to its new mode. This packet will confuse $IUT\ 2$ and gives the possibility to test inopportune behavior.

Every tester likes to know what happens on the wire, thus they always monitor the network that connects them to the other implementation. In general, "tcpdump" (or a similar tool, e.g. ethereal) is used to record the packets. Synchronization of starting and stopping the packet recorder tool and the analysis of the log files are made manually in most of the cases. We have defined a separate test component in our model to handle and to automate these tasks. *PTC M* monitors the network, records and analyzes every protocol message that the implementations or *PTC P* send. We have separated the latter two test components because their tasks are slightly different. *PTC M* only receives and analyzes packets, while *PTC P* sends protocol messages (but it can receive, as well).

The function of *MTC* is to synchronize and to coordinate the behavior of PTCs. Similarly to conformance testing, the execution of a test case is divided into 3 phases (the names of the involved PTCs are shown in parentheses):

1. Configure the implementations for the test case (*PTC R1, PTC R2*).
2. Execute the test case (*PTC R1, PTC R2, PTC P, PTC M*).
3. Restore the original configuration (*PTC R1, PTC R2*).

The *MTC* creates all the PTCs, which start the configuration phase immediately. It then waits until all the PTCs finish the first phase (the PTCs send a "Ph_1End" signal to the *MTC* when they are ready with the first phase). In the next step, *MTC* starts the execution phase by sending a "Ph2_Start" signal to every PTC. After the execution of the second phase every PTC send a "Ph2_End" signal to the *MTC*. The signaling in third phase is similar to the second one. The synchronization model can be easily extended with additional phases, if needed. If an error occurs in a PTC at any phase, it sends a "Phn_Error" signal to the MTC (where n denotes the phase number), which stops all the running PTCs and the test ends immediately. Figure 5 shows an example of the communication between *MTC, PTC R1* and *IUT 1*.

5.1 Modifications to ROHC

To perform interoperability test of ROHC one needs a source and a destination node that can generate and receive RTP packets. In real life this RTP stream is generated by an application but in case of testing we need a separate, flexible and supervised RTP packet generator. Only such a generator can ensure that the input RTP stream is syntactically and semantically correct and has controlled behavior. The packet generator must be able to generate streams with regular and irregular behavior, as well. Irregularity can be a jump in a monotonically increasing value or simply a packet loss. A destination node is also required, which must be able to check if the output RTP stream is syntactically and semantically correct, and to compare input and output streams. Thus, both nodes must be able to record RTP packets, destination node must be able to read the records of the source node and it must be also able to check whether the two streams are equivalent or not.

In order to provide a solution that better suits the need of ROHC, we have modified the original model. We have added two more test components (*PTC S*,

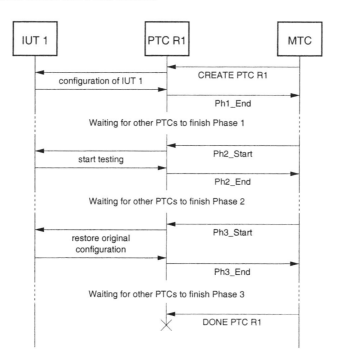

Fig. 5. Synchronization of *PTC R1* and *MTC*

PTC D) that are acting as a source and a destination node. The extended model for ROHC can be seen in Figure 6.

PTC S generates the input RTP stream for the Compressor (just like a packet generator would do) and *PTC D* receives the output RTP stream from the Decompressor. *PTC S* and *PTC D* can be configured on a test case basis and *PTC D* exactly knows the content of each packet that *PTC S* sent. The comparison of the input and the output streams became quite trivial in this case. This method eliminates the necessity of recording the packets on both sides and also the need for manual checking of the files. Thus, the whole testing process can be automated and fully controlled by using this method.

6 Advantages and Disadvantages of MAIT

In this section we describe the advantages and disadvantages of MAIT compared to ordinary conformance testing. First, let's consider the process of preparing a test suite. To write a conformance test suite, a stable description of the protocol is needed. In this sense, stable means that it doesn't change frequently, the changes doesn't affect the elementary parts of the protocol (i.e. connection establishment method or new fields in a message), etc. Then comes the development of the test purposes and the test cases. Writing test purposes usually requires the creation of some kind of a state machine with appropriate inputs, outputs and transitions.

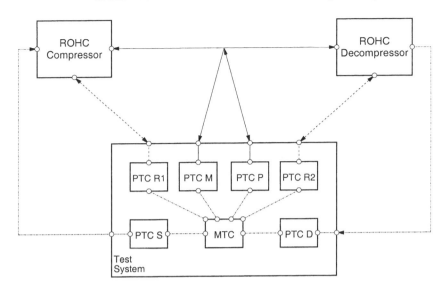

Fig. 6. Extended model of interoperability test

Test cases can be produced for the states that has definite behavior, clearly defined and reachable with the available input sequences. The exact definition of the messages (and the value of their parameters) that the test cases are sending and receiving is one of the most difficult parts of test suite writing. Finally, the test system needs to be adapted to the protocol, as well. Once the conformance test suite is ready, it can be executed against any implementation.

Naturally, a protocol description is also required in case of interoperability test, but it doesn't have to be so mature. The changes of the protocol doesn't necessarily cause trouble during the tests, because both parties' behavior is updated (since we are testing the interoperability of two different implementations of the same protocol). Let's change the packet formats in ROHC! We don't have to alter the input stream, the task of the protocol haven't changed, thus the output stream must be the same, as well. Moreover, we doesn't have to change the configuration commands. The only things that has to be updated are the type and template definitions of the protocol. It only affects *PTC M* and *PTC P* processes. To write test purposes one has to take into consideration all the possible scenarios and input sequences, just like in conformance test. Instead of creating a state machine, only the input sequences are needed in this case. This input sequence differs from the one used in conformance test. For ROHC it is a set of RTP streams, where each stream differs from the others and invokes a different action of the IUT. In other cases it can be a set of commands that establishes a connection but it is important to note that this input sequence doesn't consist of protocol messages. This nature of interoperability test make the writing of a test suite much more easier and faster.

Each test case represents a different interoperability scenario. These scenarios concentrate on the most important functionalities of the protocol. This means that the interoperability test suite is not covering all aspects of a protocol as a conformance test suite. Since MAIT defines the structure of the test cases and the implementation specific commands can be set as parameter, it provides easy-to-understand and well-structured tests. These tests describe the protocol functionalities in a clearer way than the text of the RFC. With the interoperability test the logics of the protocol is also verified.

6.1 Test Execution

If we take a look at the execution of tests, we see that Conformance testing is executed automatically. If the test suite doesn't change between two executions (which is the normal case) it produces repeatable test results. Based on these results, we can compare two implementations regarding their functionalities. Ordinary interoperability testing is usually done manually. Therefore it lacks the ability to repeat the same tests and to compare implementations based on the test results. Having the MAIT model used these disadvantages can be avoided. MAIT uses predefined test cases in the same way as conformance test uses. It also gives the possibility to automatically execute interoperability test cases and eliminates the need for human interaction (e.g. finding out if the test was successful or not, bug tracking, etc.). Since the test cases are defined in a test suite, the tests become repeatable and the results reproducible and comparable, as well.

The number of performed conformance tests grow linearly with the number of the implementations, as a test suite have to be executed on an implementation once. The reason is that a conformance test suite can also be seen as a reference implementation, by definition. In case of interoperability test, the number of test executions grow quadratically with the number of implementations because we don't have such a reference. However, we don't have a reference implementation by definition, we can appoint one or more implementations as reference if they fulfill certain requirements[1]. One requirement can be the successful interoperability test with a given number of other implementations. Others can be set easily upon common agreement of the implementors. The number of test executions decreases drastically if appointed reference implementations exist.

6.2 The Role of Testing in Protocol Design

Both conformance and interoperability testing is used to prove that products fulfill the requirements stated in the standard. Conformance testing is used usually when we have the protocol definition available as a standard. Hence, conformance testing is not part of protocol design, when the protocol itself is created but it is heavily used if the standard has a mature version and numerous implementations are available on the market.

---

[1] The authors don't intend to solve the problem in this paper. They just recommend some possible solution without giving further explanations, arguments or proofs.

In IETF the process of protocol design is iterative in this sense. Vendors implement prototypes of the protocol written in the RFC. The protocol is designed further in the IETF working-group based on the experiments of the prototyping. It is usual to find errors and ambiguities in the protocol and in its specification during interoperability test. They will be corrected, clarified in the next version of the specification or added as a separate document [14]. Either way, the specification evolves as interoperability tests are executed more and more times. It is easier to modify the interoperability test suite in case the protocol is modified, as it is smaller in size and it has a stronger dependency on the inputs than on the protocol's logic and messages. Thus, we can say that interoperability testing is an integral part of the protocol design.

6.3 Implementation Dependency and Reuse of Tests

Conformance test is implementation independent, since it uses only the standardized interfaces of implementations. This attribute makes it possible to use the same test suite for different implementations. Interoperability tests are usually performed for Internet Protocols and these protocols don't have such a detailed specification. It implies that it cannot be guaranteed that these tests will be implementation independent. We have shown in case of ROHC that implementation specific messages are used to configure the implementations, to trigger tests and to check if the required action was taken. This dependency is inevitable if we want to check all the possible interoperability test scenarios. In case of using MAIT, these implementation dependent messages can be given as parameters, so it is not necessary to recompile the Executable Test Suite. Recompilation may need tools and equipments that are usually not available at the site where interoperability testing takes place.

Nowadays, TTCN is a common language for writing conformance test suites. Its latest version [2] is general and flexible enough to describe interoperability test suites, as well. If the two different types of test suites are written in the same language (and MAIT is written in TTCN-3) then some parts of it can be reused. This situation becomes more and more likely as TTCN-3 based test tools get used widespread. The interoperability test is part of the protocol design (at least in IETF), protocol specification, implementations and tests are developed together. It also means that an interoperability test suite is written before the conformance test suite and both use the same description language. That is, results, experiences and parts of the TTCN-3 source code of the interoperability tests can be reused in conformance tests. These parts are the definitions of the protocol's data types and messages, templates and timers.

The test environment also have to be adapted to the particular protocol. This is mainly done by a software module that can transmit and transform protocol messages between the test tool and the underlying service provider. This adaptation module is reusable, too. Our experience shows that the reusable parts (data type and template definitions, adaptation module, etc.) takes 20-40% of the work needed to create an executable conformance test suite. Moreover, it is also possible to record packets during interoperability test and save their contents in order to use in conformance testing. We have already shown that the content of the protocol messages is very important in conformance testing and creation of them is one of the most difficult tasks. Storing the packets for later use can also save reasonable amount of time. In other words, writing an executable conformance test suite could take as high as 50% less time.

7 Conclusion

This paper presented the MAIT model for automated interoperability test and also defined a framework that uses this model. We gave detailed explanation on the components of the model and the role of each component. We showed the advantages and disadvantages of this model and compared it to the properties of conformance testing.

Using our method it is much more easier and faster to derive an interoperability test suite than a conformance test suite from a textual description of a protocol. It is not necessary to create a state machine, to figure out the order and the content of messages to be able to write a test case. The writer of the test suite only has to know what the IUT should do (i.e. set up a connection) in that test case and ask for the configuration commands from the implementors. It is the implementor who knows every detail about the implementation and can give the necessary information. These commands are handled as test suite parameters, so the test suite writer doesn't have to know them at all.

Unlike in case of ordinary interoperability test, the MAIT model stores the interoperability test scenarios in a form of TTCN-3 test cases. Thus, it gives the possibility to rerun the same tests on the same implementation at a different time. It ensures the ability to repeat the same test and to compare the implementations based on the interoperability test results.

Although the model uses implementation specific configuration commands, it doesn't depend on the IUT or on the protocol. It is because these commands are passed as parameters (executable conformance test suites also use similar parameters). That is, a change in these commands doesn't require the change of the test suite. In other words, an implementation can alter in a way that affects its configuration commands but the MAIT test suite can remain the very same. It also means that testing an implementation of another vendor only needs the configuration commands· of that implementation.

With the MAIT method it is easier to change the interoperability test cases if the protocol specification change. This situation happens quite often during in the design phase of a protocol. Waiting for a stable specification is not feasible in practice, because interoperability testing is also used to verify the protocol itself (IETF requires two independent inter-operating implementations to accept a protocol specification as a Proposed Standard).

Table 1 shows the needed testing types in different phases of a protocol. During protocol design phase, no stable specification is available, it can change day by day. So, it requires a testing type where it is possible to quickly adapt to changes. Protocol life-cycle phase presumes a mature standard and it gives the possibility to develop a product that is an implementation of that particular

Table 1. Testing needs in different phases of a protocol

Protocol design	prototype	Interoperability Test
Protocol life-cycle	product (protocol implemenetation)	Conformance Test

protocol. It can be seen that the need for an interoperability test suite precedes the need for a conformance test suite. But when the protocol definition reaches a stable state as a standard, conformance testing should be applied first to test the correctness and conformance of the products implementing the protocol. That is why it is important that parts of a MAIT-based interoperability test suite can be reused in a conformance test suite.

References

1. OSI - Open System Interconnection, Conformance testing methodology and framework, ISO/IEC 9646, 1997.
2. ETSI, The Testing and Test Control Notation version 3, ETSI ES 201 873, v2.2.0, May 2002.
3. S. Kang, J. Shin and M. Kim: Interoperability test suite derivation for communication protocols, The International Journal of Computer and Telecommunications Networking, Vol. 22, Num. 3, pp. 347-364, March 200.
4. N. Griffeth, R. Hao, D. Lee, R.K. Sinha: Integrated System Interoperability Testing with Applications to VoIP, Formal Methods for Distributed System Development (FORTE/PSTV 2000), Pisa, Italy, October 2000.
5. T. Kato, T. Ogishi, H. Shinbo, Y. Miyake, A. Idoue and K. Suzuki: Interoperability Testing System of TCP/IP Based Communication Systems in Operational Environment, Testing of Communicating Systems, Ottawa, Canada, September 2000.
6. J. Shin and S. Kang: Interoperability Test Suite Derivation for the ATM/B-ISDN Signaling Protocol, Testing of Communicating Systems, Tomsk, Russia, September 1998.
7. C. Bormann (ed.): RObust Header Compression (ROHC), RFC 3095, July 2001
8. R. Gecse: Conformance testing methodology of Internet protocols, Testing of Communicating Systems, Tomsk, Russia, September 1998.
9. R. Gecse, P. Krémer: Automated test of TCP congestion control algorithms, Testing of Communicating Systems, Budapest, Hungary, September 1999.
10. T. Csöndes, S. Dibuz, P. Krémer: Experiments on IPv6 testing, Testing of Communicating Systems, Ottawa, Canada, September 2000.
11. M. Degermark, B. Nordgren and S. Pink: IP Header Compression, RFC 2507, February 1999
12. S. Casner and V. Jacobson: Compressing IP/UDP/RTP Headers for Low-Speed Serial Links, RFC 2508, February 1999.
13. M. Degermark, H. Hannu, L.E. Jonsson, K. Svanbro: Evaluation of CRTP Performance over Cellular Radio Networks, IEEE Personal Communication Magazine, Volume 7, number 4, pp. 20-25, August 2000.
14. P. Krémer, L. E. Jonsson: Implementer's Guide, May 2002, http://standards.ericsson.net/kremer/draft-ietf-rohc-rtp-impguide-01.txt.

TestNet: Let's Test Together!

Ana Cavalli[1], Edgardo Montes de Oca[2], and Manuel Núñez[3]

[1] Institut National des Télécommunications GET-INT
91011 Evry Cedex, France
Ana.Cavalli@int-evry.fr
[2] Consultant Telecom, 2 Rue kuss, 75013 Paris, France
edmo@wanadoo.fr
[3] Dept. Sistemas Informáticos y Programación
Universidad Complutense de Madrid, E-28040 Madrid, Spain
mn@sip.ucm.es

Abstract. In this paper we briefly describe the main goals and organization of TestNet, a proposal for the creation of a Network of Excellence in the scope of the 6th Framework Programme of the European Community. TestNet: Integration of Testing Methodologies represents the joint effort of the different European testing communities to create a common framework to improve all the aspects of the testing process.

1 Rationale

Borrowing from [Dah95] we may say that throughout history, society has created increasingly complex systems that put themselves and their environment at risk. In fact, human-made catastrophes have increased in frequency and magnitude with industrialisation. Early designers were able to learn readily from mistakes, leading to important advances in different areas, such as transportation, architecture, and energy production. Unfortunately, the growing complexity of present-day systems and the critical application areas in which they are used have made it more difficult to avoid introducing the possibility of catastrophic failures. This is in particular true for modern software systems.

Testing, the process of checking that a system possesses a set of desired properties and behaviour, has become an integral part of innovation, production and operation of systems in order to reduce the risk of failures and to guarantee the quality and reliability of the software used. The activity of testing is already a flourishing area with the active participation of a large community of researchers and experts. Thus, there is a high level of consciousness of the importance and the impact for the future deployment and use of software and systems. There is also an awareness of an increasing need to automate the testing activity in all application areas, that this activity should be taken into consideration in all phases of the software and system life cycle, and that interoperability and standardisation rely on it. An important problem is that different testing communities use different methods. Roughly speaking, we can identify two testing communities: *Testing of software* and *testing of communication systems*. Until very recently their research had been carried out with almost no interactions. So, there is an urgent need to co-ordinate research tasks and to find a synthesis between different techniques developed in isolation by each community. In other words, the different testing communities have realised the need to unify their research efforts to define a common framework to confront the above-mentioned problems. Another relevant problem is that the existing techniques and tools for testing are not adequately applied or even known in industry. Although scientific advances are needed in several areas, even well established technologies (as an example, code coverage measures) are rarely applied as they should. Testing activities are still very expensive and often relegated to the last stages of development, so that when time and resources start to be scarce, they are often partially sacrificed. We consider this to be a crucial problem, and believe that researchers in the field should urgently take two types of measures:

- to increase awareness in the area among the software and communication industries, and
- to make the test activities more cost/effective by developing better tools and by defining techniques seamlessly integrated into the development processes.

The research on testing has been reflected in several international conferences (e.g. *IFIP-FORTE/PSTV, IFIP-TESTCOM, IEEE-ICNP, ASE, ISSTA, IEEE-COMPSAC*). Further, standards bodies (e.g. *ETSI, ISEB, ITU-T, ISO, IETF*) have also dealt with testing. Unfortunately, these instruments lack the structuring effect that would arise from the organisation of a European Network of Excellence (NoE). We propose to establish a European Network of Excellence in the area of software testing that we call TestNet. The past, on-going and necessary future activities in the area of testing, as well as the importance of it in all application domains (communication services, nuclear power, transport, aeronautics, etc), make the subject of testing an important issue to be undertaken at the European level.

The subjects related to TestNet naturally fall into the priority thematic area 1.1.2 (*Information Society technologies*) and its research priority (1.1.2.ii *Communication, computing and software technologies*). The network will mobilise researchers and experts in the field of testing "mobile communications, consumer electronics and embedded software and systems" that is essential to assure that Europe remains a leader in the field and continue to offer innovative, integrated and adapted methods and techniques to tackle the growing complexity of software and communication systems and to prevent economic failures (maintaining cost-efficiency) or social catastrophes (maintaining reliability). The current European research situation is already mature for creating consensus and standardisation. So there is an urgent need for concerted actions in the field to maintain the competitive advantage of Europe. This effort will have an important impact in the production of secure high quality software based systems, reducing social and environmental risk in the areas these systems are used. For all these reasons we feel that the financial support from the EC would result in great advantages for the European community as well as its industry and research. The creation of a NoE will serve as a catalyst that will stimulate collaborations between the main actors in the field and the achievement of the objectives described in the next section. It will have impact on the short term by stimulating the collaborations between the participants; in the medium term by reaching agreements and setting up projects and industrial platforms; and, in the long term by defining strategies and actions (standardisation, training, ...) that will extend the existence of the NoE and its impact beyond the present duration planned by this proposal. In addition, it is the objective of TestNet to encompass transversal applicability. Even though it mainly concerns the Information Society Technologies domain, it is also applicable in other priority areas such as the fields of aeronautics and nuclear energy. In fact, some of the participants are strongly involved in these areas and thus the impact of the results obtained will extend to them. We also plan to interact with other NoE's as well as with other projects working on real-time and performance aspects. In particular, we will collaborate closely with the British network *FORTEST*, the *Accompanying Measures on System Dependability*: *AMSD*, *ENCRESS* and *SECUREUBINET* (proposed networks on safety of critical systems that include testing) and the French research networks *RNRT* and *RNTL*.

Taking into account the previous considerations, we believe that for software and telecommunication systems testing the following aspects should be addressed:

- There is a great need to automate the testing procedure to reduce time to market and to improve software quality at the same time.
- There is a great need to define best practices that take into account different existing techniques in different domains that are applicable for industry.
- Testing techniques are not keeping up with software development techniques and processes. It is necessary to take up different trends in software development and study the impact of integrating testing techniques and their usability.
- Testing is not always taken into consideration as it should be. Universities and other educational institutions often do not include it in their curricula. In enterprises, the effort needed is often underestimated.

2 Objectives of TestNet

In addition to coping with the problems identified above, the main goal of TestNet is to integrate the knowledge accumulated by the most successful European teams working on testing so that the critical mass necessary to lead this area of research can be reached. Briefly, the main goals to be achieved by the network are:

1. Co-ordinate and bring together major European research groups from different testing communities and domains to strengthen and develop the collaboration in the area at the European level:
    (a) by stimulating the technological transfer and sharing of experience,
    (b) by co-ordinating research activities and defining its roadmap (medium and long term strategy),

(c) by adapting activities to integrate their competencies taking advantage of the multi-disciplinary character of participants.
2. Promote testing as a well-defined and important activity in education, research, software engineering and industry.
3. Promote the integration of different testing techniques and development tools. Stimulate the creation of open tool environments so that different test and development tools can be seamlessly connected.
4. Promote innovation and take-up of new technologies through small and medium enterprises (SMEs) and through the creation of integrated projects where industry takes a leading role.

3 Approach and Planned Activities

In order to achieve the objectives we propose the following steps and organisation.

3.1 Creation of Working Groups

The partners collaborate in working groups centred on different topics. The (tentative) list of topics is:

WG1: Interoperability and conformance testing platforms and application to heterogeneous networks (e.g., next generation communication networks, mobile and wireless communication systems).
WG2: Testing of embedded software and fault-tolerant and safety-critical systems (e.g., aeronautics and railway control).
WG3: Languages for systems and test description (e.g., UML, SDL, B, Z, MSC, TTCN).
WG4: Techniques and processes for the new paradigms of software development (e.g., component-based software testing and testing of product lines).
WG5: Theoretical foundations of testing (e.g., measures for quality of testing and quality of software).
WG6: Platforms for testing of industrial applications.

3.2 Actions

In addition to the tasks carried out by each working group, we plan the following *cross-working group actions* to accomplish the goals of the NoE:

- *Roadmap for testing*: elaboration of a strategy for research and development in all the aspects and phases of the testing process.
- *Education and training*: elaboration of new curricula to integrate testing methodologies into university degrees. Many partners already propose specific courses on testing but they would benefit from a better overall vision more adapted to industrial and practical needs. Creation of Pan-European masters as well as a common Doctoral program.

- Elaboration of *best practices* applicable globally, but also specifically for each area, for obtaining reliable software and hardware.
- To *synthesise* and define the necessary *integration of practices and techniques* deemed necessary in the development and operation of critical software and hardware.
- Participation in the *definition of standards*: *ETSI, ISO, ITU-T, ISEB, IETF,* etc.
- Promotion of *common activities with enterprises*, with special emphasis on SMEs. In particular, creation of academia-industry laboratories in the countries having the necessary critical mass.

3.3 Scientific Committee

The NoE will have a steering committee composed of the main co-ordinator of the network, the responsible person for each cross-working action, and the co-ordinators of the working groups. There will be two additional members of the steering committee: a representative of industrial partners and an academic member responsible for technological transfer. This committee will meet on a regular basis every three/four months. However, extraordinary meetings may be deemed necessary if unexpected situations arise (e.g. conflict solving). The steering committee will have the following responsibilities:

- Organisation of joint workshops (i.e. including all the working groups) as well as open workshops (i.e. including participants outside the network). It is planned to organise these two events once a year (with a six months separation).
- Organisation of summer schools covering the different aspects and communities of testing.
- Co-ordination activities regarding external projects.
- Stimulation of the development of research projects among the participating members and establishment of IPR (Intellectual Property Rights).
- Integration of new partners.

3.4 Co-ordination Activities

Co-ordination activities and collaboration among partners are essential for the deployment and relevance of the NoE and so we consider them separately. First of all, activities already carried out in the field will be continued and further stimulated by organising meetings and discussions. In addition to the co-ordination activities charged to the steering committee, partners will carry out the following activities:

- *Working Group Meetings* (approximately every 6 months). They will serve not only to promote collaborations but also to refine medium and long-term research strategies to tackle the various aspects of testing from different perspectives.

- Organisation of *training* and *exchange* of researchers and students (short and medium term).
- *Share of resources*, establishment of a common knowledge base and shared tool base. Organisation of video/audio conferencing, a web site, and other interactive working tools.
- *Promotion* of *results* and *efforts* undertaken through presentations, publications, participation in standard bodies and existing conferences, contests to stimulate the participation of young researchers and teams.
- *Stimulation* of a *two-way transfer* of technology and experience between research institutions and enterprises participating in different application areas.

3.5 Management of TestNet

The co-ordinator of the network will be in charge of its management. Given the dimension of the network, she will be helped by a full-time manager. In addition, the responsible persons for cross-working group actions together with the working group co-ordinators will participate in the management decisions. The management committee will be responsible for the administration, the financing, and the budget of the network. It guarantees that the network's terms of reference are respected. In case of problems or modifications, it is the role of this committee to react and initiate the corrective actions rapidly.

In addition to the management committee, an *external council* will be created to organise external evaluations and peer review of the research results and collaborations. In particular, they will control the quality of work and deliverables by assuring that the review and validation of documents, software and integration results is correctly carried out.

4 Partners in TestNet

In order to achieve the stated goals, TestNet will include the most renowned experts and research bodies across Europe in the fields of software testing and testing of communication systems as well as in applications to industry. All of the people have worked for at least ten years in their respective field of knowledge (in most cases this experience is even more extended) and most serve as members of the program committees of international conferences related to testing. We have intentionally limited the size of the network such as to include only partners whose experience is both solid and strongly relevant to the purpose of the network. The number of researchers will be around 180.

Acknowledgments

This paper reproduces, with few changes, the proposal submitted to the *Call for Expressions of Interest* promoted in the year 2002 as a preliminary step of the 6th

Framework Programme of the European Community. We would like to thank those partners of TestNet which actively cooperated in the final redaction of this document by sending suggestions which improved the quality of the original draft.

References

[Dah95] A.T. Dahbura. Testing through the ages. In *8th IFIP Int. Workshop on Protocol Test Systems*, pages 1–16. Chapman & Hall, 1995.

An Open Framework for Managed Regression Testing

Naina Mittal and Ira Acharya

Tata Consultancy Services, D-4, Sector 3, Noida – 201301, India
{nainam,iraa}@delhi.tcs.co.in

Abstract. In the prevailing competitive environment, companies are facing tremendous market pressures to launch defect-free products in a timely manner. This challenge is compounded when a product runs into sustenance phase because complete regression runs need to be performed for each change/enhancement made in every build/release of the product-under-test. This paper discusses an architectural framework approach to address the above challenge, thereby aiming to make regression testing a simple, repeatable and automated exercise. The framework design encapsulates hierarchical test case management, multi-user support, product version maintenance, and automated test execution and result analysis to facilitate easy testing. The open architecture of the framework allows it to augment capabilities of some other testing tools by providing adapters to them, thus, eliminating the rigidity of use of a particular tool. The paper also draws a comparison of the architectural approach with other existing frameworks and presents a cost-benefit analysis of the suggested approach.

Keywords. Test automation, managed testing, test planning, test execution, test framework, networking equipment, regression testing, black-box testing, test scripts, test bench, hierarchical test case management, test plan tree, framework deployment, test-cycle reduction, testing tool collaboration.

1 Background

In the new economy, testers face the challenge of ensuring timely, defect-free product launches in the market. In this world of ever shortening internet times, minimal time-to-market is the key for steering ahead of competition. The goal of achieving this difficult target puts a direct liability on the development and test cycle times of products-under-test. This is almost impossible without defining a streamlined process for testing and incorporating desirable automation levels at every stage. A common problem faced by most test managers is that of managing the test process itself, which is a unique challenge, requiring judgment, agility and organization. In spite of sizable investment in tools (for example, traffic generators, voice simulators, script generators) that aid in testing of the product-under-test, test managers still face the dilemma of managing testing using these disparate entities to achieve an optimal testing solution. Test management challenges include establishing a communication mechanism between these test tools, controlling test planning, test execution and test results from a central control point, tracking test cycles of multiple builds/releases of the product-under-test, arranging tests in an organized manner so as to optimize on time as well as hardware resources, automating execution of tests and much more. [1], [2]. Thus, while the investment in testing tools may appear as the apparent solution to the problem, it is an even bigger problem putting them to use effectively and preventing them from becoming shelfware.

2 Challenges Faced in a Manual Testing Scenario

A manual test system is a costly exercise in terms of time and effort. It is usually characterized by unmanaged testing, inconsistent result logging and inefficient defect-management. Moreover, there is usually no provision for tracking test cycles of various product versions which maybe undergoing testing simultaneously. In spite of dedicating a good-sized testing team for product testing, there is always the likelihood of defects left unnoticed in the shipped deliverable, which are later discovered and reported by the recipients of the deliverable. This situation is unwelcome for every product vendor, as defect detection after delivery not only has a direct impact on cost, it also implies vendor's inability to contain defects.

The need for a test management framework stems from the aforementioned problems that are associated with a manual-testing scenario. In the context of telecom/networking equipment, testing is an even more complex proposition because it involves setting up test benches and having a number of third-party test equipment in each test bench required to aid testing of different functions of the product-under-test and to simulate multiple test scenarios. This pre-test setup is followed by manually configuring the product-under-test as well as peripheral test equipment before running each test case. Finally, a test case is executed and the results as observed on the console are logged for reference. The complex nature of these tests makes their execution susceptible to errors arising due to either incorrect configuration of peripherals or incorrect test execution by testers who are likely to commit errors either due to inadequate domain knowledge or simple negligence. A mid-size system can have large number of possible test scenarios, which when tested manually take significant amount of time.

Thus, having a managed test framework to control the test management tasks, integrate the various testing tools and automate test execution resulting in better quality deliverables would make the test management process simple.

3 Addressing the Most Cumbersome Stage of Testing

Products go through multiple stages of testing in their lifetime from conception to release. Producing reliable systems requires a well-planned, comprehensive application of several techniques throughout the development life cycle, collectively known as Verification and Validation [3]. The basic stages of testing include:
- Unit Testing
- Module Testing
- Sub-system Integration Testing
- System Testing
- Regression Testing

Although it is desirable to have maximum possible automation at each stage of testing, it is also important to consider cost for introducing automation at every stage. Automation should be adopted at stages, which consumes most time and effort. This decision essentially depends on the type of product being developed. If a product is conceived with the aim of delivering it once and for all, with no liability for enhancements and maintenance support subsequent to release, then it would be intelligent to invest in automation tools at the unit and integration system testing stages. However, if the product is being developed with the intention of providing sustained support to product recipients and periodic functionality addition thereafter, then the most obvious choice for automation would be the regression testing stage.

Regression testing [4] stands for testing modified code under the same set of inputs used in previous tests. This way we can ensure that modifications in the code did not introduce new errors into the software and verify that modifications successfully eliminated existing errors. It involves executing a pre-defined battery of tests against successive builds of the product to verify that bugs are fixed and functions that were working in the previous build haven't been broken. A more precise definition [5] terms regression testing as equivalency testing, that is, rerunning a suite of tests to assure that the current version behaves identically to the previous version except in such areas as are known to have been changed. Regression testing is an essential part of testing, but is very repetitive and can become tedious when manually executed build after build. Moreover, while running the entire regression test set manually, testers tend to be selective in test case execution due to stringent time constraints, again leaving scope for the final deliverable containing a few unnoticed defects. To further optimize the regression testing, a number of test selection techniques have been proposed by many. [6], [7], [8], [9].

A typical test scenario requires some basic essential steps for an automated black-box testing exercise. These include:

- Test Plan Building
- Test and User Management
- Test Scheduling and Automated Execution
- Result Analysis and Logging
- Report Generation
- Defect Tracking

An ideal test framework should encapsulate all these functions to automate the test management and execution tasks in an effective way. It should also implement a mechanism whereby all test cases of the test plan of the product-under-test can be classified and arranged in the most optimal manner to facilitate organized testing subsequently. Moreover, the framework should be able to store scripts and test parameters and use the appropriate scripts at the time of testing.

4 An Open Framework Approach

Keeping regression testing as the target for automation, this paper describes an open architectural framework for regression testing, which intends to address some of its bottlenecks and irritants. Though the framework specializes in automating the system (black box) testing of networking software, (such as that of devices like routers, switches, integrated access devices and others of similar nature), it is fairly generic in nature and can be extended for the testing of other software as well. The framework eliminates manual intervention by mimicking the actions of a tester through its scripts. With this framework, the commands or instructions that a tester would otherwise issue manually (for example, through a PC), come from a simulation of the keyboard rather than the keyboard itself. To validate the results of the test, the framework captures a portion of the terminal output, compares it to expected results, and then decides whether the test case succeeded or failed.

Overview

The salient features of the framework are discussed in this section. Essentially, it is designed to provide for black box testing of software (for example, networking software), given the test plan of the product under test. It supports a multi-threaded, multi-user environment and is responsible for maintaining sessions with all the users who have logged on.

It provides a GUI which captures all information related to regression tests such as test script parameters, hardware test bed(s)[1], test plan for one or more builds/releases of the product, users of the system, test schemes[2], etc. At the time of test execution, it allows (through its GUI) users to Start, Stop, Pause or Cancel the current execution of test cases at any given point of time.

It also facilitates the test cases to be easily designed and organized in order to streamline the testing process by grouping optimally the functional features as well as hardware resources required for testing. This is achieved by introducing the concept of a dependency tree in the framework design. This feature gives the users the flexibility to run any test case or a group of cases of their choice, and also serves to utilize the hardware resources optimally. At the end of a test execution run, a number of reports are generated for user reference.

The framework has an interface with advanced automation scripts giving the user complete control over test execution and provides for quick, consistent, reliable and repeatable automated tests. Simulating the presence of a "virtual" user, the scripts execute pre-defined streams of actions, and compare the output to valid states to determine whether or not the test was successful.

Framework Architecture and Components

The framework comprises of two main components, namely:
- Engine component
- Script component

The engine component is the test case manager and test script driver. It is the central controlling entity that controls the concurrent execution of scripts on various test beds and handles multiple simultaneous sessions with testers who are using the framework for testing. The engine further has three sub-components, namely:

---

[1] A group of equipment physically connected in a certain configuration, which is used to test a set of functional features of the product under test.
[2] Those sub-sections of the test plan of the system under test, for which the user wishes to run test cases at a particular instance.

- Engine client (GUI)
- Engine server
- Database repository

The script component consists of intelligent scripts that actually simulate the presence of a user performing a device configuration followed by a test case execution. The scripts also have the ability to determine whether a test case has executed successfully or failed. The scripts, after completion, return the result of the test case to the engine component which is responsible for logging the same in its database and reflecting the result on the GUI so that the tester is constantly updated with the latest status.

The architecture of the framework is depicted in Fig.1. The various sub-components of the engine are discussed in the ensuing sections.

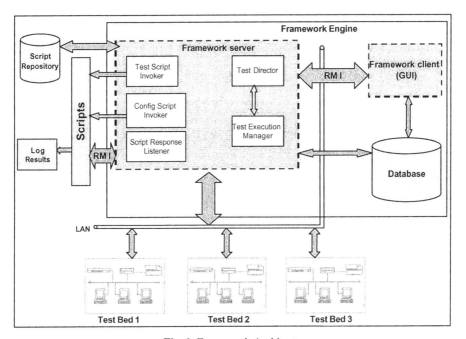

Fig. 1. Framework Architecture

Engine Client (GUI). The engine client provides a convenient interface to the user for entering the data for storage into the framework and for viewing results of the test execution. There are a number of screens, which facilitate this data entry. The kind of information that has to be entered into the framework engine pertains to the release/build of the product-under-test, the details of equipment that constitute a test bed, the number of such test beds, the test plan tree of the product-under-test, test scripts and their expected parameters, etc. If required, more information can be added (such as additional test cases defined for a feature) or existing data updated (such as test scripts, script parameters, etc.) at a later time. Once all the information discussed above is registered with the framework, the users just need to define their individual test schemes and perform tasks such as Start Testing, Stop Testing, and Verify Test Setup for each run of the test cases.

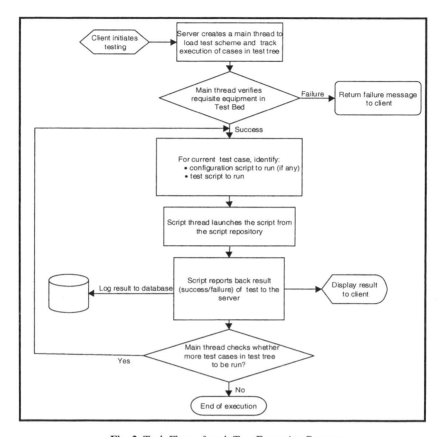

Fig. 2. Task Flow of each Test Execution Request

At the end of a test execution run, progress bars on the main screen give an at-a-glance view of the percentage of the test cases that have succeeded/failed. Detailed results can also be viewed through the GUI itself. A 'History' screen is provided for the purpose of viewing results of any past execution. The results view of historical runs can be filtered by specifying sorting criteria. Summary reports of the test execution results can also be generated through the GUI.

Engine Server. The interaction of the server with other entities, that is, client, database repository and scripts is illustrated in Fig.1 and the task flow of each test execution request as processed by the server is represented in the flowchart in Fig.2. At all times, only one instance of the server runs and it allows multiple clients to connect to it concurrently. The server allocates a unique identifier to each test execution request from a client and tracks that request to completion maintaining this identifier as the key. While multiple clients can execute their schemes concurrently, the execution of test cases within a test scheme is sequential. The server is multi-threaded with a thread allocated to each script, and is responsible for launching the correct script and waiting (for a pre-defined finite time period) for it to return execution results. To maintain the integrity of each client's test execution, the server ensures that each test execution occurs on a separate test bed so as to prevent interference.

Once a script reports its test execution result to the server, the server stores the result in the database, with the unique test execution identifier as the key. Java RMI is used for communication between client, server and scripts. Taking into consideration that all entities in the test bed need not necessarily support the same operating system, the framework has been developed in Java so as to provide platform independence and hence greater flexibility.

Database Repository. The framework uses a Relational Database as a repository for persistent storage of information (that is, products, test plans, test cases, scripts, parameters, test beds, test results). Most of this information is defined in a one-time data entry by the administrator through the GUI and can be reused for later executions. The framework currently provides support for Oracle 8i and MySQL [10] as the underlying databases.

Scripts. The scripts in the framework are primarily written in Expect/TCL, but it also has adapters to some other scripting languages such as Perl, Python, C, etc (Refer Fig.4). Expect [11] is a non-proprietary scripting language, which automates interactive testing by imitating the manual test steps performed by an individual. With the use of Expect, it is possible to simulate a virtual user and perform actions that a user would carry out manually to conduct a test. The scripts in the framework are written based on the test case procedures in the system test plan of the product-under-test. This test plan is pre-defined by the testers and contains all possible test cases to test the various functional features of the product. The scripts are written in such as manner so as to encapsulate the functionality defined by each test case. Thus, the scripts have the intelligence to implement the logic of each functional feature of the product-under-test and of each test case within that feature. The engine controls the execution of the scripts in the correct order and handles the results returned by these scripts.

In case the functionality of the product-under-test undergoes modifications/enhancements, it is likely to impact the test cases for that functionality. Since the scripts corresponding to the changed test cases in turn also get impacted, therefore, it is required to modify the existing scripts or add new scripts in order to reflect the changed functionality.

The new scripts can easily be integrated into the framework engine and subsequently run in an automated fashion as part of the regression test suite for that product.

Hierarchical Test Case Management

A key feature of the framework is the introduction of a dependency tree concept in its design for effective test case organization and management (Fig.3). The grouping of test cases is done with a two-fold objective:

- To group test cases in functional clusters starting at the highest level functionality right down to the test case level, so as to facilitate testers to start testing from any intermediate node in the hierarchy and test all the test cases that lie below that node.
- To group test cases in such a manner that the hardware resources are utilized optimally and clusters are formed so that test cases using similar hardware configurations are placed in the same group. This would reduce manual intervention to a minimum.

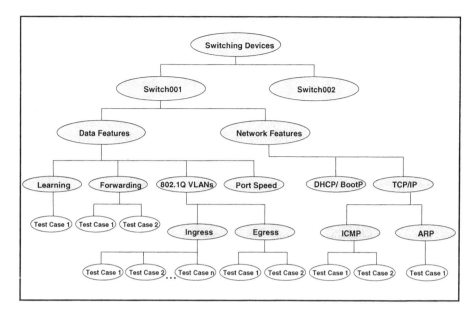

Fig. 3. Hierarchical Test Case Management

The hierarchical test case management approach:
- allows the association of multiple test cases/scenarios to a particular node.
- allows the association of a particular node to a hardware setup. All cases lying below that node in the hierarchy will use the same hardware test setup. For example, if we define a test bed setup at node "Data Features" (Fig.3.), then all branches of the tree lying below this node in the hierarchy would be able to be tested without altering the physical configuration of the defined test bed.
- streamlines the functionality based on scenarios.
- gives the user the ability to perform testing from a particular node/point.
- covers all the test cases under that node.
- bypasses all cases under a node, in case of failure at that node.
- uses the existing structure of the test plan as a baseline, in case a new release is introduced.
- allows testing of a new product, by reusing the entire tree of an existing product for the new product (provided the new product is similar in nature to the existing one and has the same test plan structure).

5 Collaboration with Other Testing Tools

With the growing need for automation, there is a wide spectrum of testing tools available in the market. One of the limitations of most of the testing tool from various vendors is the proprietary scripting language or APIs that come as a package with the tool and hence enforce the binding for a user to use it as it is. Our framework endeavors to do away with this limitation by providing adapters to other tools and scripting languages, so that the strengths of both can be leveraged and synergized to provide an optimal testing solution. The test case management capabilities of the framework can be integrated with an underlying proprietary or non-proprietary scripting tool, so as to benefit from the advantages provided by both. Envisaging this, adapters have been developed in the framework for integrating with underlying scripting languages such Expect, TCL, Perl, Python and 4Test (A proprietary scripting language of the SilkTest® Tool [12]).

Moreover, since various tests can be automated best in a particular scripting language, (for example, '4Test' for a GUI automation; 'Expect' for Command Line Automation), the framework can interleave running of these scripts, by invoking the appropriate adapter at runtime. The above feature of collaboration with other tools is illustrated in Fig.4.

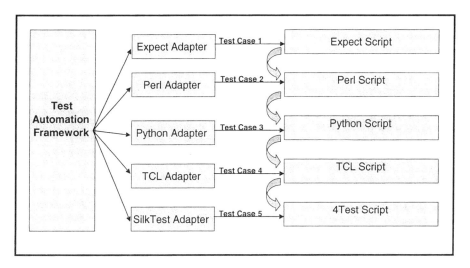

Fig. 4. Collaboration with Testing Tools

6 Framework Deployment

A typical instance of the framework deployment is depicted in Fig.5, with reference to which, the step-wise explanation of how a test execution request is handled by the framework is discussed as follows:

1. The user at the framework client PC chooses a test scheme (that is, test cases to be run in one cycle) from the framework GUI and triggers the testing by pressing the 'Start Testing' button.
2. Once the request reaches the framework engine server, the server extracts information from the database pertaining to:
   - The hardware test bed on which the tests are to be run. (Note: Concurrent testing on more than one test bed is supported by the framework, so it is essential to determine for which test bed a certain execution is targeted.)
   - The scripts corresponding to the test cases in the test scheme.
   - The correct parameters that are to be passed to each script for it to execute the intended functionality of each test case in the test scheme.

3. The framework engine server launches the scripts one by one from the script repository in the correct order onto the hardware test bed. Since the framework is on the same LAN as the test bed, the scripts are able to communicate with all the equipment within the test bed. The scripts are launched in the order in which the test tree is defined in the test scheme. The execution commences with the execution of 'verification' scripts that poll requisite equipment in test bed to ensure their presence for subsequent testing. This is followed by running of 'configuration' scripts that execute the pre-condition to a test case by configuring all the equipment in the test bed with the desired configuration parameter values. Finally, once the configuration is successful, the 'test script' is run. In case the configuration for a particular test case is not successful, then the framework bypasses the test case execution and moves onto the next one.
4. Each script, after completing its execution, reports its result (success/failure), to the framework server. The success or failure result is also accompanied by a brief remark regarding the reason of success/failure.
5. The server logs these results into the database. Simultaneously, the result is updated on the framework client GUI in the form of a progress bar for user reference.
6. Once the entire test scheme has been executed, the same is conveyed to the user by a prompt on the GUI. The user can now choose to view or print the various summary reports, which provide details of the test execution results. The script failures can then be investigated and resolved by the user.

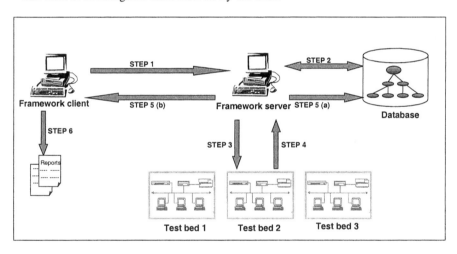

Fig. 5. Framework Deployment Scenario

7 Other Applications of the Framework

Though in this paper, we have taken networking equipment testing as the test scenario, the open architecture of the framework allows it to test a number of other applications and does not limit its ability to testing of networking equipment alone. We have deployed it for the testing of a certain operating system abstraction software, which provided abstraction to an embedded software developer from the details of any underlying real-time operating system. This operating system abstraction software [13] was required to be regression tested with each port to a different operating system in order to validate its functionality across various operating systems. The system test cases were to remain the same for regression runs across different operating systems and hence the test automation framework provided a befitting solution for the regression testing of the software.

8 A Comparison with Other Frameworks

The framework under discussion is comparable to some of the others available in the market, intending to do similar tasks. Some such equivalent frameworks include TestDirector from Mercury Interactive [14], STAF (Software Test Automation Framework) [14] and Telcordia™ Mynah System [16].

A brief parallel with each of these is drawn in the following discussion:

TestDirector

TestDirector is Mercury Interactive's software test management tool. TestDirector helps organize and manage various phases of the software testing process, including planning tests, executing tests, and tracking defects. TestDirector is Windows-based and helps create a framework, which acts as a foundation for a testing workflow.

If a parallel is drawn between our framework and TestDirector, it will be observed that they both perform a number of similar tasks, but in a different manner. These include management of test plans and test cases for each product, supporting automated test execution (local or remote) with the help of an underlying scripting language/tool, result logging and report generation for a test set executed, and multi-user support.

TestDirector has additional components like Version Control Manager (for configuration management of test scripts), Defect Tracker (for logging, assigning, fixing and tracking defects to closure), Test Plan import facility (for directly importing test plans from a word document or an Excel spreadsheet), reordering facility (for rearranging test cases before execution), integrated support for WinRunner, [17] (script generation tool from Mercury Interactive, which has its proprietary scripting language-TSL) and third-party tool integration facility through its Visual API.

Our framework has differentiating features like an intelligent test plan tree (for hierarchical management of test plans based on product functionality), facility for specification of configuration and test parameters (which are stored persistently for subsequent regression runs), platform independence (as it works on Windows as well as Linux platforms), and facility to run interleaved tests through different underlying scripting languages. Our framework is focused on the test process of network equipment as against TestDirector, which is a general framework for any enterprise application software testing.

STAF

The Software Testing Automation Framework (STAF) is a framework designed to improve the level of reuse and automation in test cases and test environments. The goal of STAF is to provide a complete end-to-end automation solution for testers. This framework is open source, and is available under GNU license. STAF can run locally or remotely, provided STAF is installed on the machine. It operates in a peer-to-peer environment; in other words, there is no client-server hierarchy among machines running STAF. A number of processes can be launched simultaneously provided an equal number of resources are available for STAF to be able to handle their concurrent execution. Back-end applications can be developed in Java, C, C++, REXX, Perl and TCL.

When a parallel was drawn between STAF and our framework, it was observed that both support console environment, both have a Java client (for specification of test data and triggering execution) which can work across platforms, both support a multi-user environment, amongst other features.

What STAF does not apparently have is a persistent data storage repository, a report generation facility and managed testing of test scenarios.

All in all, the STAF framework exhibits a lot of resemblance to our framework as is evident from the comparison drawn above and can be considered as its subset as far as its features and capabilities are concerned.

MYNAH

The MYNAH System provides a fast and reliable way to leverage data stored in legacy systems by using scripts to automate functions. It performs the data-intensive work that traditionally required manual intervention. The MYNAH System can be used for:

- Data reconciliation
- High-volume data entry
- Database and system conversions
- Proactive system monitoring
- Rapid application development
- Software testing

Like our framework, the MYNAH System supports TCL scripts at the backend for automation of test cases. It also provides a Graphical User Interface, through which:

- Test scripts can be captured using the automated script builder (which allows scripts to be recorded).
- Test scripts can be modified and enhanced using TCL language package commands.
- Test results can be processed using test management features.

MYNAH System can also be customized for specific testing requirements and apparently supports Unix platform.

9 Return on Investment

In order to arrive at the return on investment achievable with the use of this framework, a pilot was performed to automate the regression test process of an integrated access device (IAD). This device under test integrated voice, data, video and Internet services on up to two broadband connections over TDM and Frame Relay or DSL or ATM. It combined a channel bank, CSU/DSU, IP router, DHCP server, and firewall in one small footprint. In order to test its features in a black box environment, a number of support equipment such as CSU/DSU, DAC, Router, Abacus (third-party tool for voice generation), and PCs were required to be present in the test bed. Various parameters of these support equipment were required to be configured with different values before actually running individual test cases. A sample of 15 data features of varying complexity was picked for the purpose of the pilot automation. The features of the device that were tested are illustrated in the functional hierarchy in Figure 3.

These were first executed manually to estimate the time taken for one regression test run without any automation aid. These were then organized in a functional hierarchy and automated by writing scripts. They were then integrated into the framework and executed in an automated fashion. The exercise claimed an investment of 3 months, but the benefits derived from it subsequently made it a worthwhile activity. The summary of the manual vs. automated tally obtained at the end of the pilot is presented in the following table.

Table 1. Manual vs. Automated Execution Results

No. of Test Cases in Scope	Manual		Automated	
	Execution Time (hh:mm)	Elapse Days	Execution Time (hh:mm)	Elapse Days
15 data features consisting of 125 test cases	70 hours 45 min.	9 days	16 hours 30 min.	1 day (unattended)

$T_{SE} = [(Time_{manual} - Time_{automated})/ Time_{manual}] * 100 = 76.6\%$
$T_{SD} = [(Elapse_{manual} - Elapse_{automated}) / Elapse_{manual}] * 100 = 89\%$

where:
T_{SE} = Time saving in terms of execution time
T_{SD} = Time saving in terms of elapse days

10 Conclusion

In today's environment of plummeting cycle times, test automation has become an increasingly critical and strategic necessity. It should be adopted as an integral part of the test cycle because few investment opportunities offer a more tangible or reliable return on investment than test automation. A number of frameworks have been conceived with the intention of facilitating test automation at various testing stages. [18], [19], [20], [21]. The previous sections discuss one such framework which attempts to alleviate testers from the tedious task of manual testing.

The framework has been put to test in a real test environment and has proved to be extremely cost-effective in cutting down the test cycle time, as well as meeting other objectives of test automation.

Future challenges lie in enhancing its capabilities to cater to a wider segment of applications/systems to test. In its current state, it can help in the functional/regression test automation of networking equipment, thereby aiding networking equipment vendors in making their test process efficient and easy. Moreover, it can be integrated with some of the other testing tools available in the market.

The usage of the framework for testing more general applications, where a request-response mechanism through a CLI is used to test the application, is being looked at. The framework in such cases would be used to store and manage the test cases, execute the automation scripts in the correct sequence and generate summary reports. Another value addition that is being planned is the building of ready-made test suites for the conformance testing of various protocols.

References

1. Rex Black: Managing the Testing Process: Practical Tools and Techniques for Managing Hardware and Software Testing. Wiley, John & Sons, Incorporated
2. Edward Kit: Software Testing in the Real World (STRW). Addison-Wesley. (1995).
3. Wallace, D. and Fujii, R.: Software Verification and Validation: An Overview. IEEE Software. (May 1989). 10-17
4. Bret Pettichord: Seven Steps to Test Automation Success. STAR West. (November 1999).
5. Boris Beizer: Black-Box Testing: Techniques for Functional Testing of Software and Systems. John Wiley & Sons. (1995).
6. Agrawal, H., Horgan, J., Krauser, E. and London, S.: Incremental regression testing. In Proceedings of the Conference on Software Maintenance (September 1993). 348–357.
7. Chen, Y. -F., Rosenblum, D.S., and Vo, K. -P.: TestTube: A system for selective regression testing. In Proceedings of the 16th International Conference on Software Engineering (ICSE '94), Sorrento, Italy. (May 1994). 211–220.
8. Todd L. Graves, Mary Jean Harrold, Jung-Min Kim and Adam Porter, Gregg Rothermel: An Empirical Study of Regression Test Selection Techniques. ACM Transactions on Software Engineering and Methodology, Vol. 10, No. 2. (April 2001). 184–208.
9. John Bible and Gregg Rothermel: A Comparative Study of Coarse- and Fine-Grained Safe Regression Test-Selection Techniques. ACM Transactions on Software Engineering and Methodology, Vol. 10, No. 2. (April 2001). 149–183
10. MySQL (http://www.mysql.com/)
11. Expect Home Page. (http://expect.nist.gov/)
12. Segue Solutions (http://www.segue.com/html/s_solutions/s_silktest/s_silktest_toc.htm)
13. mOSAic Design Document , Tata Consultancy Services.
14. TestDirector (http://www-svca.mercuryinteractive.com/products/testdirector/)
15. STAF (http://sourceforge.net/projects/staf/)
16. Mynah System (http://www.telcordia.com/ADAPTX/mynahbft.html)
17. WinRunner (http://www-svca.mercuryinteractive.com/products/winrunner/)
18. Hoffman, D. and Strooper, P.: Automated Module Testing in Prolog. IEEE Transactions on Software Engineering, Vol. 17, No. 9. (September 1991).
19. Chang Liu: Platform-Independent And Tool-Neutral Test Descriptions For Automated Software Testing. In Proceedings of the 22nd international conference on Software Engineering, Limerick, Ireland. (2000). 713 - 715
20. Usha Santhanam: Automating Software Module Testing for FAA Certification. In Proceedings of the annual conference on ACM SIGAda annual international conference (SIGAda 2001), Bloomington, MN. (2001)
21. Schot C.A., Sim M.N. and Kist P.M.: ANT - A Test Harness for the NELSIS CAD System. In Proceedings of the conference on European Design Automation, Congress Centrum Hamburg, Hamburg, Germany. (November 1992)

TUB-TCI
An Architecture for Dynamic Deployment of Test Components

Markus Lepper, Baltasar Trancón y Widemann, and Jacob Wieland

Technische Universität Berlin, Fakultät IV, ÜBB
Institut für Softwaretechnik und Theoretische Informatik, Sekr. FR 5–13
Franklinstr. 28/29, D–10587 Berlin
{lepper,bt,ugh}@cs.tu-berlin.de

Abstract. The test definition language TTCN-3 is currently under standardization by ETSI/ITU-T. Its intended field of application is testing and performance measurement of communication hard- and software. TTCN-3 does include mechanisms for specifying remote hardware access and for distributed execution of testing code components.
But for running compiled TTCN-3 code distributed onto distinct nodes of different vendors an architecture is needed which offers standardized means for dynamic, program controlled deployment, configuration and status inquiry of active and passive resources.
TUB-TCI is a proposal for such an architecture, characterized by (1) totally generic definition of component classes, (2) coexistence of standardized (XML based) and specialized (high-speed) communication channels, (3) a model-based, strictly formal definition given in Z and (4) a simple, minimized but powerful execution model.
Especially because of the integration of non-standard, high speed data channels TUB-TCI seems applicable for dynamical routing of real-time signals in general, beyond the field of test execution.

Keywords: Dynamic Deployment, Conformance Testing, XML based configuration.

1 Design Principles and Features of TUB-TCI

1.1 The TCI Problem in General

TCI stands for "Test Configuration Infrastructure" and is a topic discussed by a recently installed working group of ETSI, continuing the work presented in [6]. Protocol conformance testing and performance measurement — as far as communication technology is concerned, but possibly also in other fields — will increasingly have to deal with heterogenous ensembles of hardware nodes, in which tests are performed by co-operating pieces of code *distributed* to distinct hardware nodes, each of them may be based on very different technologies.

Any compiler, when translating a so called "abstract test suite" ("ATS") written e.g. in TTCN-3 (cf. [8]) to a collection of pieces of executable code ("executable test suite", "ETS"), can easily do this for different execution platforms.

But to really make these code objects co-operate, the pure language definitions and their semantics do not suffice. Indeed an infrastructure is needed which provides additional information concerning the concrete set-up, and which allows user data and control information to be passed between different nodes in a transparent way.

This arises from the following basic contradiction:

- Advanced concepts of testing, like in TTCN-3, provide the means for *dynamic*, program-controlled creating, deleting, configuring and linking test components. This feature will become more and more important, especially for automated, batch driven "over-night" tests, as well as for periodically scheduled, automated in-field "online-tests" in dynamically changing topologies.
  So the actual test code execution needs some information concerning (1) the specific capabilities and API definitions of the involved hardware nodes, their current topology and the currently valid set-up of addressing and routing, and (2) about the creation and configuration commands for certain hardware resources (timers, ports, local I/O-devices) and the required parameterization information, which is specific for the type and vendor of the hardware device.
- Both requirements conflict with the requirement of *abstractness*, i.e. that the same abstract test suite specification and the corresponding compiled code should be able to run in different hardware settings of divergent topologies: neither (1) the deployment strategies, which need information on the current hardware configuration, nor (2) the driver specific parameters should (or even: can) be contained in the source text of the test program.
- Consequently, there has to be a kind of "merging of semantics": The ETS-code will provide certain pieces of information, concerning e.g. "abstract timers", "abstract ports", etc., which are complete in the sense of the semantics of the abstract test suite, but only more or less sufficient w.r.t. the needs of the addressed particular hardware drivers. They must be completed by the TCI infrastructure, using some strategic knowledge concerning the concrete set-up, and passed to the different "drivers" of the selected hardware node, which are the only subsystems able to really perform the allocation and configuration.

Currently all these issues are addressed by the *hand-coding* of so called "test adapters", pieces of code which realize the mapping from the abstract semantics of e.g. TTCN-3 to the concrete interfaces of the concrete hardware setting. These code objects can be re-used only in a limited way, so the work has to be done from scratch for each family of test situations. This work is a source of further possible *errors* in the test process, and can hardly be considered intellectually challenging engineering.

A central intention of TUB-TCI is to relieve the implementor of the test adapter by some ability of *self-organization* of the underlying infrastructure, establishing means for declaring the additionally needed pieces of information in a generic way.

1.2 Basic Paradigms and Current State of TUB-TCI

These are the issues which must be addressed by any TCI, and TUB-TCI is a possible solution proposed by the authors. Its central design goals are

- Versatility,
- Robustness, and
- Precise Specification.

It may further be characterized by

- Co-existence of different paradigms, e.g. central and distributed knowledge, unified and specialized communication channels.
- Being equally well suited for interactive and batch driven testing.
- Preserving referential integrity.

TUB-TCI currently exists as a complete model-based specification (cf. [3]). In contrast to the official TCI specification by ETSI (cf. [4]), which concentrates on defining APIs callable by the executable test suite, our approach also defines the *behaviour* of the single subsystems and the rules of their co-operation, and thus specifies the behaviour of the total system.

There are severe advantages of such a model based specification which uses some abstract but (potentially) executable language:

- It allows to concentrate on the *intended functionality* and its realization by state transitions and message interchange operations. In contrast, when using a concrete programming language for modeling, i.e. "implementing", more than half of the text would have to deal with the specific idiosyncratics of this language, e.g. encoding of type sums, process scheduling, memory management etc.
  So the specification text turns out to be much shorter and clearer than the text of a conventional implementation, but nevertheless *is* a kind of implementation. So e.g. the grade of robustness of the rules could be tested by paper-and-pencil evaluations, in which human intention simulated the behaviour of "malicious" test components.
  The final translation of the mathematically defined state transitions (and the pure functional auxiliaries) into a conventional programming language can in most cases be done in a rather straight-forward way. Here any programming language and any operating system can be chosen, since the (non time-critical) configuration information is exchanged as standardized XML fragments.
- A model in an abstract, mathematically founded language should allow a wider and deeper *discussion* in the community, since basic mathematics are a language known to every engineer and project manager, — independent of her/his experiences with distinct programming languages.
- If the appropriate tools for the chosen modeling language are available, execution ("simulation") and tests can be performed on the mathematical model

immediately. Even *automated verification* of certain properties of the specification are possible, using model-checking techniques, which is practically impossible with general-purpose programming languages[1].

TUB-TCI is intended as a base for *discussion*. While being complete, consistent and ready to work, principal decisions do — of course — have some alternatives, and we do expect revisions when creating the first implementation.

The project has already proved that it is possible to define the behaviour of a complex and generic architecture *completely* through all system layers and all execution phases by precise mathematical means, provided that the level of *abstraction* is chosen for each system layer appropriately.

1.3 Principles of the Architecture

TUB-TCI assumes distributed testing to be performed by collaboration of distinct *subsystems*, hosted on the same or on different hardware nodes. Each test session is an alternating sequence of preparation phases (TPrep), in which components are installed and hardware resources allocated, and test execution phases (TRun), in which the behaviour of the total system is under control of the running "executable test suite". The main issue of any TCI design is that the state of the whole system is *dynamic*, that is, allocation, installation and control of components may happen under program control, i.e. in TRun.

TUB-TCI does *not* specify any strategic knowledge needed for deployment decisions, nor does it define any concrete classes of devices. However, it precisely defines the generic means for "installing" these pieces of information into a system, — the former by run-time services requested or received by a "test manager", the latter by a formalism for declaring new device classes.

TUB-TCI is specified by giving its operational semantics as a collection of transition rules. Most of these involve two subsystems, thereby defining the possible *message passing*. These rules and the data space they operate on are given as Z formulae[2]. Table 1 completely lists all services offered by all categories of subsystems.

All message formats are given as Z schemata, too. As soon as message passing crosses the boundaries of hardware nodes, it is implemented as an exchange of XML fragments. For implementation or practical standardization some canonical representation for Z schemata as XML schemata will have to be chosen. The Z, and not the XML representation is first-class resident in the TUB-TCI specification, because the dynamic semantics can only be formulated in the former.

---

[1] Due to the limitation of resources, these goals could not be addressed in our project and are left to future work.

[2] The notation used is indeed a *derivation* from Z: Strict requirements needed for the mathematical foundation of the Z semantics could be discarded, so that some shorthand notations (e.g. for type sums, step semantics, parallelism, reflection, i.e. converting schema *definitions* from/to data *values* describing schemas, and call of external driver functions) could be added for the sake of readability.

W.r.t. the work of the authors, the appropriate structure and semantics of the meta-language is indeed a central topic of *research* on its own.

TUB-TCI is *minimal*, as it defines the minimal required, specialized subset of the functionality of an "object broker", a "routing mechanism" and a "communication protocol". So TUB-TCI indeed does contain a "micro-CORBA", "micro-IP" and a "micro-TCP". This does in no case imply that we want to re-invent existing technologies, — contrarily: the state transitions specified on these layers of TUB-TCI can easily be mapped on existing implementations coming from "large", general purpose implementations (TCP/IP, CORBA), or can be implemented in few lines of code from scratch.

Because of this minimality, the specification of TUB-TCI does span the whole range of architectural layers, from top-level declaration of "semantic types" down to the lowest level containing the (loosely specified) primitive bus driver communication handshakes.

The architecture consists of the following layers:

- Basic node initialization and communication mechanism.
- Meta-Services for defining new actor classes and their access modes.
- Runtime services for creating, configuring and destroying active and passive entities.
- Runtime services for user data transmission.

2 Subsystems Forming a TUB-TCI Setting

In TCI it is assumed that a multitude of hardware nodes co-operate. These nodes are connected by communication channels of widely divergent technologies. On each node there is a "testing sandbox": Parallel to all other activities running on a node, this sandbox contains all testing or measuring code, which is controlled by TCI.

All activities in TUB-TCI happen as *Service Requests* between two *subsystems*, which are implemented by message exchange. We define different categories of such subsystems, the most important of which are characterized as follows:

2.1 TM = Test Manager

Even in case of distributed execution, testing in practice will always be controlled by one central instance, which launches, starts and stops the diverse test phases and calculates the intermediate and final verdicts and results. This is true for batch driven as well as for interactive test execution.

Such a central instance is modeled by one single instance of the "Test Manager" (TM) category of subsystems. A TM must be "external" to each TCI concept, as its behaviour should not and cannot be specified therein.

Contrarily, the currently active TM is an arbitrary program, of which only two aspects are specified:

(1) Each TM is an active component issuing service requests for installing and controlling other active or passive components (i.e. it is a subclass of the Actor subsystem, and *must* fulfill the role of an RBD, see below, section 2.4.

Additionally it *may* act as signal drain, i.e. an RBQ, see below section 3.2.2).
More precisely: The TM is the *only initially active* component, and as such the
source of all other activities in a TCl system.

(2) A TM *offers* only one single service called "decideNode()", which, given
an indication for the class of which a new component shall be created, together
with an appropriate parameter set, delivers an indication for the hosting node
of this new component and a (perhaps) modified parameter setting.

So TM is specified only as being the subsystem containing all deployment
strategies and the total routing information, — the implementation of both is
(currently) outside the scope of each TCl.

2.2 CAS = Central Access Server

As it is with TM, there is always only one single instance of CAS in each TCl
setting. Contrarily to TM, the behaviour of CAS is totally specified by TUB-TCl.

The (single instance of) CAS caches all deployment information calculated
by TM, calculates unique IDs for all newly created objects of global scope and
holds a catalog of these for answering the corresponding inquiries, controls the
sequentialization of reset() commands, keeps track of the initialization state of
all nodes and gathers all verdict and error messages for passing them to TM.

In contrast to TM the CAS must be hosted on a node which is *reachable* from
all other nodes (c.f. section 3.1)[3].

2.3 NodeS = Node Server and Factories

On each node of the TCl setting there is exactly one Node Server running. The
NodeS has total control of the testing sandbox of its hosting node, in which all
test components will be living. It is also responsible for bootstrapping of the
underlying basic communication layer (Trans, see section 2.5) and its routing
information. Furthermore, it is the recipient for all DOcreate() and DOdelete()
commands, which install new subsystems on this node.

For really performing the creation, the NodeS delegates DOcreate requests
for all new subsystems of a certain class to the corresponding Factory. There
is one factory running on each node for each class of subsystems which can be
created on this node.

2.4 Actors = Dynamically Created Active and Passive Subsystems (Components)

All subsystems which can be dynamically created or allocated are subsumed as
"Actors". This includes hardware resources (IO ports, timers or their respective
drivers), code images loaded onto a distinct node, and also running code, e.g.
"jobs" or "threads", created by instantiating parts of these images.

---

[3] Of course the node hosting the CAS must also be able to reach the node hosting the TM.

TUB-TCI. An Architecture for Dynamic Deployment of Test Components   285

Actors realized by active running code can act in two different roles, an active and a passive one, called RBD and RBQ resp.[4].

RBDs are the only "user code" running in a TUB-TCI environment and the only sources of activity.

The passive RBQ interface is offered by all those components which can act as signal *drains*, that is they can consume a data stream generated by some other hardware Actor or by some running RBD.

2.5   Trans = Inter-Node Communication and Node Initialization

As induced by the TTCN-3 language, on "user data level" any TCI architecture must support synchronous as well as asynchronous communication. Furthermore, on "system level" communication necessary for component control and configuration must be realized.

Both levels of communication are mapped to a basic layer called Trans, which realizes the (small) necessary portions of "data link", "transport" and "network" layer of the ISO-OSI model.

We support two primitive communication acts: unsolicited, "UDP-like" datagrams, and solicited datagrams, i.e. one request and one single response.

When sending an unsolicited datagram, control returns to the sending component code immediately. When sending a solicited datagram, the communication is synchronous. A time out duration value has to be given and control returns to the sending code either with the return message or with a timeout indication. A central feature yielding the *robustness* of TUB-TCI is that there is only one single source for time-out generation: the Trans subsystem of the node hosting the client.

3   Scenarios of Dynamic Behaviour

3.1   Node Topology, Node Initialization and Routing

The hardware nodes forming a TCI setting must be connected by some communication channels. It is *not* required that each such channel is bi-directional. It *is* required, however, that each node has at least one input and one output channel, and that the node hosting the CAS is initial and final, i.e. can reach any other node and is reachable by any other node.

When powering up (the testing sandbox internal to) a distinct node, this node is in an uninitialized state. The only function it performs is listening on dedicated ports for a loadRouting() service request. This message contains (1) the assignment of one globally unique identifier for this node (*NodeId*), (2) a routing table indicating for each *NodeId* the bus driver and bus address, if the corresponding node is directly reachable, or the *NodeId* of the node which is to

---

[4] The names RBD and RBQ are historically determined, cf. [6], and stand for "Runtime Behaviour / Dynamic" resp. "Runtime Behaviour / Queue".

Table 1. Offered Services

TM	called by CAS :	decideNode()
CAS	called by TM :	CASreset()
		bootNodes()
	called by RBD :	getGlobalParameter()
		create()
		delete()
		setverdict()
	called by Trans :	lookup()
	called by Factory :	registersubactors()
		deleted()
		desmudge()
NodeS	called by CAS :	reset()
		loadRouting()
		sendHdwStateInfo()
		startsession()
		stopsession()
		DOcreate()
		DOdelete()
		HScreatelink()
		HSregisteroutlink()
Factory	called by NodeS :	DOcreate()
		DOdelete()
Actor	called by RBD :	setconfigparams()
		getconfigparams()
		RToperation()
RBQ	called by QAS :	RToperation (putQ())
		HSputQ()
QAS	called by RBD :	subscribe()
		unsubscribe()
		HSsubscribe()
		HSunsubscribe()
	called by Actor(-demon/-irqHdl) :	value_event()
		HSvalue_event()
Trans	called by Trans (= interface from/to *Bus Adapters*) :	
		deliver()
	called by *Client* (=all but RBQ) :	req()
	delivered to *Server* :	service()
	called by *Server* :	reply()
	delivered to *Client* :	answer()
back door	called by *client* :	BDopen()
interface	called by hosting node of *server* :	BDregisterServer()
(≈ TCP)	called by *client* and *server* :	BDread()
		BDwrite()
		BDclose()

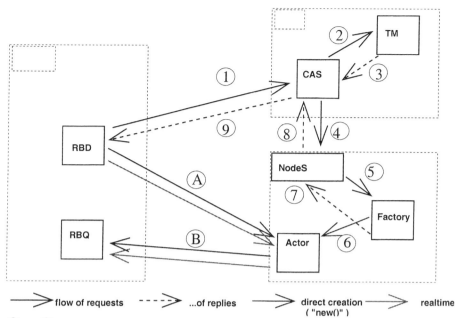

Fig. 1. Subsystems and Collaboration Diagram for creating a new actor

① – ⑨ : Sequence of Requests/Replies for creating an Actor.
Ⓐ : Run-Time Control on Actors
Ⓑ : Real-Time Signal Flow from Spontaneous Signal Sources.

be used as a gateway, and (3) a routing table assigning a set of *NodeId*s to each *broadcast group*.

The initialization and re-initialization procedures can be rather critical and complicated, depending on the topology of the network: Direct confirmation of successful initialization or reset may not be possible until third nodes are initialized, which are needed as gateway to reach the CAS node. Figure 2 just wants to give an impression how complicated the correct schedule of init-commands (upper sequence in the figure) and reset-commands (lower sequence) can turn out, and that in a complicated topology both schedules are neither identical nor just simply inverse.

Furthermore the initialization sequence may have to consider latency requirements. The concrete schedule of loadRouting() and reset() messages requires strategic knowledge and is left to the TM, external to the TUB-TCI specification. The specification *does* require, that every reset() must be distributed to *all* nodes in the setting, so that only a *total* reset is legal for sake of robustness.

3.2 Inter-component Communication

On *application* level, i.e. from the viewpoint of a compiled TTCN-3 code and its runtime library, runtime communication happens between Actors, i.e. dynamically created components.

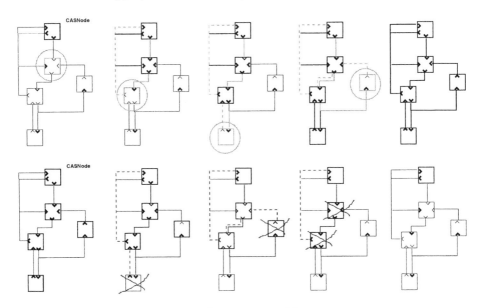

Fig. 2. Sequences of Initialization (○) and Reset (X) in a non-trivial Network Topology

We distinguish two totally disjoint flavors of communication, each of which can be realized by two totally different mechanisms:

3.2.1 Single-Drain Communication. Single-drain communication channels are related to one single actor which is the *target* of all messages. These messages are operation control commands issued by a parallel running component. Examples are: start,stop and launch commands of timers, write commands to outgoing data ports etc.

The corresponding messages can be sent solicited or unsolicited.

The timing of events in these channels is typically irregular: Multiple active components (RBDs) independent from each other can send messages to one single drain spontaneously.

3.2.2 Single-Source Communication. Single-source channels are related to one single actor which is the *source* of all communication. They are mainly used for realizing a stream of user data, e.g. ticks generated by a timer or incoming messages received by a port.

The corresponding messages can only be sent unsolicited.

The frequency of these messages is often *regular*. More than one listener can *subscribe* a distinct source, and will from then on be notified of each event by their "RBQ" interface, until they *unsubscribe* the channel.

Both kinds of communication can be realized in two ways:

3.2.3 User Level Communication by Service Requests.

The first possibility of runtime user level communication is realized by mapping it to the standard service request mechanism:

In the single-drain case an RToperation() message is sent by the emitting RBD to the target Actor which shall be controlled or configured. The argument of this message is a further schema representing the distinct action to perform. All possible real time control messages and their formats are defined with the class of the target Actor, see section 3.3.

In the case of single-source the RBD of the Actor sends a subscribe() or unsubscribe()message to a subsystem called QAS(= Queue Access Server)[5]. This message is parameterized with the id of the Actor the outgoing data stream of which shall be consumed.

The driver of this (passive) component will simply generate a single value_event message to QAS each time it generates or receives a data event.

On the other side the Actor the active code of which performs such a subscribing must also implement the passive RBQ interface. This means that the among the set of RToperation() it understands there must be the putQ() command, which pushes the data into its data input queue.

The scheduling and dispatching of all these messages is done automatically and totally dynamic by the operational semantics of TUB-TCI. The signal flow will always be minimized in so far as each node hosting a subscriber and/or needed as a gateway will only receive one single copy of each data event.

All data in this flavor of communication is encoded as an XML object, which is derived from the corresponding schema defined with the Actor Class. So there is unlimited compatibility: Each RBQ can subscribe each source, and one single RBQ can subscribe multiple sources simultaneously, de-multiplexing the incoming data using the tags contained therein.

3.2.4 User Level Communication by High Speed Channels.

Secondly there is the possibility for a given node and its factories to offer specialized *high speed channels* (HS-channels).

These are communication channels which bypass the Trans layer, but are defined directly on "driver level". In contrast to the "normal" communication they may be based on specialized hardware links between nodes (e.g. time slot busses), may use specialized addressing protocols and transmit *binary coded* data, maybe in a proprietary format. Both flavors of communication (single-drain component control or single-source data stream consumption) can be realized by HS-channels.

Their intended purposes are (1) the monitoring of data with high bandwidth generated on a external node, and (2) the transmission of single events which should reach the target with minimal latency.

---

[5] For sake of this paper, QAS can be considered to be one single virtual subsystem, which is directly reachable for any RBD. Indeed it will be *implemented* on each node as part of the node's NodeS.

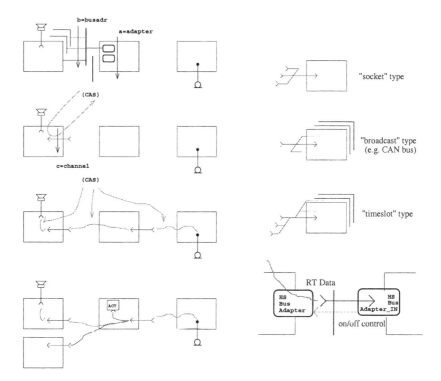

Fig. 3. Creation of HsLinks / Variants of HsLinks determined by the underlying Technology

The principle of total dynamic configuration and scheduling naturally contradicts these purposes. So here we have chosen a different strategy: All HS-channels can (and should) be allocated and configured by the TM *in advance*. When linking and loading the compiled code it has to be parameterized with the IDs of the HS-channels corresponding to the source level communication statements.

Only in those cases where the HS-channel is bi-directional, the HSsubscribe() and HSunsubscribe() messages issued by the consumer do have an effect: This effect is just "switching on and off" the data stream from the sending node, as soon as the first consumer subscribes or the last consumer unsubscribes. This back channel should of course be implemented also on driver level for sake of optimal performance and lowest latency.

Figure 3 tries to depict that the main achievement of TUB-TCI lies in the abstraction from the different flavours of low-level communication (many-to-one, one-to-many, etc., as depicted in the right column of the figure), and from the different way of addressing the nodes. The left column shows the typical two-stage approach for establishing HS-channels by the CAS (controlled by strategic knowledge of the TM): First the "logical in-ports" are allocated, thereby resolving all specific requirements of the lower communication layer. So in a second step the concrete connections into these ports can be established in a uniform way.

3.3 Actor Class Declaration and Instantiation

As mentioned above, the declaration of the classes of which Actors can be created on a given node is outside the scope of TUB-TCI. Even most basic foundation classes like "active code" or "thread" are unknown to this layer of architecture. In fact only the mechanisms for defining and deriving classes is part of the TUB-TCI.

Each Actor Class is given as a Z schema, and declares for each instance (1) the collection of configuration parameters and (2) the runtime operations applicable. The latter is just a *free type* describing the possible control messages the Actors of this class understand during runtime. The former gives for each parameter a basic type, arbitrarily chosen constraints on the possible values, and the "update allowance", which describes if this parameter cannot be set at all (CO and RO), can be written once when creating the Actor(WI) or re-written during setup time (RW) or can be changed even at runtime (RWRT).

3.4 Example

Let us illustrate this mechanism by the example of a "timer" Actor Class:

First we define the configuration parameters and the real-time operations separately[6]:

$$\begin{array}{|l}
\hline
\ PI\_Timer \\
\hline
minResolution : Duration \\
curResolution\ \ : Duration \\
maxValue\ \ \ \ \ \ \ : Duration \\
\hline
\end{array}$$

$$\begin{array}{|l}
RT\_Timer\ ::=\ \ \texttt{start} \\
\ \ \ \ \ \ \ \ \ \ \ \ \ \ \ |\ \texttt{stop} \\
\ \ \ \ \ \ \ \ \ \ \ \ \ \ \ |\ \texttt{reset}\langle\!\langle Duration \rangle\!\rangle \\
\ \ \ \ \ \ \ \ \ \ \ \ \ \ \ |\ \texttt{read}\ \Rightarrow\ \texttt{timerval}\langle\!\langle Duration \rangle\!\rangle
\end{array}$$

No we use these both data types and build a schema which includes the schema *ActorClassDescription*:

$$\begin{array}{|l}
\hline
\ AC\_Timer \\
\hline
ActorClassDescription \\
\hline
PI\ \subset\ unpack\ PI\_Timer \\
RT\ =\ RT\_Timer \\
PI\ (\texttt{"minResolution"})\ .mode\ =\ \texttt{CO} \\
PI\ (\texttt{"maxDuration"})\ .mode\ \ \ =\ \texttt{CO} \\
\hline
\end{array}$$

---

[6] Please note the slight enhancements and weakenings of the Z notation, used for purpose of readability: The notation $\alpha \Rightarrow \beta$ simultaneously defines one case of a free type used as service request (α) together with the case (β) of another free type, representing the *corresponding* reply. This correspondence is considered informal, i.e. only for reader's information. The operation *"unpack"* belongs to our reflection tools: it takes a schema and lifts it to a data structure.

Now we can derive the definition of a special timer, offering additional operations and requiring more parameters:

$$
\begin{array}{|l}
PI\_Timer\_4711 == [maxMarkerCount : \mathbb{N}] \\
RT\_Timer\_4711 ::= \quad \texttt{setMark}(\mathbb{N}) \\
\qquad\qquad\qquad | \ \ \texttt{readMark}(\mathbb{N}) \Rrightarrow \texttt{timestamp}(\mathit{Time}) \\
AC\_Timer\_4711 == [\ extend(AC\_Timer, PI\_Timer\_4711, RT\_Timer\_4711) \\
\qquad\qquad\qquad | \ PI\ (\texttt{"maxMarkerCount"}).mode\ =\ \texttt{WI} \\
\qquad\qquad\qquad]
\end{array}
$$

At last we declare a **factory**, i.e. a concrete "driver", which probably will be supported only on nodes of dedicated node classes. We simply build a new schema which combines (1) the schema of the most specific **actor class** for which the **factory** is an implementation, (2) the generic *Factory* schema (cf. figure 4), and (3) further constraints on the configuration parameters.

$$
\begin{array}{|l}
\text{—— } FACT\_timer\_x4711\_2.2\_runs\_on\_Tektronix\_0815 \text{ ——————} \\[4pt]
AC\_Timer\_4711 \\
Factory \\
\hline
\\
V\ \texttt{minResolution} \qquad\qquad\ \ =\ sec(0.001) \\
V\ \texttt{maxDuration} \qquad\qquad\quad <\ sec(3600.00) \\
PI(\texttt{"curResolution"}).mode\ \ =\ \texttt{RW} \\
PI(\texttt{"curResolution"}).default\ =\ sec(0.01) \\
sec(0.001) \leq V\ \texttt{curResolution} \leq sec(0.1) \\
\exists n : \mathbb{Z}\ \bullet\ V\ \texttt{curResolution} = n * sec(0.001) \\
V\ \texttt{maxDuration}\ /\ V\ \texttt{curResolution} \leq 2^{31} - 1
\end{array}
$$

Please note that this schema imposes real *dynamic constraints* on the configuration parameters' values, since the value of `maxDuration` varies depending on the value of `curResolution`.

4 Related and Future Work

4.1 Related Work

Our work on TCI is based on early industrial approaches as found in [2], and did have some influence to the recent development of TRI ([6]), a communication protocol which can be interpreted as one fixed instance of a TCI, but without formal definitions of behaviour.

There are numerous significant theoretical works concerning the algebraic aspects of distributed testing ([1,5,9]).

$ParameterUpdateAllowance ::=$ CO | RO | WI | RW | RWRT

CO < RO < WI < RW < RWRT

─── *ParameterDescription* ───────────────────────────
$type \qquad : \_TYPE\_$
$updatemode : ParameterUpdateAllowance$
$value \qquad : type$
$allowed \qquad : \mathbb{P}\ type$
$default \qquad : type\ \cup\ \{\bot\}$
───────────────────────────
$default \in allowed\ \vee\ default = \bot$

| $ParameterDescriptionList\ ==\ IDENT \nrightarrow ParameterDescription$

─── *ActorClassDescription* ───────────────────────────
$PI\ : ParameterDescriptionList$
$RT\ : \_FREE\_TYPE\_\_$

─── *Actor_Factory* ───────────────────────────

$ActorClassDescription$

|→| create   : $LUID \times pack\ PI \to GR$
|→| delete   : $LUID \times LUID \to GR$
|→| changeParamas   : $LUID \times pack\ PI \to GR$
|→| dumpParamas   : $LUID \to pack\ PI$
|→| executeRtOperation : $LUID \times RT \to \_DATA\_\_$

// GR = General Result data type, $LUID$ = "local unique id" = a unique id for a dynamically created Actor.

Fig. 4. Basic definitions for Defining Actor Classes

On the other side there is, to the best of our knowledge, only one published approach concerning practical implementation: the industrial implementation of TSP1 [7], as part of the TINA project.

Our approach is somehow in the middle, — it does not cope with reasoning as the former, but it does contain precise specification of operational semantics and genericity, both lacking in the latter.

4.2 Next Steps

As mentioned above, currently TUB-TCI is a concept meant as a base for discussion. A first implementation is planned as part of a research project, and will certainly lead to modifications.

Additionally there are higher, lower or parallel layers of a possible TCI architecture, which still have to be specified:

- The Factories corresponding to these classes are currently assumed to be "hard wired" into the Node Server of the hardware node.
  This corresponds to the fact, that our specification does not yet model nodes, node classes and node vendors: in our declaration example on page 292 the corresponding dependencies are only given informally by choosing the "human readable name" of the factory to be "$FACT\_timer\_x4711\_2.2\_runs\_on\_Tektronix\_0815$".
  An ubiquitous, standardized semantic basis for modeling and implementing (!) these relations probably will need the allocation of "$ASN.1$ *information objects*", cf. ISO/IEC 8824-2, i.e. entities of world-wide unique meaning assigned to vendors by a standardization board.
- This also applies to the *binary* data encoding used for the HS-Channels. Currently the data encoding is totally unspecified and the correct "routing" from sources to sinks is left to the informal knowledge of the TM. Different encodings, even vendor-specific, could be indicated by "information objects", so that Actors realizing *encoding converters* could be inserted automatically and correctly w.r.t. typing.

References

1. Mohammed Bennattou, Leo Cacciari, Régis Pasini, and Omar Rafiq. Principles and tools for testing open distributed systems. In *IFIP TC5 12th International Workshop on Testing Communicating Systems*. Kluwer Academic Publishers, 1999.
2. *Generic Compiler/Interpreter Interface*. INTOOL CGI / NPL 038 (v.2.2), december 1996.
3. Markus Lepper. TUB-TCI — A Generic Architecture for Distributed Test Execution. Technical report, Berlin, September 2002. http://uebb.cs.tu-berlin.de/papers/published/TR02-08.ps.
4. TTCN-3 Control Interface (TCI). Technical report, ETSI ES 201 837-5 , to appear 2003.
5. Maria Törö. Decision on tester configuration for multiparty testing. In *IFIP TC5 12th International Workshop on Testing Communicating Systems*. Kluwer Academic Publishers, 1999.
6. TRI — the TTCN-3 Runtime Interface. Technical report, ETSI TR 102 043 V1.1.1, Sofia-Antipolis, April 2002.
7. Test Synchronization Protocol 1 Plus (TSP1+) Specification. Technical report, ETSI TC-MTS, ETSI Standard ES 201 770, Sofia-Antipolis, Jan 1997.
8. Methods for Testing and Specification (MTS); Part 1: TTCN-3 Core Language. Technical report, ETSI ES 201 837-1 (V1.0.11), Sofia-Antipolis, May 2001.
9. Andreas Ulrich and Hartmut König. Architectures for testing distributed systems. In *IFIP TC5 12th International Workshop on Testing Communicating Systems*. Kluwer Academic Publishers, 1999.

Fast Testing of Critical Properties through Passive Testing*

José Antonio Arnedo[1], Ana Cavalli[1], and Manuel Núñez[2]

[1] Institut National des Télécommunications GET-INT
91011 Evry Cedex, France
{Jose-Antonio.Arnedo-Rodriguez,Ana.Cavalli}@int-evry.fr
[2] Dept. Sistemas Informáticos y Programación
Universidad Complutense de Madrid, E-28040 Madrid, Spain
mn@sip.ucm.es

Abstract. We present a novel methodology to perform *passive testing*. The usual approach consists in recording the trace produced by the implementation under test and trying to find a fault by comparing this trace with the specification. We propose a more *active* approach to passive testing where the minimum set of (critical) properties required to a correct implementation may be explicitly indicated. In short, an invariant expresses that each time that the implementation under test performs a given sequence of input/output actions, then it must show a behavior reflected in the invariant. By using an adaptation of the classical pattern matching algorithms on strings, we obtain that the complexity of checking whether an invariant is fulfilled by the observed trace is in $\mathcal{O}(n \cdot m)$, where n and m are the lengths of the trace and the invariant, respectively. If the length of the invariant is much smaller than the length of the trace then this complexity is *almost linear* with respect to the length of the trace. Actually, this is usually the case for most practical examples. In addition to our methodology, we present the case study that was the driving force for the development of our theory: The Wireless Application Protocol (WAP). We present a test architecture for WAP as well as the experimental results obtained from the application of our passive testing with invariants approach.

1 Introduction

The main purpose of testing is to find out whether an implementation presents the behavior that was indicated by the corresponding specification. In order to perform this task, several techniques, algorithms, and semantic frameworks have been introduced in the literature (see e.g. [LY96,Lai02] for two overviews on the topic). Most of the proposals for testing are based on the so-called *active* testing. Intuitively, the tester sends an input to the implementation and waits

---
* Research supported in part by the Spanish *Ministerio de Ciencia y Tecnología* projects MASTER and AMEVA. This research was carried out while the third author was visiting the GET-INT.

for an output. If the output belongs to the expected ones, according to the specification, then the process continues; otherwise, a fault has been detected in the implementation. This kind of testing is called active because the tester has total control over the inputs provided to the implementation under test. On the contrary, *passive testing* (see e.g. [AAD79,LNS+97,Mil98,TC99,TCI99]) does not involve the presence of an *active* tester. In passive testing the implementation under test is allowed to run independently without any interaction with a tester. However, the trace that the implementation is executing is observed so that it can be analyzed. By comparing the obtained trace with the specification we may detect some faults in the implementation.

Unfortunately, passive testing is less powerful than active testing. This is so because active testing allows a closer control of the implementation under test. For example, depending on the received output, we may choose among a set of inputs to be applied to the implementation. In passive testing this capability is lost. Thus, it can happen that faults that could be detected by choosing an appropriate input are not found because the implementation does not take that path. Nevertheless, passive testing presents some important advantages. First, active testing is in general more costly because the testing process has to be closely controlled. Second, there are situations where active testing is not even feasible. For instance, passive testing is attracting a lot of study in the field of network management where the testing process has to be minimized in order to reduce the use of the network (see [MA01a,MA01b,WZY01] for some recent works on the topic).

Even though there is ongoing work on passive testing[1] most of these proposals share a common pattern. Usually, the trace is taken and the specification is traversed in order to detect a fault. A drawback of this approach is that it presents a low performance (in terms of complexity in the worst case) if non-deterministic specifications are considered. A notable exception is presented in [CGP01] where the specification is (partially) set apart. Actually, they extract some test sequences from the specification, called *invariants*, and they check whether the trace obtained from the implementation under test is correct with respect to them. So, once these test are extracted, the specification plays no role in the testing process. They consider three kinds of invariants. However, the most useful for practical purposes are the so-called *output* invariants. Intuitively, an (output) invariant as $i_1/o_1, \ldots, i_{n-1}/o_{n-1}, i_n/o_n$ must be interpreted as "*each time the implementation performs the sequence $i_1/o_1, \ldots, i_{n-1}/o_{n-1}, i_n$ the next observed output must be o_n*".

This paper extends and improves [CGP01] in several ways. Actually, our methodology represents a real example of theory guided by practice. In fact, we came out with our different notions of invariants when we tried to apply [CGP01] to a real protocol. By doing so we found some shortcomings. First, their invariants were not expressive enough for our purposes. For example, the complete sequence has to be indicated. That is, properties as

---

[1] Nevertheless, the effort spent in passive testing is not yet comparable with the one for the study of active testing techniques.

Each time that a user asks for connection and the connection is granted, if after performing some operations the user asks for disconnection then he is disconnected.

cannot be easily represented by using their invariants, because all the possible sequences of actions expressing the idea of *some operations* must be explicitly written. So, we have added the possibility of specifying wild-card characters in invariants. Besides, we now allow a set of outputs (instead of a single output) as termination of the invariant. Thus, we may specify properties as

Each time that a user asks for a resource (e.g. a web page) either the resource is obtained or an error is produced.

Finally, their invariants were automatically extracted from the specification. This approach presents two drawbacks. First, *interesting* invariants cannot be distinguished from *trailing* invariants. Second, and more important, the complexity of extracting invariants exponentially increases with their length. That is, the complexity of extracting the invariants of length n is in $\mathcal{O}(|Tr|^n)$, where $|Tr|$ is equal to the number of transitions in the specification. We claim that invariants should be supplied by the specifier/tester. In this case, the first step must be to check that the invariant is in fact correct with respect to the specification. We provide an algorithm that checks this correctness in linear time, with respect to the number of transitions, if the invariant does not contain the wild-card character $*$; this complexity is quadratic if the symbol $*$ appears in the invariant.

Once we have a set of (correct) invariants, our approach to passive testing proceeds as follows: We observe the trace produced by the implementation under test and we decide whether this trace respects the invariants. In order to do so, we have implemented a simple variant of the classical algorithms for pattern matching on strings (see e.g. [BM77,KMP77]). Our algorithm works, in the worst case, in time $\mathcal{O}(m \cdot n)$, where n and m are the length of the trace and the length of the invariant, respectively. Let us remark that in most practical cases the length of the invariant is several orders of magnitude smaller than the length of the trace. Thus, we may consider that the complexity is almost linear with respect to the length of the trace.

In addition to the formal framework, in this paper we also report our experiments on the WAP (Wireless Application Protocol). This protocol is an open global specification that empowers mobile users with wireless devices to easily access and interact with information and services instantly. It is worth to point out that this protocol represents a typical example where active testing cannot be applied. In general, there is no direct access to the interfaces between the different layers. Thus, the tester cannot control how internal communications are established. However, in our experiments we have used a software free protocol stack, namely Kannel, and we have the possibility of installing points of observation, in short POs, between the different layers. Moreover, a platform and a test architecture capable to deal with passive testing in a mobile phone environment (WAP, GPRS, UMTS) have been defined. The platform and the architecture are

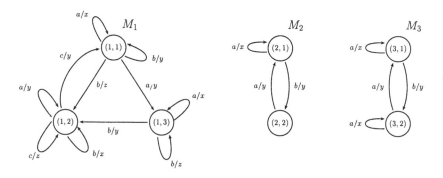

Fig. 1. Examples of FSMs.

used to apply our passive testing with invariants approach. We take the output of the platform, in the form of log files, and we apply the appropriate invariants to the obtained trace. This experiment represents an original contribution because such a study has been never performed in a systematic way.

The rest of the paper is organized as follows. In Section 2 we present our notion of invariants, that we call *simple invariants*. These invariants are able to express properties as "after x has happened then we must have that y happens". We give algorithms to decide the correctness of our invariants with respect to a given specification and we explain how our invariants are *applied* to the observed trace. In Section 3 we present the WAP and we briefly comment on the performed experiments by using our notion of passive testing. Finally, in Section 4 we give our conclusions and some lines for future work.

2 Invariants and Passive Testing

In this section we introduce our invariants and the corresponding algorithms to decide whether they are correct with respect to specifications. We consider that specifications are represented as Finite State Machines. However, we will comment on how our invariants can be extended to deal with data.

Definition 1. A *Finite State Machine*, in the following FSM, is a tuple $M = (S, \mathcal{I}, \mathcal{O}, Tr, s_{in})$ where S is a finite set of states, $\mathcal{I}$ is the set of input actions, $\mathcal{O}$ is the set of output actions, Tr is the set of transitions, and s_{in} is the initial state.

Each transition $t \in Tr$ is a tuple $t = (s, s', i, o)$ where $s, s' \in S$ are the initial and final states of the transition, respectively, and $i \in \mathcal{I}$, $o \in \mathcal{O}$ are the input and output actions, respectively.

Let $s, s' \in S$ be states and $tr = i_1/o_1, \ldots, i_n/o_n$, for $n \geq 1$, be a sequence of pairs such that for any $1 \leq j \leq n$ we have $i_j \in \mathcal{I}$ and $o_j \in \mathcal{O}$. We write $s \xrightarrow{tr} s'$ if either $tr = \epsilon$ and $s = s'$ or there exist n transitions $t_1, \ldots, t_n \in Tr$ and states $s_1, \ldots, s_{n-1} \in S$ such that $t_1 = (s, s_1, i_1, o_1)$, $t_n = (s_{n-1}, s', i_n, o_n)$, and for any $1 < j < n$ we have $t_j = (s_{j-1}, s_j, i_j, o_j)$. □

First, let us note that this notion of FSM does not restrict specifications to be deterministic, so that we work with a general notion of FSM. Intuitively, a transition $t = (s, s', i, o)$ indicates that if the machine is in state s and receives the input i then the machine emits the output o and the current state becomes s'. We will sometimes write $s \xrightarrow{i/o} s'$ to denote that we have the transition $(s, s', i, o) \in Tr$. We can extend the notion of single transition to a sequence of transitions. Thus, $s \xrightarrow{tr} s'$ simply denotes that we can traverse from the state s to the state s' by following transitions containing the corresponding pairs i/o appearing in tr. In Figure 1 we present some simple examples of Finite State Machines.

Once we have a FSM we may extend its set of transitions so that the wild-card characters ? and ∗ can be taken into account. In particular, these special symbols can appear as part of a sequence of transitions.

Definition 2. Let $M = (S, \mathcal{I}, \mathcal{O}, Tr, s_{in})$ be an FSM. We write $s \xrightarrow{?/o} s'$ (resp. $s \xrightarrow{i/?} s'$) if there exists $i \in \mathcal{I}$ (resp. $o \in \mathcal{O}$) such that $s \xrightarrow{i/o} s'$. Besides, we write $s \xrightarrow{?/?} s'$ if there exist $i \in \mathcal{I}$ and $o \in \mathcal{O}$ such that $s \xrightarrow{i/o} s'$. We write $s \xrightarrow{*} s'$ if there exists a sequence of input/output pairs tr such that $s \xrightarrow{tr} s'$. □

2.1 Introducing Simple Invariants

Next we present our notion of invariant. We call them *simple invariants*, or just *invariants*. Let $M = (S, \mathcal{I}, \mathcal{O}, Tr, s_{in})$ be an FSM. Intuitively, a trace as $i_1/o_1, \ldots, i_{n-1}/o_{n-1}, i_n/O$ is a simple invariant for M if each time that the trace $i_1/o_1, \ldots, i_{n-1}/o_{n-1}$ is observed, if we obtain the input i_n then we necessarily get an output belonging to O, where $O \subseteq \mathcal{O}$. In addition to sequences of input/output symbols, we will allow the *wildcard* characters ? and ∗. In our framework, the meaning of ? is the standard one in the pattern matching community (that is, to replace any symbol). However, we will slightly modify the usual meaning of ∗. For example, the intuitive meaning of an invariant as $i/o, *, i'/O$ is that if we detect the transition i/o then the first occurrence of the input symbol i' is followed by an output belonging to the set O. In other words, ∗ replaces any sequence of symbols not containing the input symbol i'.

Definition 3. Let $M = (S, \mathcal{I}, \mathcal{O}, Tr, s_{in})$ be an FSM. We say that the sequence tr is a (simple) *invariant* for M if the following two conditions hold:

1. tr is defined according to the EBNF $tr ::= i/O \,|\, *, tr \,|\, a/z, tr$. In the previous expression we consider $i \in \mathcal{I}$, $a \in \mathcal{I} \cup \{?\}$, $z \in \mathcal{O} \cup \{?\}$, and $O \subseteq \mathcal{O}$.
2. tr is *correct* with respect to M.

We denote the set of (simple) invariants for M by Inv_M. □

In Figures 2 and 3 we introduce two algorithms to decide whether an invariant is *correct* with respect to a specification. First, we present some examples of invariants to show what kind of critical properties can be tested as well as how our invariants work.

Example 1. Our notion of invariant allows us to express several interesting properties. For example, we can test that when a user requests a disconnection then he is in fact disconnected by using the invariant

$$I_1 = req\_disconnect/\{disconnected\}$$

The idea is that each time that the symbol *req_disconnect* appears in the trace then it is followed by the output symbol *disconnected*. For instance, this invariant has the same distinguishing power as the invariant

$$I'_1 = *, req\_disconnect/\{disconnected\}$$

We can specify a more complex property by taking into account that we are interested in disconnections only if a connection was requested. In this case we have

$$I_2 = req\_connect/?, *, req\_disconnect/\{disconnected\}$$

We can refine the previous invariant if we only consider the cases where the connection was granted

$$I_3 = req\_connect/granted\_connection, *, req\_disconnect/\{disconnected\}$$

For example, a trace is correct with respect to I_3 if each time that we find a (sub)sequence starting with the pair *req_connect/granted_connection* then the first occurrence of the input symbol *req_disconnect* is paired with the output symbol *disconnected*. Let us remark that it can be the case that the pair *req_connect/granted_connection* appears in the observed trace but that the input *req_disconnect* is not detected afterwards in the corresponding trace. In such a situation we cannot conclude that the implementation fails: It may happen that we have stopped *too soon* observing the behavior of the implementation. Finally, an invariant as

$$I_4 = req\_connect/\{granted\_connection, error\}$$

indicates that after requesting a connection we either are granted with it or an error is produced. □

We could adapt to our framework the algorithm given in [CGP01] to extract their invariants, up to a length n, for a specification M. However, as we explained in the introduction of the paper, this process presents several drawbacks. In particular, the complexity exponentially increases with the length of invariants. On the contrary, we advocate that invariants should be indicated by the specifier/tester as the set of critical properties that the implementation must fulfill. Fortunately, we have found an algorithm that detects in linear time (with respect to the number of transitions in the FSM) whether a sequence of symbols not containing the character $*$ is in fact an invariant for a specification. Obviously, before we try to find out whether the trace observed from the behavior of the implementation is *correct* with respect to an invariant, we should assure

Input: $M = (S, \mathcal{I}, \mathcal{O}, Tr, s_{in})$, $I = i_1/o_1, \ldots, i_n/O$, $\forall\, 1 \leq j \leq n : i_j \neq * \wedge o_j \neq *$
Ouput: true/false $\equiv$ invariant I correct/incorrect

$j := 1;\ S' := S;$
while $j < n$ and $S' \neq \emptyset$ do begin
   $T := Tr;\ S'' := \emptyset;$
   while $T \neq \emptyset$ do begin
      choose $t \in T;\ \{t = (s, s', a, z)\};$
      $T := T - \{t\}$
      if $s \in S'$ and $a = i_j$ and $z = o_j$ then $S'' := S'' \cup \{s'\}$
   end;
   $S' := S'';\ j := j + 1$
end;
{last pair of the invariant or empty set S' of current states}
if $S' = \emptyset$ then return(false)
    else begin
        $error :=$ false; $T := Tr;$
        while $T \neq \emptyset$ and not $error$ do begin
           choose $t \in T;\ \{t = (s, s', a, z)\};$
           $T := T - \{t\};$
           if $s \in S'$ and $a = i_n$ and $z \notin \mathcal{O}$ then $error :=$ true
        end;
        return (not $error$)
    end

We consider that both $i = ?$ and $o = ?$ hold.

Fig. 2. Checking correctness of Invariants (1/2).

that the invariant is in fact *correct* with respect to the specification. In order to facilitate the reading, we first present an algorithm (see Figure 2) deciding correctness of invariants without occurrences of the * wild-card character. In Figure 3 we extend this algorithm to deal with invariants where the symbol * can appear.

The algorithm given in Figure 2 works as follows. The first *while-loop* computes those states $s \in S$ such that they can be reached from one of the states in S by following the sequence $i_1/o_1, \ldots, i_{n-1}/o_{n-1}$. Let us remark that if one of the symbols in the sequence is the wild-card character ? then any symbol can be used. Besides, it may happen that after some steps we find that there do not exist two states connected by the analyzed sub-sequence. In this case, S' becomes empty. Let us note that for each execution of the loop we perform a number of operations proportional to the number of transitions in the corresponding specification. So, in the worst case we perform a number of operations proportional to the number of transitions times the length of the sequence, that is, n. The second *while-loop* analyzes the last pair of the invariant. If the auxiliar set of states S' is empty then the invariant is incorrect. Actually, this means that there does not exist a state in the specification such that the sequence of pairs forming the invariant can be performed from it. Thus, we should not consider that this *candidate* represents any property of the specification[2]. If that set is not empty then we check that for any transition labelled by the input i_n we receive an output belonging to O. Again, the complexity of this last loop is given by the number of transitions. Besides, we need $|S|+|Tr|$ additional space. Let us note that if the graph induced by the corresponding FSM is connected then $|S| \leq |Tr|+1$ (otherwise we can discard those states and transitions not reachable from the initial state and the result holds for the new sets of states and transitions).

Proposition 1. *Let $M = (S, \mathcal{I}, \mathcal{O}, Tr, s_{in})$ be an FSM and $I = i_1/o_1, \ldots, i_n/O$ be an invariant such that for any $1 \leq j \leq n$ we have $i_j \neq * \wedge o_j \neq *$. The worst case of the algorithm given in Figure 2 checks the correctness of the invariant I with respect to M in time $\mathcal{O}(n \cdot |Tr|)$ and space $\mathcal{O}(|Tr|)$.* □

Next, we present some examples of correct/incorrect invariants for a given specification.

Example 2. Let us consider the FSMs presented in Figure 1. For example, the following invariants are correct for M_1:

$$I_1 = a/\{x, y\} \qquad I_2 = a/?, c/z, b/\{x\}$$

Let us remark that I_1 is also correct for both M_2 and M_3. On the contrary, I_2 is incorrect for them since the sequence $a/?, c/z$ cannot be performed from any state belonging either to M_2 or M_3. If we consider the invariant

$$I_3 = b/y, a/\{y\}$$

we have that I_3 is incorrect for M_1. For instance, we have a transition labelled by b/y outgoing from the state $(1, 1)$ and reaching the same state $(1, 1)$. Then, we have a transition labelled by a/x from the state $(1, 1)$. So, there exists a state (in this case $(1, 1)$) such that the sequence of transitions b/y can be performed in such a way that the reached state (i.e. $(1, 1)$) may perform a transition whose input action is a but the corresponding output action does not belong to the set $\{y\}$. Besides, this invariant is correct for M_2 while it is incorrect for M_3. □

In Figure 3 we extend the previous algorithm to deal with invariants containing the wild-card character $*$. As in the previous case, we traverse the invariant from left to right. We also have that the external *while-loop* has as termination condition that either the remaining sequence has length one or that the current set of states is empty. However, instead of advancing by incrementing a counter we consider two auxiliar functions: **head**(I) returns the first element of I and **tail**(I) removes the first element from I. If the first element of the remaining invariant is a pair i/o where $i \in \mathcal{I} \cup \{?\}$ and $o \in \mathcal{O} \cup \{?\}$ then the algorithm

---

[2] Another possibility would be to consider the usual meaning of the logical implication. Thus, the predicate "*each time* that the prefix is performed then *something* happens" would hold since the premise is false.

Input: $M = (S, \mathcal{I}, \mathcal{O}, Tr, s_{in})$, $I = i_1/o_1, \ldots, i_n/O$
Ouput: true/false $\equiv$ invariant I correct/incorrect

$\quad I' := I;\ S' := S;$
$\quad$ **while** $I' \neq b/O$ **and** $S' \neq \emptyset$ **do begin**
$\quad\quad\quad first := \mathbf{head}(I');\ I' := \mathbf{tail}(I');$
$\quad\quad\quad$ **if** $first \neq *$ **then begin** $\{first := i/o\}$
$\quad\quad\quad\quad\quad T := Tr;\ S'' := \emptyset;$
$\quad\quad\quad\quad\quad$ **while** $T \neq \emptyset$ **do begin**
$\quad\quad\quad\quad\quad\quad\quad$ choose $t \in T;\ \{t = (s, s', a, z)\};$
$\quad\quad\quad\quad\quad\quad\quad T := T - \{t\};$
$\quad\quad\quad\quad\quad\quad\quad$ **if** $s \in S'$ **and** $i = a$ **and** $o = z$ **then** $S'' := S'' \cup \{s'\}$
$\quad\quad\quad\quad\quad$ **end**;
$\quad\quad\quad\quad\quad S' := S''$
$\quad\quad\quad$ **end**
$\quad\quad\quad$ **else begin** $\{first = *\}$
$\quad\quad\quad\quad\quad$ **while** $\mathbf{head}(I') = *$ **do** $I' := \mathbf{tail}(I');$ {skip a sequence of *'s}
$\quad\quad\quad\quad\quad first := \mathbf{head}(I');\ \{first := i/o\}$
$\quad\quad\quad\quad\quad S' := \{s \in S \mid \exists s' \in S' : path(s', s, i)\}$
$\quad\quad\quad$ **end**
$\quad$ **end**;
$\quad$ {last pair of the invariant or empty set S' of current states}

See Algorithm in Figure 2 for dealing with the last pair of the invariant.

We consider that both $i = ?$ and $o = ?$ hold.

Fig. 3. Checking correctness of Invariants (2/2).

proceeds as the algorithm presented in Figure 2. If we have that the first element is * then we skip all consecutive *'s. Afterwards, we consider the first element of the remaining trace. Let us remark that this element must be a pair i/o. We compute those states s connected with one of the states $s' \in S'$ by a path that does not contain the symbol i. These paths are computed by the predicate $path(s', s, i)$. Formally, $path(s', s, i)$ if there exists a sequence of input/output pairs $tr = a_1/z_1, \ldots a_r/z_r$ such that $s' \xrightarrow{tr} s$ and for any $1 \leq j \leq r$ we have $a_i \neq i$. As a special case, if i is equal to ? then $path(s', s, ?)$ holds if $s' \xrightarrow{*} s$. Let us remark that the complexity in time in the worst case for computing this new set of states is in $\mathcal{O}(|S| \cdot |Tr|)$. This is so because we only need to compute a breath-first-search (the complexity of this operation is in $\mathcal{O}(|Tr|)$) for each of the states belonging to S' (at most $|S|$ states). Again, if we consider that the induced graph is connected then we have that the previous complexity is bounded by $\mathcal{O}(|Tr|^2)$. Besides, we need $|S| + |Tr|$ additional space. Finally, the last element of the sequence is treated as in the algorithm given in Figure 2.

The next result indicates the complexity of the previous algorithm. We consider that there are no trailing occurrences of the wild-card character * in invariants, that is, no consecutive occurrences of *.

Proposition 2. Let $M = (S, \mathcal{I}, \mathcal{O}, Tr, s_{in})$ be a FSM and $I = i_1/o_1, \ldots, i_n/O$ be an invariant without trailing occurrences of $*$. The worst case of the algorithm given in Figure 3 checks the correctness of the invariant I with respect to M in time $\mathcal{O}(k \cdot |Tr|^2 + (n-k) \cdot |Tr|)$, where k is equal to the number of $*$'s in I. Besides, the needed extra space is in $\mathcal{O}(|Tr|)$. □

Example 3. If we consider again the FSMs depicted in Figure 1 we have that the invariant $a/x, *, b/\{y, z\}$ is correct for all of the specifications. □

Next, we have to determine whether the trace obtained from the implementation satisfies the properties indicated by the invariants that we are interested in. Let us remark a very important difference with respect to previous proposals for passive testing. That is, a homing state phase is not needed for this kind of invariants. This is so because invariants have to be fulfilled at any point of the implementation. Thus, it is not relevant the state where the machine was placed when we started to observe the trace. In order to test the trace we perform a pattern matching strategy. We have implemented a simple adaptation of the classical algorithms for pattern matching on strings (e.g. [BM77,KMP77]). The inclusion of wild-card characters is easy. In addition, for an invariant of length n we have to consider all the occurrences of the first $n-1$ elements in the trace and then if we find a pair i/o such that $i_n = i$ (let us remind that if $i_n = ?$ then this equality holds) then we have to check that $o \in O$. We can say that we have found a mismatch (that is, a fault) if this last condition does not hold. Regarding the complexity of our pattern matching strategy, in the worst case we obtain $\mathcal{O}(m \cdot n)$. Let us remark that even though *good* algorithms for pattern matching on strings perform in $\mathcal{O}(m)$ (after the *pre-processing* phase) we cannot achieve this complexity because we must check all the occurrences of the pattern in the trace. However, as we commented before, if we consider that the length of the invariant is *much smaller* than the length of the trace, as it is usually the case, we have that this complexity is almost linear with respect to the length of the trace.

We finish this section by presenting some relations between different invariants and their correctness with respect to a given specification. The proofs of these results are easy (but tedious) with respect to the algorithm given in Figure 3.

Lemma 1. Let $M = (S, \mathcal{I}, \mathcal{O}, Tr, s_{in})$ be an FSM. The following properties hold:

- The invariant $*, i_1/o_1, \ldots, i_n/O$ is correct for M iff $i_1/o_1, \ldots, i_n/O$ is correct for M.
- If $i_1/o_1, \ldots, i_n/O$ is correct for M and $O \subseteq O'$ then $i_1/o_1, \ldots, i_n/O'$ is correct for M.
- Let $I = i_1/o_1, \ldots, ?/o_j, \ldots, i_n/O$ be a correct invariant for M. Then, for any $I' = i_1/o_1, \ldots, i/o_j, \ldots, i_n/O$, with $i \in \mathcal{I}$, such that $\exists\, s, s' \in S : s \xrightarrow{I'} s'$ we have that I' is correct for M.
- Let $i_1/o_1, \ldots, i_j/?, \ldots, i_n/O$ be a correct invariant for M. Then, for any $I' = i_1/o_1, \ldots, i_j/o, \ldots, i_n/O$, with $o \in \mathcal{O}$, such that $\exists\, s, s' \in S : s \xrightarrow{I'} s'$ we have that I' is correct for M.

– Let I be a correct invariant for M. If we consider the invariant I' where any occurrence of $*$ in I is replaced by a sequence of symbols $i_1/o_1, \ldots, i_j/o_j$ such that $\exists\, s, s' : s \xrightarrow{I'} s'$ we have that I' is correct for M. □

Let us note that the condition $s \xrightarrow{I'} s'$ appearing in the last three cases indicates that there exists (at least) a pair of states such that they are connected by the sequence I'. Moreover, the reverse implication of the last four results do not hold.

2.2 Extending Invariants to Deal with EFSMs

The extension of our framework to deal with EFSMs is far from trivial. In particular, we have the problem that the values of the variables cannot be, in general, observed. As we will comment in the conclusions of the paper, the work recently reported in [LCH+02] opens a new perspective for passive testing to cope with data values. Nevertheless, it is very easy to adapt our formalism to deal with invariants containing only constant data. Actually, this small inclusion is rather useful when dealing with real protocols as the WAP. For instance, we may use an invariant as

$$req\_connect(Peter)/?, *, req\_disconnect(Peter)/\{disconnected(Peter)\}$$

to check that the disconnection of the service performed by $Peter$ is linked to a request of connection made by Peter himself.

3 Applying Passive Testing with Invariants

In this section we report our experiments with the WAP (Wireless Application Protocol) when using our notion of passive testing with invariants. We briefly present this protocol, we explain how our observation points are placed, and we discuss on the invariants that we have used.

3.1 The Wireless Application Protocol

The WAP is the standard protocol conceived to provide Internet content and advanced telephony services to wireless terminals. The Wireless Application Environment, in short WAE, is the main interface to the client device. It includes the content to be displayed (that is, a WML page). The WSP is a stateful binary protocol used in conjunction with WTP and WDP to provide session oriented services, or directly with WDP to provide connectionless service. It supports sessions initiation, suspension, and resumption. A session is initiated by a WAP client and is maintained until it is explicitly disconnected. WTP, is a confirmed transaction protocol, a light weighted version of TCP. There are three classes: A non-confirmed simple flow of information in one direction (class 0), a simple send-acknowledge exchange (class 1) and a class 2 for a three-way handshake.

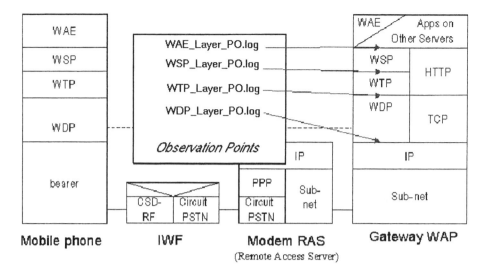

Fig. 4. A Passive Test Architecture for the WAP.

WTP also has an optional capability to segment and reassemble data. WDP, is a datagram oriented, network layer protocol. Its main purpose consists in making lower layers transparent to higher ones. It makes no delivery confirmation, packet retransmission or error correction. WTLS is a session oriented, secure protocol layer conceived after the Secure Session Layer (SSL) and Transaction Layer Security (TLS) protocols. This is optional and independent of the other parts. A schematic presentation of the protocol stacks is given in Figure 4.

We have developed an architecture capable of dealing with passive testing in a mobile phone environment (GSM-WAP). In order to do so, we deployed a platform that behaves as a normal WAP gateway[3]. In addition, we have included observation points, in short POs. These POs are placed in every layer of the WAP stack to show the flow of information in real time. So, whenever a communication between a mobile phone and a gateway exists, we have access to the involved messages and the information contained inside them. It is important to note that a layer is able to interpret only data belonging to the layer itself. This means that embedded data (i.e. from an upper layer) is not visible. The current state of every layer is also shown. These POs are entirely programmed in C. In order to have a closer control, an HTML interface with several PHP and CGI routines has also been developed.

3.2 Experimental Results on the WAP

For the sake of simplicity, in this paper we will only consider the PO that has been placed in the uppermost WAP layer, that is the session layer WSP. Thus, we can

---

[3] That is, a network component that works as interface between the mobile phone side (wireless communication) and the Internet. The gateway we took is called Kannel. It is a free-software and it can be downloaded from http://www.kannel.org.

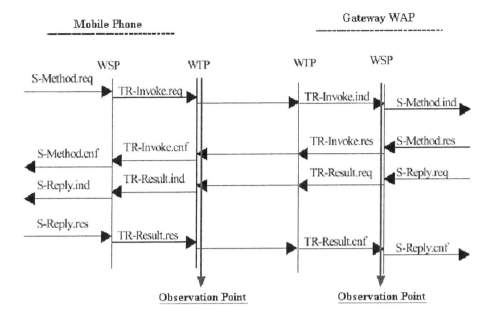

Fig. 5. Messages between layers and observation point.

observe those messages that are sent to or received from the lower layer WTP. The relation between these two layers, with respect to the transmitted messages, can be seen in Figure 5. The set of interesting events is given by: *TR-Invoke*, *TR-Result*, and *TR-Abort*. Each of these events may have one of the following attributes: *req*, *ind*, *res*, and *cnf*. Intuitively, a client C sends and *invoke* message. This message is considered in the protocol as *TR-Invoke.ind* and it is received by the gateway G. Afterwards, an acknowledge *TR-Invoke.res* is sent. Then, G tries to get the page requested by the user. Once G gets the corresponding WML page, it sends a message to the client: *TR-Result.req*. The client receives this message as a *TR-Result.ind* event. Then, it sends an acknowledge, denoted by the event *TR-Result.res*. Finally, G receives the event *TR-Result.cnf* denoting that the client received the requested information.

In order to test the protocol, we tried several properties extracted from the Wap Forum specification (see http://www.wapforum.org). The whole protocol was running autonomously and we were observing traces of 300 input/output pairs. Next we comment on the most relevant invariants that we were considering. First, we present a *misleading* example.

$$I_1 = TR\text{-}Invoke.req/?, *, TR\text{-}Result.res/\{TR\text{-}Result.cnf\}$$

This invariant is useful to check that whenever the client (mobile phone side) asks to download a WAP page, it is successfully received. More precisely, this is the way used by the WTP class 2 to acknowledge messages. Our experiments indicated that the traces were correct with respect to this invariant. However, we knew that this *invariant* was in fact not correct! Actually, an *abort* event can appear if the operation cannot be completed. For instance, this is the case if the requested web page is not available. So, we *removed* some of the requested web pages and we found that the new observed trace did not respect the invariant. In fact, the correct invariant is

$$I_2 = TR\text{-}Invoke.req/?, *, TR\text{-}Result.res/\{TR\text{-}Result.cnf, TR\text{-}Abort.ind\}$$

If we work within WTP class 0, no acknowledges are sent after a message is received. In this case, in order to check that the gateway sends the data to the cell phone we have to consider the event *TR-Result.req*. Thus, in WTP class 0 we have the invariant

$$I_3 = TR\text{-}Invoke.req/?, *, TR\text{-}Result.req/\{TR\text{-}Result.ind, TR\text{-}Abort.ind\}$$

Let us note that this invariant also holds in WTP class 2.

Our log files include a field called *handle* that uniquely identifies a given communication (between a cell phone and the gateway). Actually, any user of the protocol is uniquely identified. For instance, in GSM the identification is dynamically assigned each time the cell phone connects to the gateway: The modem assigns a dynamic IP. So, we may use our invariants to check some privacy aspects of the protocol. For example,

$$I_4 = ?/TR\text{-}Invoke.ind(\text{user1}), *, TR\text{-}Result.res(\text{user1})/O$$

where $O = \{TR\text{-}Result.cnf(\text{user1}), TR\text{-}Abort.ind\}$.

In the next table we summarize the main data related to the application of our invariants to the observed traces. We have normalized all the times with respect to the ones for I_1. In the first row we show the (normalized) times taken to check whether the invariants were correct. The second row indicates whether an error was found. Finally, the third row shows the (average and normalized) times obtained by matching the traces with the corresponding invariants. Note that the time associated with the first invariant is (much) smaller in this case because the process is stopped once a fault is detected

Inv	I_1	I_2	I_3	I_4
Time-Correct	1	1.01	1.01	1.01
Error?	Yes	No	No	No
Time-Trace	1	8.72	8.46	9.22

Finally, we would like to briefly comment on implementation details. We have developed a package where all the algorithms related to this paper are implemented. Besides, suitable interfaces between the corresponding modules have been implemented so that the process can be completely automatized. The code was written in C (on Linux) so that the whole test platform can be as portable as possible. As we have explained in the bulk of the paper, we have implemented a modification of the classical pattern matching algorithms. Specifically, the pattern matching algorithm for simple invariants is an adaptation of [KMP77] where pattern matching is performed from left to right. The algorithms for deciding whether invariants are correct with respect to a specification are completely original. The whole code for all of our algorithms and interfaces is around 1000 lines long.

4 Conclusions and Future Work

We have presented a new methodology for passive testing. Our approach to passive testing includes the definition of a new concept of invariant as well as the corresponding algorithms to deal with them. In addition, we give a new architecture to test properties on a given trace of the implementation. This methodology has been applied to the passive testing of the WAP protocol. Several properties have been tested and the results of experimentations are very promising.

Passive testing has very large domains of application. It can be used as a monitoring technique to detect and report errors (that is the use we consider in this paper). Other applications of passive testing are in network management to detect configuration problems, resource provisioning, etc. It can be also used to study the feasibility of new features: Classes of services, network security, congestion control, etc. We plan to continue our work on passive testing with invariants to the detection of errors in critical systems where active testing is not feasible. Another field of application that we are exploring is network security. We consider that our techniques are well adapted to detect anomalies, attacks and intrusions, but this point has to be further investigated. Finally, we would like to extend our invariants with capabilities so that not only the control part of protocols but also the data part can be taken into account. In this line, [LCH+02] represents a step forward in the passive testing methodology because data is formally considered. However, we still need to perform a more thorough study of this work in order to find out whether our invariants can be adapted to this new framework.

References

[AAD79] J.M. Ayache, P. Azema, and M. Diaz. Observer: A concept for on-line detection of control errors in concurrent systems. In *9th Symposium on Fault-Tolerant Computing*, 1979.

[BM77] R.S. Boyer and J.S. Moore. A fast string searching algorithm. *Communications of the ACM*, 20:762–772, 1977.

[CGP01] A. Cavalli, C. Gervy, and S. Prokopenko. New approaches for passive testing using an extended finite state machine specification. In *Concordia Prestigious Workshop on Communication Software Engineering*, pages 225–250, 2001.

[KMP77] D.E. Knuth, J.H. Morris, and V.R. Pratt. Fast pattern matching in strings. *SIAM Journal on Computing*, 6(1):323–350, 1977.

[Lai02] R. Lai. A survey of communication protocol testing. *Journal of Systems and Software*, 62:21–46, 2002.

[LCH+02] D. Lee, D. Chen, R. Hao, R. Miller, J. Wu, and X. Yin. A formal approach for passive testing of protocol data portions. In *10th IEEE Int. Conf. on Network Protocols, ICNP'02*, pages 122–131. IEEE Computer Society Press, 2002.

[LNS+97] D. Lee, A.N. Netravali, K.K. Sabnani, B. Sugla, and A. John. Passive testing and applications to network management. In *5th IEEE Int. Conf. on Network Protocols, ICNP'97*, pages 113–122. IEEE Computer Society Press, 1997.

[LY96] D. Lee and M. Yannakakis. Principles and methods of testing finite state machines: A survey. *Proceedings of the IEEE*, 84(8):1090–1123, 1996.

[MA01a] R.E. Miller and K.A. Arisha. Fault coverage in networks by passive testing. In *International Conference on Internet Computing 2001, IC'2001*, pages 413–419. CSREA Press, 2001.

[MA01b] R.E. Miller and K.A. Arisha. Fault identification in networks by passive testing. In *34th Simulation Symposium, SS'01*, pages 277–284. IEEE Computer Society Press, 2001.

[Mil98] R.E. Miller. Passive testing of networks using a CFSM specification. In *IEEE Int. Performance Computing and Communications Conference*, pages 111–116. IEEE Computer Society Press, 1998.

[TC99] M. Tabourier and A. Cavalli. Passive testing and application to the GSM-MAP protocol. *Journal of Information and Software Technology*, 41:813–821, 1999.

[TCI99] M. Tabourier, A. Cavalli, and M. Ionescu. A GSM-MAP protocol experiment using passive testing. In *World Congress on Formal Methods in the Development of Computing Systems, FM'99, LNCS 1708*, pages 915–934. Springer, 1999.

[WZY01] J. Wu, Y. Zhao, and X. Yin. From active to passive: Progress in testing of internet routing protocols. In *FORTE 2001*, pages 101–116. Kluwer Academic Publishers, 2001.

Author Index

Acharya, Ira 20, 265
Arnedo, José Antonio 295

Bergengruen, Olaf 10
Bochmann, Gregor v. 197
Boroday, Sergiy 180

Cavalli, Ana 258, 295
Cousin, Philippe 128

Dai, Zhen Ru 79, 110
Deruelle, Olivier 33
Dibuz, Sarolta 63, 243
Dssouli, Rachida 211

El-Fakih, Khaled 197
En-Nouaary, Abdeslam 211

Frey, Maximilian 163

Gecse, Roland 63
Grabowski, Jens 79, 110

Hallal, Hesham 180
Huo, Jia Le 129

Kim, Myungchul 226
Krémer, Péter 243
Kumar Singh, Hemendra 20

Lepper, Markus 279
Li, Zhongjie 49

Mittal, Naina 265
Montes de Oca, Edgardo 258
Montgomery, Doug 33

Neukirchen, Helmut 110
Núñez, Manuel 258, 295

Petrenko, Alexandre 129, 180
Prokopenko, Svetlana 197

Ranganathan, M. 33
Rennoch, Axel 79

Schieferdecker, Ina 79, 95
Schlingloff, Bernd-Holger 163
Seol, Soonuk 226

Trancón y Widemann, Baltasar 279
Trenkaev, Vadim 226

Ulrich, Andreas 180
Ural, Hasan 146

Vassiliou-Gioles, Theofanis 95

Weyuker, Elaine J. 1
Wieland, Jacob 279
Williams, Craig 146
Wu, Jianping 49

Yevtushenko, Nina 129, 197
Yin, Xia 49

IFIP - The International Federation for Information Processing

IFIP was founded in 1960 under the auspices of UNESCO, following the First World Computer Congress held in Paris the previous year. An umbrella organization for societes working in information processing, IFIP's aim is two-fold: to support information processing within its member countries and to encourage technology transfer to developing nations. As its mission statement clearly states,

IFIP's mission is to be the leading, truly international, apolitical organization which encourages and assists in the development, exploitation and application of information technology for the benefit of all people.

IFIP is a non-profitmaking organization, run almost solely by 2500 volunteers. It operates through a number of technical committees, which organize events and publications. IFIP's events range from an international congress to local seminars, but the most important are:

- The IFIP World Computer Congress, held every second year
- open conferences
- working conferences

The flagship event is the IFIP World Computer Congress, at which both invited and contributed papers are presented. Contributed papers are rigorously refereed and the rejection rate is high.

As with the Congress, participation in the open conferences is open to all and papers may be invited or submitted. Again, submitted papers are stringently refereed.

The working conferences are structured differently. They are usually run by a working group and attendance is small and by invitation only. Their purpose is to create an atmosphere conducive to innovation and development. Refereeing is less rigorous and papers are subjected to extensive group discussion.

Publications arising from IFIP events vary. The papers presented at the IFIP World Computer Congress and at open conferences are published as conference proceedings, while the results of the working conferences are often published as collections of selected and edited papers.

Any national society whose primary activity is in information may apply to become a full member of IFIP, although full membership is restricted to one society per country. Full members are entitled to vote at the annual General Assembly, National societies preferring a less committed involvement may apply for associate or corresponding membership. Associate members enjoy the same benefits as full members, but without voting rights. Corresponding members are not represented in IFIP bodies. Affiliated membership is open to non-national societies, and individual and honorary membership schemes are also offered.

Lecture Notes in Computer Science

For information about Vols. 1–2561

please contact your bookseller or Springer-Verlag

Vol. 2562: V. Dahl, P. Wadler (Eds.), Practical Aspects of Declarative Languages. Proceedings, 2003. X, 315 pages. 2002.

Vol. 2563: Y. Manolopoulos, S, Evripidou, A.C. Kakas (Eds.), Advances in Informatics. Proceedings, 2001. XI, 498 pages. 2002.

Vol. 2565: J.M.L.M. Palma, J. Dongarra, V. Hernández, A. Augusto Sousa (Eds.), High Performance Computing for Computational Science – VECPAR 2002. Proceedings, 2002. XVII, 732 pages. 2003.

Vol. 2566: T.Æ. Mogensen, D.A. Schmidt, I.H. Sudborough (Eds.), The Essence of Computation. XIV, 473 pages. 2002.

Vol. 2567: Y.G. Desmedt (Ed.), Public Key Cryptography – PKC 2003. Proceedings, 2003. XI, 365 pages. 2002.

Vol. 2568: M. Hagiya, A. Ohuchi (Eds.), DNA Computing. Proceedings, 2002. XI, 338 pages. 2003.

Vol. 2569: D. Gollmann, G. Karjoth, M. Waidner (Eds.), Computer Security – ESORICS 2002. Proceedings, 2002. XIII, 648 pages. 2002. (Subseries LNAI).

Vol. 2570: M. Jünger, G. Reinelt, G. Rinaldi (Eds.), Combinatorial Optimization – Eureka, You Shrink!. Proceedings, 2001. X, 209 pages. 2003.

Vol. 2571: S.K. Das, S. Bhattacharya (Eds.), Distributed Computing. Proceedings, 2002. XIV, 354 pages. 2002.

Vol. 2572: D. Calvanese, M. Lenzerini, R. Motwani (Eds.), Database Theory – ICDT 2003. Proceedings, 2003. XI, 455 pages. 2002.

Vol. 2574: M.-S. Chen, P.K. Chrysanthis, M. Sloman, A. Zaslavsky (Eds.), Mobile Data Management. Proceedings, 2003. XII, 414 pages. 2003.

Vol. 2575: L.D. Zuck, P.C. Attie, A. Cortesi, S. Mukhopadhyay (Eds.), Verification, Model Checking, and Abstract Interpretation. Proceedings, 2003. XI, 325 pages. 2003.

Vol. 2576: S. Cimato, C. Galdi, G. Persiano (Eds.), Security in Communication Networks. Proceedings, 2002. IX, 365 pages. 2003.

Vol. 2577: P. Petta, R. Tolksdorf, F. Zambonelli (Eds.), Engineering Societies in the Agents World III. Proceedings, 2002. X, 285 pages. 2003. (Subseries LNAI).

Vol. 2578: F.A.P. Petitcolas (Ed.), Information Hiding. Proceedings, 2002. IX, 427 pages. 2003.

Vol. 2580: H. Erdogmus, T. Weng (Eds.), COTS-Based Software Systems. Proceedings, 2003. XVIII, 261 pages. 2003.

Vol. 2581: J.S. Sichman, F. Bousquet, P. Davidsson (Eds.), Multi-Agent-Based Simulation II. Proceedings, 2002. X, 195 pages. 2003. (Subseries LNAI).

Vol. 2582: L. Bertossi, G.O.H. Katona, K.-D. Schewe, B. Thalheim (Eds.), Semantics in Databases. Proceedings, 2001. IX, 229 pages. 2003.

Vol. 2583: S. Matwin, C. Sammut (Eds.), Inductive Logic Programming. Proceedings, 2002. X, 351 pages. 2003. (Subseries LNAI).

Vol. 2584: A. Schiper, A.A. Shvartsman, H. Weatherspoon, B.Y. Zhao (Eds.), Future Directions in Distributed Computing. X, 219 pages. 2003.

Vol. 2585: F. Giunchiglia, J. Odell, G. Weiß (Eds.), Agent-Oriented Software Engineering III. Proceedings, 2002. X, 229 pages. 2003.

Vol. 2586: M. Klusch, S. Bergamaschi, P. Edwards, P. Petta (Eds.), Intelligent Information Agents. VI, 275 pages. 2003. (Subseries LNAI).

Vol. 2587: P.J. Lee, C.H. Lim (Eds.), Information Security and Cryptology – ICISC 2002. Proceedings, 2002. XI, 536 pages. 2003.

Vol. 2588: A. Gelbukh (Ed.), Computational Linguistics and Intelligent Text Processing. Proceedings, 2003. XV, 648 pages. 2003.

Vol. 2589: E. Börger, A. Gargantini, E. Riccobene (Eds.), Abstract State Machines 2003. Proceedings, 2003. XI, 427 pages. 2003.

Vol. 2590: S. Bressan, A.B. Chaudhri, M.L. Lee, J.X. Yu, Z. Lacroix (Eds.), Efficiency and Effectiveness of XML Tools and Techniques and Data Integration over the Web. Proceedings, 2002. X, 259 pages. 2003.

Vol. 2591: M. Aksit, M. Mezini, R. Unland (Eds.), Objects, Components, Architectures, Services, and Applications for a Networked World. Proceedings, 2002. XI, 431 pages. 2003.

Vol. 2592: R. Kowalczyk, J.P. Müller, H. Tianfield, R. Unland (Eds.), Agent Technologies, Infrastructures, Tools, and Applications for E-Services. Proceedings, 2002. XVII, 371 pages. 2003. (Subseries LNAI).

Vol. 2593: A.B. Chaudhri, M. Jeckle, E. Rahm, R. Unland (Eds.), Web, Web-Services, and Database Systems. Proceedings, 2002. XI, 311 pages. 2003.

Vol. 2594: A. Asperti, B. Buchberger, J.H. Davenport (Eds.), Mathematical Knowledge Management. Proceedings, 2003. X, 225 pages. 2003.

Vol. 2595: K. Nyberg, H. Heys (Eds.), Selected Areas in Cryptography. Proceedings, 2002. XI, 405 pages. 2003.

Vol. 2596: A. Coen-Porisini, A. van der Hoek (Eds.), Software Engineering and Middleware. Proceedings, 2002. XII, 239 pages. 2003.

Vol. 2597: G. Păun, G. Rozenberg, A. Salomaa, C. Zandron (Eds.), Membrane Computing. Proceedings, 2002. VIII, 423 pages. 2003.

Vol. 2598: R. Klein, H.-W. Six, L. Wegner (Eds.), Computer Science in Perspective. X, 357 pages. 2003.

Vol. 2599: E. Sherratt (Ed.), Telecommunications and beyond: The Broader Applicability of SDL and MSC. Proceedings, 2002. X, 253 pages. 2003.

Vol. 2600: S. Mendelson, A.J. Smola, Advanced Lectures on Machine Learning. Proceedings, 2002. IX, 259 pages. 2003. (Subseries LNAI).

Vol. 2601: M. Ajmone Marsan, G. Corazza, M. Listanti, A. Roveri (Eds.) Quality of Service in Multiservice IP Networks. Proceedings, 2003. XV, 759 pages. 2003.

Vol. 2602: C. Priami (Ed.), Computational Methods in Systems Biology. Proceedings, 2003. IX, 214 pages. 2003.

Vol. 2603: A. Garcia, C. Lucena, F. Zambonelli, A. Omicini, J. Castro (Eds.), Software Engineering for Large-Scale Multi-Agent Systems. XIV, 285 pages. 2003.

Vol. 2604: N. Guelfi, E. Astesiano, G. Reggio (Eds.), Scientific Engineering for Distributed Java Applications. Proceedings, 2002. X, 205 pages. 2003.

Vol. 2606: A.M. Tyrrell, P.C. Haddow, J. Torresen (Eds.), Evolvable Systems: From Biology to Hardware. Proceedings, 2003. XIV, 468 pages. 2003.

Vol. 2607: H. Alt, M. Habib (Eds.), STACS 2003. Proceedings, 2003. XVII, 700 pages. 2003.

Vol. 2609: M. Okada, B. Pierce, A. Scedrov, H. Tokuda, A. Yonezawa (Eds.), Software Security – Theories and Systems. Proceedings, 2002. XI, 471 pages. 2003.

Vol. 2610: C. Ryan, T. Soule, M. Keijzer, E. Tsang, R. Poli, E. Costa (Eds.), Genetic Programming. Proceedings, 2003. XII, 486 pages. 2003.

Vol. 2611: S. Cagnoni, J.J. Romero Cardalda, D.W. Corne, J. Gottlieb, A. Guillot, E. Hart, C.G. Johnson, E. Marchiori, J.-A. Meyer, M. Middendorf, G.R. Raidl (Eds.), Applications of Evolutionary Computing. Proceedings, 2003. XXI, 708 pages. 2003.

Vol. 2612: M. Joye (Ed.), Topics in Cryptology – CT-RSA 2003. Proceedings, 2003. XI, 417 pages. 2003.

Vol. 2613: F.A.P. Petitcolas, H.J. Kim (Eds.), Digital Watermarking. Proceedings, 2002. XI, 265 pages. 2003.

Vol. 2614: R. Laddaga, P. Robertson, H. Shrobe (Eds.), Self-Adaptive Software: Applications. Proceedings, 2001. VIII, 291 pages. 2003.

Vol. 2615: N. Carbonell, C. Stephanidis (Eds.), Universal Access. Proceedings, 2002. XIV, 534 pages. 2003.

Vol. 2616: T. Asano, R. Klette, C. Ronse (Eds.), Geometry, Morphology, and Computational Imaging. Proceedings, 2002. X, 437 pages. 2003.

Vol. 2617: H.A. Reijers (Eds.), Design and Control of Workflow Processes. Proceedings, 2002. XV, 624 pages. 2003.

Vol. 2618: P. Degano (Ed.), Programming Languages and Systems. Proceedings, 2003. XV, 415 pages. 2003.

Vol. 2619: H. Garavel, J. Hatcliff (Eds.), Tools and Algorithms for the Construction and Analysis of Systems. Proceedings, 2003. XVI, 604 pages. 2003.

Vol. 2620: A.D. Gordon (Ed.), Foundations of Software Science and Computation Structures. Proceedings, 2003. XII, 441 pages. 2003.

Vol. 2621: M. Pezzè (Ed.), Fundamental Approaches to Software Engineering. Proceedings, 2003. XIV, 403 pages. 2003.

Vol. 2622: G. Hedin (Ed.), Compiler Construction. Proceedings, 2003. XII, 335 pages. 2003.

Vol. 2623: O. Maler, A. Pnueli (Eds.), Hybrid Systems: Computation and Control. Proceedings, 2003. XII, 558 pages. 2003.

Vol. 2625: U. Meyer, P. Sanders, J. Sibeyn (Eds.), Algorithms for Memory Hierarchies. Proceedings, 2003. XVIII, 428 pages. 2003.

Vol. 2626: J.L. Crowley, J.H. Piater, M. Vincze, L. Paletta (Eds.), Computer Vision Systems. Proceedings, 2003. XIII, 546 pages. 2003.

Vol. 2627: B. O'Sullivan (Ed.), Recent Advances in Constraints. Proceedings, 2002. X, 201 pages. 2003. (Subseries LNAI).

Vol. 2628: T. Fahringer, B. Scholz, Advanced Symbolic Analysis for Compilers. XII, 129 pages. 2003.

Vol. 2631: R. Falcone, S. Barber, L. Korba, M. Singh (Eds.), Trust, Reputation, and Security: Theories and Practice. Proceedings, 2002. X, 235 pages. 2003. (Subseries LNAI).

Vol. 2632: C.M. Fonseca, P.J. Fleming, E. Zitzler, K. Deb, L. Thiele (Eds.), Evolutionary Multi-Criterion Optimization. Proceedings, 2003. XV, 812 pages. 2003.

Vol. 2633: F. Sebastiani (Ed.), Advances in Information Retrieval. Proceedings, 2003. XIII, 546 pages. 2003.

Vol. 2634: F. Zhao, L. Guibas (Eds.), Information Processing in Sensor Networks. Proceedings, 2003. XII, 692 pages. 2003.

Vol. 2636: E. Alonso, D, Kudenko, D. Kazakov (Eds.), Adaptive Agents and Multi-Agent Systems. XIV, 323 pages. 2003. (Subseries LNAI).

Vol. 2637: K.-Y. Whang, J. Jeon, K. Shim, J. Srivastava (Eds.), Advances in Knowledge Discovery and Data Mining. Proceedings, 2003. XVIII, 610 pages. 2003. (Subseries LNAI).

Vol. 2639: G. Wang, Q. Liu, Y. Yao, A. Skowron (Eds.), Rough Sets, Fuzzy Sets, Data Mining, and Granular Computing. Proceedings, 2003. XVII, 741 pages. 2003. (Subseries LNAI).

Vol. 2642: X. Zhou, Y. Zhang, M.E. Orlowska (Eds.), Web Technologies and Applications. Proceedings, 2003. XIII, 608 pages. 2003.

Vol. 2643: M. Fossorier, T. Høholdt, A. Poli (Eds.), Applied Algebra, Algebraic Algorithms and Error-Correcting Codes. Proceedings, 2003. X, 256 pages. 2003.

Vol. 2644: D. Hogrefe, A. Wiles (Eds.), Testing of Communicating Systems. Proceedings, 2003. XII, 311 pages. 2003.

Vol. 2646: H. Geuvers, F, Wiedijk (Eds.), Types for Proofs and Programs. Proceedings, 2002. VIII, 331 pages. 2003.

Vol. 2648: T. Ball, S.K. Rajamani (Eds.), Model Checking Software. Proceedings, 2003. VIII, 241 pages. 2003.

Vol. 2649: B. Westfechtel, A. van der Hoek (Eds.), Software Configuration Management. Proceedings, 2003. VIII, 241 pages. 2003.

Vol. 2656: E. Biham (Ed.), Advances in Cryptology – EUROCRPYT 2003. Proceedings, 2003. XIV, 649 pages. 2003.

Vol. 2663: E. Menasalvas, J. Segovia, P.S. Szczepaniak (Eds.), Advances in Web Intelligence. Proceedings, 2003. XII, 350 pages. 2003. (Subseries LNAI).